Birkhäuser

Reinhard Racke

Lectures on Nonlinear Evolution Equations

Initial Value Problems

Second Edition

Birkhäuser

Reinhard Racke
Department of Mathematics and Statistics
University of Konstanz
Konstanz, Germany

ISBN 978-3-319-21872-4 ISBN 978-3-319-21873-1 (eBook)
DOI 10.1007/978-3-319-21873-1

Library of Congress Control Number: 2015948864

Mathematics Subject Classification (2010): 35B40, 35K05, 35K55, 35L05, 35L45, 35L70, 35Q99, 36Q60, 35Q74, 74F05

Springer Cham Heidelberg New York Dordrecht London
First edition published by Vieweg, Wiesbaden, 1992

Cover design: deblik, Berlin

Printed on acid-free paper

Springer International Publishing AG Switzerland is part of Springer Science+Business Media (www.birkhauser-science.com)

Preface

This book is the second edition of the book *Lectures on nonlinear evolution equations. Initial value problems* [150] from 1992. Additionally, it now includes a new Chapter 13 on *initial-boundary* value problems for waveguides, addressing more advanced students and researchers.

Several people contributed helpful comments on the first edition and on the new Chapter 13. In particular I would like to thank Dipl.-Math. Karin Borgmeyer, Dr. Michael Pokojovy, Dipl.-Math. Marco Ritter, and Dipl.-Math. Alexander Schöwe. For typing Chapter 13 I thank Gerda Baumann. I am obliged to Birkhäuser, in particular to Clemens Heine, for the interest in publishing this book.

Konstanz, April 2015 Reinhard Racke

Preface to the first edition:

The book in hand is based on lectures which were given at the University of Bonn in the winter semesters of 1989/90 and 1990/91. The aim of the lectures was to present an elementary, self-contained introduction into some important aspects of the theory of global, small, smooth solutions to initial value problems for nonlinear evolution equations. The addressed audience included graduate students of both mathematics and physics who were only assumed to have a basic knowledge of linear partial differential equations. Thus, in the spirit of the underlying series, this book is intended to serve as a detailed basis for lectures on the subject as well as for self-studies for students or for other newcomers to this field.

The presentation of the theory is made using the classical method of continuation of local solutions with the help of a priori estimates obtained for small data. The corresponding global existence theorems have been proved mainly in the last decade, focussing on fully nonlinear systems. Related questions concerning large data problems, the existence of weak solutions or the analysis of shock waves are not discussed. Also the question of optimal regularity assumptions on the coefficients is beyond the scope of the book and is touched only in part and exemplarily.

Most of the material presented here has only been previously published in original papers, and some of the material has never been published until now. Therefore, I hope that both the interested beginner in the field and the expert will benefit from reading the book. In addition, a long list of references has been included, although it is not intended to be exhaustive. Of course the selection of the material follows personal interests and tastes.

Several colleagues and students helped me with their comments on earlier versions of this book. In particular I would like to thank R. Arlt, S. Jiang, S. Noelle, P. P. Schirmer, R. P. Spindler, M. Stoth and F. Willems. Special thanks are due to R. Leis who also suggested writing first lecture notes in 1989 (SFB 256 Vorlesungsreihe Nr. 13, Universität Bonn (1990), in German). I am obliged to the Verlag Vieweg and to the editor of the "Aspects of Mathematics", K. Diederich, for including the book in this series. The major part of typing the manuscript was done by R. Müller and A. Thiedemann whom I thank for their expert work. Last, but not least, I would like to thank the Deutsche Forschungsgemeinschaft, Sonderforschungsbereich 256, for generous and continuous support.

Bonn, August 1991 Reinhard Racke

Contents

Introduction

Many problems arising in the applied sciences lead to nonlinear initial value problems (nonlinear Cauchy problems) of the following type

$$V_t + AV = F(V, \ldots, \nabla^\beta V), \quad V(t=0) = V^0.$$

Here $V = V(t,x)$ is a vector-valued function taking values in $\mathbb{R}^k$ (or $\mathbb{C}^k$), where $t \geq 0$, $x \in \mathbb{R}^n$, and A is a given linear differential operator of order m with $k, n, m \in \mathbb{N}$. F is a given nonlinear function of V and its derivatives up to order $|\beta| \leq m$, and ∇ denotes the gradient with respect to x, while V^0 is a given initial value. In particular the case $|\beta| = m$, i.e. the case of fully nonlinear initial value problems, is of interest.
An important example from mathematical physics is the wave equation describing an infinite vibrating string (membrane, sound wave, respectively) in $\mathbb{R}^1$ ($\mathbb{R}^2$, $\mathbb{R}^3$, respectively; generalized: $\mathbb{R}^n$). The second-order differential equation for the elongation $y = y(t,x)$ at time t and position x is the following:

$$y_{tt} - \nabla' \frac{\nabla y}{\sqrt{1+|\nabla y|^2}} = 0,$$

where ∇' denotes the divergence. This can also be written as

$$y_{tt} - \Delta y = \nabla' \frac{\nabla y}{\sqrt{1+|\nabla y|^2}} - \Delta y =: f(\nabla y, \nabla^2 y).$$

We notice that f has the following property:

$$f(W) = \mathcal{O}(|W|^3) \quad \text{as } |W| \to 0.$$

Additionally one has prescribed initial values

$$y(t=0) = y_0, \quad y_t(t=0) = y_1.$$

The transformation defined by $V := (y_t, \nabla y)$ turns the nonlinear wave equation for y into a first-order system for V as described above. The investigation of such nonlinear evolution equations has found an increasing interest in the last years, in particular because of their application to the typical partial differential equations arising in mathematical physics.
We are interested in the existence and uniqueness of global solutions, i.e. solutions $V = V(t,x)$ which are defined for all values of the time parameter t. The solutions will be smooth solutions, e.g. C^1-functions with respect to t taking values in Sobolev spaces of sufficiently high order of differentiability. In particular they will be classical solutions. Moreover we wish to describe the asymptotic behavior of the solutions as $t \to \infty$.

It is well known for the nonlinear wave equation, the first example above, that in general one cannot expect to obtain a global smooth solution. That is to say, the solution may develop singularities in finite time, no matter how smooth or how small the initial data are. This phenomenon is known for more general nonlinear hyperbolic systems and also for many other systems from mathematical physics, biology, etc., including the systems which are mentioned below. Therefore, a general global existence theorem can only be proved under special assumptions on the nonlinearity and on the initial data. The result will be a theorem which is applicable for small initial data, assuming a certain degree of vanishing of the nonlinearity near zero. The necessary degree depends on the space dimension, being a weaker assumption for higher dimensions. This is strongly connected with the asymptotic behavior of solutions to the associated linearized system ($F \equiv 0$ resp. $f \equiv 0$ in the example above) as $t \to \infty$, which gives a first insight into the means used for the proof.

Further examples of nonlinear evolution equations which can be written in the general first-order form after a suitable transformation are the following. They will be discussed in more detail in Chapter 11.

- Equations of elasticity:

$$\partial_t^2 U_i = \sum_{m,j,k=1}^{n} C_{imjk}(\nabla U)\partial_m \partial_k U_j, \quad i = 1, \dots, n,$$

$$U(t=0) = U^0, \quad U_t(t=0) = U^1.$$

  We shall discuss the homogeneous, initially isotropic case for $n = 3$ and the homogeneous, initially cubic case for $n = 2$.

- Heat equations:

$$u_t - \Delta u = F(u, \nabla u, \nabla^2 u), \quad u(t=0) = u_0.$$

- Equations of thermoelasticity:

$$\partial_t^2 U_i = \sum_{m,j,k=1}^{n} C_{imjk}(\nabla U, \theta)\partial_m \partial_k U_j + \tilde{C}_{im}(\nabla U, \theta)\partial_m \theta, \quad i = 1, 2, 3,$$

$$(\theta + T_0)a(\nabla U, \theta)\partial_t \theta = \nabla' q(\nabla U, \theta, \nabla\theta) + tr\{\tilde{C}_{km}(\nabla U, \theta)'_{km} \cdot (\partial_t \partial_s U_r)_{rs}\}(\theta + T_0),$$

$$U(t=0) = U^0, \quad U_t(t=0) = U^1, \quad \theta(t=0) = \theta^0.$$

  The homogeneous, initially isotropic case will be discussed here.

- Schrödinger equations:

$$u_t - \mathrm{i}\Delta u = F(u, \nabla u), \quad u(t=0) = u_0.$$

- Klein–Gordon equations:

$$y_{tt} - \Delta y + my = f(y, y_t, \nabla y, \nabla y_t, \nabla^2 y), \quad m > 0,$$

$$y(t = 0) = y_0, \quad y_t(t = 0) = y_1.$$

- Maxwell equations:

$$\begin{aligned} D_t - \nabla \times H &= 0, \\ B_t + \nabla \times E &= 0, \end{aligned}$$

$$\begin{aligned} D(t = 0) = D^0, &\quad B(t = 0) = B^0, \\ \nabla' D = 0, &\quad \nabla' B = 0, \\ D = \varepsilon(E), &\quad B = \mu(H). \end{aligned}$$

- Plate equations:

$$y_{tt} + \Delta^2 y = f(y_t, \nabla^2 y) + \sum_{i=1}^{n} b_i(y_t, \nabla^2 y)\partial_i y_t,$$

$$y(t = 0) = y_0, \quad y_t(t = 0) = y_1.$$

In order to obtain existence theorems to these systems, we shall apply the classical method of continuing local solutions (local with respect to t), provided a priori estimates are known. The proof of the a priori estimates represents the non-classical part of the approach. It requires ideas and techniques which mainly have been developed in the last years, in particular the idea of using the decay of solutions to the associated linearized problems. These new techniques were essential to overcome the difficulties in the study of *fully* nonlinear systems, i.e. systems where the nonlinearity involves the highest derivatives appearing on the linear left-hand side. We remark that in this sense the Schrödinger equations and the plate equations above are not fully nonlinear. The highest derivatives that appear in the nonlinearity can still directly be dominated by the linear part in the energy estimates, see Chapter 11.
The general method by which all the systems mentioned before can be dealt with (cum grano salis) is described by the following scheme.
We discuss the system

$$V_t + AV = F(V, \dots, \nabla^\beta V), \quad V(t = 0) = V^0,$$

where F is assumed to be smooth and to satisfy

$$F(W) = \mathcal{O}(|W|^{\alpha+1}) \quad \text{as } |W| \to 0, \quad \text{for some } \alpha \in \mathbb{N}.$$

The larger α is, the smaller is the impact that the nonlinearity will have for small values of $|W|$, i.e. the linear behavior will dominate for some time and there is some hope that it will lead to global solutions for sufficiently small data if the linear decay is strong enough. This will depend on the space dimension.
The general scheme consists of the following Steps **A–E**.

A: Decay of solutions to the linearized system:

A solution V to the associated linearized problem

$$V_t + AV = 0, \quad V(t=0) = V^0,$$

satisfies

$$\|V(t)\|_q \leq c(1+t)^{-d}\|V^0\|_{N,p},$$

where $2 \leq q \leq \infty$ (or $2 \leq q < \infty$), $1/p + 1/q = 1$; $c, d > 0$ and $N \in \mathbb{N}$ are functions of q and of the space dimension n. (E.g. for the wave equation above: $d = \frac{n-1}{2}(1-\frac{2}{q})$.) This is usually proved by using explicit representation formulae and/or the representation via the Fourier transform.

B: Local existence and uniqueness:

There is a local solution V to the nonlinear system on some time interval $[0, T]$, $T > 0$, with the following regularity:

$$V \in C^0([0,T], W^{s,2}) \cap C^1([0,T], W^{\tilde{s},2}),$$

where $s, \tilde{s} \in \mathbb{N}$ are sufficiently large to guarantee a classical solution. The proof of a local existence theorem is always a problem itself. We shall present the proof of the corresponding theorem for the wave equation in detail.

C: High energy estimates:

The local solution V satisfies

$$\|V(t)\|_{s,2} \leq C\|V^0\|_{s,2} \cdot \exp\left\{C\int_0^t \|V(r)\|_{b,\infty}^{\alpha}\, dr\right\}, \quad t \in [0,T].$$

C only depends on s, not on T or V^0. b is independent of s, that is, the exponential term does not involve higher derivatives in the L^∞-norm (which allows to close the circle in Step **E**). This inequality is proved using general inequalities for composite functions (see Chapter 4).

D: Weighted a priori estimates:

The local solution satisfies

$$\sup_{0\leq t\leq T}(1+t)^{d_1}\|V(t)\|_{s_1,q_1} \leq M_0 < \infty,$$

where M_0 is independent of T, s_1 is sufficiently large, $q_1 = q_1(\alpha)$ is chosen appropriately for each problem and $d_1 = d(q_1, n)$ according to **A**, provided V^0 is sufficiently small (in a sense to be made precise later; roughly, high Sobolev norms of V^0 are small).

In this step the information obtained in **A** is exploited with the help of the classical formula

$$V(t) = \mathrm{e}^{-tA}V^0 + \int_0^t \mathrm{e}^{-(t-r)A}F(V, \dots, \nabla^\beta V)(r)dr,$$

where $\mathrm{e}^{-tA}V^0$ symbolically stands for the solution to the linearized problem with initial value V^0.

E: Final energy estimate:

The results in **C** and **D** easily lead to the following a priori bound:

$$\|V(t)\|_{s,2} \leq K\|V^0\|_{s,2}, \quad 0 \leq t \leq T,$$

$s \in \mathbb{N}$ being sufficiently large, V^0 being sufficiently small and K being independent of T. This a priori estimate allows us to apply now the standard continuation argument and to continue the local solution obtained in Step **B** to a solution defined for all $t \in [0, \infty)$.

The method described above immediately provides information on the asymptotic behavior of the global solution as $t \to \infty$ in Step **D** and in Step **E**.

This general scheme applies to all the above systems mutatis mutandis; for example, there may appear certain derivatives with respect to t of V in the integrand of the exponential in Step **C**. Moreover the nonlinearity may depend on t and x explicitly. Nevertheless, difficult questions can arise in the discussion of the details for each specific system. Particularly interesting are the necessary modifications that have to be made for the equations of thermoelasticity. This system cannot directly be put into the framework just described because it consists of different types of differential equations (hyperbolic, parabolic), and also different types of nonlinearities appear which exclude for example a uniform sharp estimate as in Step **A**. Instead different components of V have to be dealt with in different ways. Altogether however, global existence theorems will again be proved in the spirit of the Steps **A**–**E**.

This underlines the generality of the approach. Of course, this generality prevents the results from being optimal in some cases. We shall discuss this in detail for the following general wave equation:

$$y_{tt} - \Delta y = f(y_t, \nabla y, \nabla y_t, \nabla^2 y),$$

$$y(t=0) = y_0, \quad y_t(t=0) = y_1.$$

For this we shall go through the Steps **A**–**E** in Chapters 1–8. Moreover, a more or less optimal result is presented, the proof of which uses invariance properties of the d'Alembert operator $\partial_t^2 - \Delta$ under the generators of the Lorentz group. The other examples will be studied in Chapter 11. In several of the cases there, these subtle invariances are not available.

To underline the necessity of studying conditions under which small data problems allow global solutions we shall shortly describe some blow-up results — results on the development of singularities in finite time even for small data — in Chapter 10. In Chapter 9 a few other methods are briefly mentioned and Chapter 12 tries to outline some recent developments and future projects going beyond the main line of this book.

The scheme described above can be found in [94]. Similar ideas are present in [117, 119, 158, 178].

One may think of the global existence results as a kind of stability result for small perturbations of the associated linear problems. Of course it is of great interest to study solutions for large data but this is beyond the scope of this book. We refer the interested reader to the literature [138, 179, 180, 186]. We also remark that there are much more results on semilinear systems. The emphasis in this book lies on fully nonlinear systems.

In the second edition, we shall treat in the new Chapter 13 linear and nonlinear initial-boundary value problems in waveguides, giving insight into the impact of the geometry of domains with boundaries, and, simultaneously, demonstrating that following the steps **A**–**E** also here applies, mutatis mutandis.

1 Global solutions to wave equations — existence theorems

We shall start with the formulation of a global existence theorem for solutions of a class of nonlinear wave equations. The first theorem, Theorem 1.1, is typical for the kind of existence theorems that will be obtained for other evolution equations in Chapter 11. The second theorem, Theorem 1.2, optimizes in some sense the result for wave equations. We shall conclude this section with giving a few examples characterizing the behavior of solutions to nonlinear wave equations in general, thus pointing out the crucial parts of the assumptions in the existence theorems.

The nonlinear wave equations which shall be considered here are

$$y_{tt} - \Delta y = f\left(y_t, \nabla y, \nabla y_t, \nabla^2 y\right) \equiv f\left(Dy, \nabla Dy\right), \tag{1.1}$$

with prescribed initial data

$$y(t=0) = y_0, \quad y_t(t=0) = y_1. \tag{1.2}$$

The following notation is used:

$y = y(t,x) \in \mathbb{R}$, $t \geq 0$, $x \in \mathbb{R}^n$, $n \in \mathbb{N}$ arbitrary.

$\Delta = \sum\limits_{i=1}^{n} \partial_i^2$, $\partial_i = \partial/\partial x_i$, $i = 1, \ldots, n$, $y_t = \partial_t y$, $y_{tt} = \partial_t^2 y$ $\partial_t = \partial/\partial_t$,

$D = \begin{pmatrix} \partial_t \\ \nabla \end{pmatrix}$, $\nabla = (\partial_1, \ldots, \partial_n)'$.

Let

$$\begin{aligned} u &:= Dy = (\partial_t y, \partial_1 y, \ldots, \partial_n y), \quad \nabla u = \nabla D y = (\partial_1 u, \ldots, \partial_n u), \\ u_0 &:= (y_1, \nabla y_0), \quad \text{(as column-vectors)}. \end{aligned}$$

We assume that the nonlinear function f satisfies

$$\left.\begin{aligned} &f \in C^\infty\left(\mathbb{R}^{(n+1)^2}, \mathbb{R}\right), \\ &\exists \alpha \in \mathbb{N}: \quad f(u, \nabla u) = \mathcal{O}\left((|u| + |\nabla u|)^{\alpha+1}\right) \quad \text{as} \quad |u| + |\nabla u| \to 0, \end{aligned}\right\} \tag{1.3}$$

where $C^\infty\left(\mathbb{R}^m, \mathbb{R}^k\right)$ is the space of infinitely differentiable functions from $\mathbb{R}^m$ into $\mathbb{R}^k$, $m, k \in \mathbb{N}$. Let us introduce some more notation:

$W^{m,p} := W^{m,p}(\mathbb{R}^n)$: usual Sobolev spaces, $m \in \mathbb{N}_0$, $1 \leq p \leq \infty$, with norm $\|\cdot\|_{m,p}$, (see R.A. Adams [1]) (*Sergei L'vovich Sobolev*, 6.10.1908 – 3.1.1989).

$L^p := W^{0,p}$ with norm $\|\cdot\|_p$, $1 \leq p \leq \infty$.

$C^k(I, E) :=$ space of k times continuously differentiable functions from an interval

$I \subset \mathbb{R}$ into a Banach space E, $k \in \mathbb{N}_0$ (*Stefan Banach*, 30.3.1892 – 31.8.1945).
Now we are ready to formulate the first existence theorem.

Theorem 1.1 *We assume (1.3) with* $\frac{1}{\alpha}\left(1+\frac{1}{\alpha}\right) < \frac{n-1}{2}$. *Then there exist an integer* $s_0 > \frac{n}{2}+1$ *and a* $\delta > 0$ *such that the following holds:*
If $u_0 = (y_1, \nabla y_0)$ *belongs to* $W^{s,2} \cap W^{s,p}$ *with* $s \geq s_0$ *and* $p = \frac{2\alpha+2}{2\alpha+1}$ *and*

$$\|u_0\|_{s,2} + \|u_0\|_{s,p} < \delta,$$

then there is a unique solution y *of the initial value problem to the nonlinear wave equation (1.1), (1.2) with*

$$(y_t, \nabla y) \in C^0\left([0,\infty), W^{s,2}\right) \cap C^1\left([0,\infty), W^{s-1,2}\right).$$

Moreover, we have

$$\begin{aligned} \|(y_t, \nabla y)(t)\|_\infty + \|(y_t, \nabla y)(t)\|_{2\alpha+2} &= \mathcal{O}\left(t^{-\frac{n-1}{2}\frac{\alpha}{\alpha+1}}\right), \\ \|(y_t, \nabla y)(t)\|_{s,2} &= \mathcal{O}(1) \quad \text{as} \quad t \to \infty. \end{aligned}$$

The proof of Theorem 1.1 will be presented in Chapter 8 based on results that will be obtained in Chapters 2–7.
By Sobolev's imbedding theorem the solution y obtained in Theorem 1.1 is a classical solution:

$$y \in C^2\left([0,\infty) \times \mathbb{R}^n\right).$$

The $L^{2\alpha+2}$-decay rate given above is optimal, but the L^∞-decay rate is not optimal. This results from the decay rate for sufficiently many derivatives of $(y_t, \nabla y)$ in the $L^{2\alpha+2}$-norm just by Sobolev's imbedding theorem (see Chapter 7). The optimal decay rate for the L^∞-norm is $\frac{n-1}{2}$ (instead of $\frac{n-1}{2}\frac{\alpha}{\alpha+1}$), see Theorem 1.2 below.
As far as the regularity assumption on f is concerned, we remark that the C^∞-assumption can be weakened, cf. the remarks in Chapters 5 and 8.
Theorem 1.1 was given by Klainerman & Ponce in [94]. It provides sufficient conditions for the global existence of small, smooth solutions to the nonlinear wave equation (1.1). Moreover, the asymptotic behavior of the solution as $t \to \infty$ is described with decay rates.
The condition

$$\frac{1}{\alpha}\left(1+\frac{1}{\alpha}\right) < \frac{n-1}{2} \tag{1.4}$$

obviously connects the space dimension and the degree of vanishing of the nonlinearity near zero. The larger α and/or n are, the better the situation is.

For the example from the introduction,

$$y_{tt} - \nabla' \frac{\nabla y}{\sqrt{1+|\nabla y|^2}} = 0$$

or

$$y_{tt} - \Delta y = \nabla' \frac{\nabla y}{\sqrt{1+|\nabla y|^2}} - \Delta y$$

we have

$$\begin{aligned} f(Dy, \nabla Dy) &= \Delta y \left(\frac{1}{\sqrt{1+|\nabla y|^2}} - 1 \right) + \nabla y \nabla' \frac{1}{\sqrt{1+|\nabla y|^2}} \\ &= \mathcal{O}\left(\left(|\nabla y| + |\nabla^2 y| \right)^3 \right) \quad \text{as} \quad |\nabla y| + |\nabla^2 y| \to 0, \end{aligned}$$

i.e. we have $\alpha = 2$ and the condition (1.4) turns into: $n > 5/2$, i.e. $n \geq 3$.
In general we can express the relation between α and n as given in Table 1.1.

$\alpha =$	1	2	3, 4, ...
$n \geq$	6	3	2

Table 1.1: Sufficient conditions in Theorem 1.1

Quadratic nonlinearities ($\alpha = 1$) require n to be at least 6. This is not optimal. Since the method leading to Theorem 1.1 is very general, being applicable to hyperbolic, parabolic and many other equations, it is not surprising that it is not sharp in all cases — although it is sharp in many cases! The optimal condition here being necessary is $n \geq 4$ for quadratic nonlinearities. To prove this result one has to use rather special properties of the operator $\partial_t^2 - \Delta$. The corresponding result is stated in the next theorem. It is optimal in the sense that quadratic nonlinearities in $\mathbb{R}^3$ in general tend to develop singularities in finite time, see below and Chapter 10.
Let the initial data y_0, y_1 be given in the form

$$y(t=0) = y_0 = \varepsilon\varphi, \quad y_t(t=0) = y_1 = \varepsilon\psi, \tag{1.5}$$

where $\varphi, \psi \in C_0^\infty \equiv C_0^\infty(\mathbb{R}^n)$ (test functions) and $\varepsilon > 0$ is a (small) parameter. Let $T_\infty(\varepsilon)$ denote the *life span* of a solution to the initial value problem (1.1), (1.5), i.e. $T_\infty(\varepsilon)$ equals the supremum of all times $T > 0$ for which there exists a C^∞-solution to (1.1), (1.5) for all $x \in \mathbb{R}^n$, $0 \leq t < T$.
We assume that f satisfies (1.3) with $\alpha = 1$. Then we have

Theorem 1.2 **(i)** *Let* $n > 3$. *Then there is an* $\varepsilon_0 > 0$ *such that for all* ε *with* $0 < \varepsilon \le \varepsilon_0$ *we have*

$$T_\infty(\varepsilon) = \infty,$$

where ε_0 *depends on at most* $(2n+3)$ *derivatives of* f *and* ψ *and on at most* $(2n+4)$ *derivatives of* φ. *The global solution* y *satisfies*

$$(y_t, \nabla y) \in C^0\left([0,\infty), W^{2n+3,2}\right) \cap C^1\left([0,\infty), W^{2n+2,2}\right).$$

Moreover, we have

$$\begin{aligned} \|(y_t, \nabla y)(t)\|_\infty &= \mathcal{O}\left(t^{-\frac{n-1}{2}}\right), \\ \|(y_t, \nabla y)(t)\|_{2n+3,2} &= \mathcal{O}(1) \quad \text{as} \quad t \to \infty. \end{aligned}$$

(ii) *Let* $n = 3$. *There exist an* $\varepsilon_0 > 0$ *and an* $A > 0$ *such that for all* ε *with* $0 < \varepsilon \le \varepsilon_0$ *we have*

$$T_\infty(\varepsilon) \ge \mathrm{e}^{A/\varepsilon},$$

where ε_0 *and* A *depend on at most* 9 *derivatives of* f *and* ψ *and on at most* 10 *derivatives of* φ.

If f vanishes of order $\alpha + 1$ near zero, $\alpha \ge 1$, then the proof of Theorem 1.2 in Chapter 8 will show that the following condition would replace the condition (1.4):

$$\frac{1}{\alpha} < \frac{n-1}{2}. \tag{1.6}$$

Since this only changes the value for n in Table 1.1 if $\alpha = 1$, Theorem 1.2 has been formulated for this case. The reason for having the improved relation (1.6) is that a precise analysis of the invariance properties of $\partial_t^2 - \Delta$ allows to replace an L^1–L^∞-estimate for solutions to the linear wave equation by an L^2–L^∞-estimate of a general Sobolev type with the same decay rate $\frac{n-1}{2}$; see Chapter 8.

The L^∞-decay rate of $(y_t, \nabla y)$ given for the global solution y in Theorem 1.2, (i) is optimal.

Theorem 1.2 was given by S. Klainerman in [88]. The result is optimal with respect to the relation between α and n in the following sense. It is known that quadratic nonlinearities ($\alpha = 1$) in $\mathbb{R}^3$ in general tend to develop singularities in finite time. Examples have been given by F. John (*Fritz John*, 14.6.1910 – 10.2.1994) in [68, 70]; see Chapter 10.

The following two examples illustrate typical situations. The first example is a special quadratic nonlinearity in $\mathbb{R}^3$. It shows that global solutions may exist but also singularities may develop depending on the size of the data.

Example 1: (cf. [86, pp. 45–46])

$$y_{tt} - \Delta y = |\nabla y|^2 - y_t^2, \tag{1.7}$$

$$y(t=0)=0, \quad y_t(t=0)=h\in C^2(\mathbb{R}^3), \tag{1.8}$$

where $y=y(t,x)$, $t\in\mathbb{R}$, $x\in\mathbb{R}^3$; further conditions on h are given below. If y is a solution to (1.7), (1.8), $y\in C^2(\mathbb{R}\times\mathbb{R}^3)$, then v defined by $v:=e^y$ satisfies

$$v_t=vy_t, \quad v_{tt}=vy_{tt}+vy_t^2, \quad \nabla v=v\nabla y, \quad \Delta v=v\Delta y+v|\nabla y|^2$$

which implies

$$\begin{aligned} v_{tt}-\Delta v &= v\left(y_{tt}-\Delta y+y_t^2-|\nabla y|^2\right)=0, \\ v(t=0) &= 1, \quad v_t(t=0)=h. \end{aligned}$$

Thus v is explicitly given by

$$v(t,x)=1+\frac{t}{4\pi}\int_{S^2}h(x+t\xi)d\xi,$$

S^2 being the unit sphere in $\mathbb{R}^3$, (see Chapter 2 for this formula).
If $v>0$ we obtain y as

$$y(t,x)=\log\left(1+\frac{t}{4\pi}\int_{S^2}h(x+t\xi)d\xi\right). \tag{1.9}$$

Hence it is always possible to find a function $h\in C_0^\infty(\mathbb{R}^3)$ such that the corresponding solution y develops a singularity not later than at time t_0 at the position x_0, where t_0 and x_0 are arbitrary. h only has to satisfy the following relation:

$$\frac{t_0}{4\pi}\int_{S^2}h(x_0+t_0\xi)d\xi=-1.$$

On the other hand we can find conditions on h such that y is defined globally. For this let

$$\begin{gathered} h(z)=\mathcal{O}\left(|z|^{-1}\right) \quad \text{as} \quad |z|\to\infty, \\ \|\nabla h\|_1<4\pi, \quad \|h\|_\infty<1. \end{gathered}$$

Then we have for $t>1$:

$$\begin{aligned} \left|t\int_{S^2}h(x+t\xi)d\xi\right| &= t\left|\int_{S^2}\int_t^\infty\frac{d}{ds}h(x+s\xi)dsd\xi\right| \\ &\le t\int_{S^2}\int_t^\infty\frac{s^2}{t^2}|(\nabla h)(x+s\xi)|dsd\xi \\ &\le \frac{1}{t}\|\nabla h\|_1 \\ &< \frac{4\pi}{t}. \end{aligned}$$

This implies

$$\left| \frac{t}{4\pi} \int_{S^2} h(x+t\xi)d\xi \right| < 1 \quad \text{for} \quad t > 1.$$

Analogously for $t < -1$. For $|t| \leq 1$ we have

$$\left| \frac{t}{4\pi} \int_{S^2} h(x+t\xi)d\xi \right| \leq |t| \; \|h\|_\infty < 1.$$

Therefore, $v(t,x) > 0$ for all $(t,x) \in \mathbb{R} \times \mathbb{R}^3$ and y is defined globally in (1.9).

Remark: The nonlinear wave equation (1.7) is a special case of the differential equation

$$y_{tt} - \Delta y = a|\nabla y|^2 + by_t^2, \quad a, b \in \mathbb{R} \quad \text{fixed}.$$

For this the following holds: Global solutions always exist for sufficiently small data if and only if $a + b = 0$. The if-part has just been shown by the example above (taking $v := e^{ay}$ in general). The only-if-part was proved by Hanouzet & Joly in [43]. In the case $a + b = 0$ the nonlinearity satisfies the so-called *null condition* which is a sufficient condition for quadratic nonlinearities in $\mathbb{R}^3$ to allow small, global solutions, see [90] and Chapter 9.

The second example is an example in one space dimension which shows that the solution itself and its first derivatives may stay bounded but that second derivatives may develop singularities in finite time. This is also a typical phenomenon observed for nonlinear wave equations.

Example 2: (cf. [65, pp. 649–650])

$$y_{tt} = (1 + y_x)^2 y_{xx} \tag{1.10}$$

or equivalently

$$y_{tt} - y_{xx} = 2y_x y_{xx} + y_x^2 y_{xx},$$

$$y(t=0) = H, \quad y_t(t=0) = -\tfrac{1}{2}(H')^2 - H', \tag{1.11}$$

where $y = y(t,x)$, $t \geq 0$, $x \in \mathbb{R}$, $y_{xx} = \frac{\partial^2}{\partial x^2} y$. H is a given function with $H \in C_0^\infty(\mathbb{R})$ and

$$h := \min_{x \in \mathbb{R}} H''(x) < 0$$

$\left(H''(x) = \frac{d^2}{dx^2} H(x)\right)$.

We construct a solution $y \in C^2\left([0, -1/h) \times \mathbb{R}\right)$ which becomes singular as $t \to -1/h$, more precisely:

$$y_{xx}(t_n, x_0) \to -\infty$$

for a sequence $(t_n)_n \subset [0, -1/h)$, $t_n \to -1/h$, and for some $x_0 \in \mathbb{R}$.

For this purpose let $\theta \in C^1\left([0,-1/h) \times \mathbb{R}\right)$ be implicitly defined by

$$\theta(t,x) = H'\left(x-(1+\theta(t,x))t\right).$$

This is possible by the implicit function theorem because for $t \in [0,-1/h)$ there holds

$$H''\left(x-(1+\theta)t\right)(-t) - 1 \neq 0$$

which implies

$$\frac{d}{d\theta}\left(H'(x-(1+\theta)t) - \theta\right) \neq 0.$$

Let y be defined by

$$y(t,x) := \frac{t}{2}\theta^2(t,x) + H\left(x-(1+\theta(t,x))t\right).$$

CLAIM: y solves (1.10), (1.11) (for $(t,x) \in [0,-1/h) \times \mathbb{R}$).
PROOF: $y(t=0) = H$ is obvious.

$$\begin{aligned} y_t(t,x) &= \frac{1}{2}\theta^2(t,x) + t\theta(t,x)\theta_t(t,x) + H'\left(x-(1+\theta(t,x))t\right)\left(-1-\theta(t,x)-t\theta_t(t,x)\right) \\ &= -\frac{1}{2}\theta^2(t,x) - \theta(t,x). \end{aligned}$$

This implies

$$y_t(t=0) = -\frac{1}{2}(H')^2 - H'$$

and

$$y_{tt}(t,x) = -\theta_t(t,x)\theta(t,x) - \theta_t(t,x) = -\theta_t(t,x)\left(1+\theta(t,x)\right).$$

Moreover

$$\begin{aligned} y_x(t,x) &= t\theta(t,x)\theta_x(t,x) + H'\left(x-(1+\theta(t,x))t\right)\left(1-t\theta_x(t,x)\right) \\ &= \theta(t,x) \end{aligned}$$

which implies

$$y_{xx}(t,x) = \theta_x(t,x).$$

On the other hand we have

$$\theta_x(t,x) = H''\left(x-(1+\theta(t,x))t\right)\left(1-t\theta_x(t,x)\right)$$

which yields

$$\theta_x(t,x) = \frac{H''\left(x-(1+\theta(t,x))t\right)}{1+tH''\left(x-(1+\theta(t,x))t\right)}.$$

Analogously we obtain

$$\theta_t(t,x) = \frac{-H''(x-(1+\theta(t,x))t)\left(1+\theta(t,x)\right)}{1+tH''\left(x-(1+\theta(t,x))t\right)}.$$

This implies

$$\theta_t(t,x) = -\,(1+\theta(t,x))\,\theta_x(t,x)$$

and finally

$$y_{tt}(t,x) = (1+\theta(t,x))^2\,\theta_x(t,x) = (1+y_x(t,x))^2\,y_{xx}(t,x).$$

Q.E.D.

CLAIM: There are sequences $(t_n)_n \subset [0,-1/h)$ with $\lim_{n\to\infty} t_n = -1/h$, and $(x_n)_n \subset \mathbb{R}$ with $\lim_{n\to\infty} x_n = x_0$ for some $x_0 \in \mathbb{R}$ such that

$$\lim_{n\to\infty} y_{xx}(t_n,x_n) = -\infty.$$

PROOF: Let $\xi \in \mathbb{R}$ with $H''(\xi) = h$, $(t_n)_n \subset [0,-\frac{1}{h})$, arbitrary with $\lim_{n\to\infty} t_n = -1/h$. Since θ is bounded we conclude

$$\forall n \in \mathbb{N} \quad \exists x_n \in \mathbb{R}: \quad x_n - (1+\theta(t_n,x_n))\,t_n = \xi.$$

Hence there is a subsequence which converges to some $x_0 \in \mathbb{R}$. We obtain

$$y_{xx}(t_n,x_n) = \theta_x(t_n,x_n) = \frac{H''(\xi)}{1+t_n H''(\xi)} \longrightarrow -\infty \quad \text{as } n\to\infty.$$

Q.E.D.

By the definition of y and the derived formulae for y_t and y_x it is obvious that y, y_t and y_x stay bounded in $[0,-1/h)\times\mathbb{R}$.

More examples will be given in Chapter 10.

2 L^p–L^q-decay estimates for the linear wave equation

For the proof of Theorem 1.1 simple decay properties of solutions to the linear wave equation play an important role (see Chapter 7). The decay rates of L^q-norms are typically of polynomial order in $\mathbb{R}^n$ depending on the space dimension n and on q.
We consider the solution of the linear initial value problem

$$y_{tt} - \Delta y = 0, \tag{2.1}$$

$$y(t=0) = 0, \quad y_t(t=0) = g, \tag{2.2}$$

where $y = y(t,x)$ is a real-valued function, $t \geq 0, x \in \mathbb{R}^n$ and g is assumed to be smooth for the moment.
Let the operator $w(t)$ be defined through

$$(w(t)g)(x) := y(t,x).$$

Remark: The assumption $y(t=0) = 0$ is made without loss of generality because the function y_1 defined by

$$y_1(t,x) := \partial_t(w(t)g)(x)$$

solves the initial value problem

$$\partial_t^2 y_1 - \Delta y_1 = 0$$

$$y_1(t=0) = g, \quad \partial_t y_1(t=0) = \partial_t^2(w(t)g)(t=0) = \Delta w(t=0)g = 0.$$

(Cf. the representation of solutions in Chapter 7 and the considerations in Section 11.5.)

Theorem 2.1 *$\exists c = c(n) > 0 \quad \forall g \in C_0^\infty \quad \forall t \geq 0$:*

(i) $\|Dw(t)g\|_2 \;=\; \|g\|_2,$

(ii) $\|Dw(t)g\|_\infty \;\leq\; c(1+t)^{-\frac{n-1}{2}}\|g\|_{n,1}.$

PROOF: Let $g \in C_0^\infty$. Then $y = w(\cdot)g \in C^\infty([0,\infty) \times \mathbb{R}^n)$ and $D^\alpha w \in C^0([0,\infty), L^2)$ for $\alpha \in \mathbb{N}_0^n$. (Cf. Chapter 3 or the book of R. Leis [98].) c will denote various positive constants at most depending on n.
Multiplying both sides of (2.1) with $y_t(t,\cdot)$ in L^2 (inner product denoted by $\langle\cdot,\cdot\rangle$) and dropping the parameter t, we obtain

$$\begin{aligned} 0 &= \langle y_{tt}, y_t\rangle + \langle \nabla y, \nabla y_t\rangle \\ &= \frac{1}{2}\frac{d}{dt}(\|y_t\|_2^2 + \|\nabla y\|_2^2) \\ &= \frac{1}{2}\frac{d}{dt}\|Dw(t)g\|_2^2. \end{aligned}$$

This proves (i).
(ii) will be proved here for $n = 1$ and $n = 3$ to give some main ideas. For odd space dimensions $n \geq 3$ or even space dimensions see Section 11.5 and the paper of W. von Wahl [187], respectively.
$n = 1$: The solution y is given by d'Alembert's formula:

$$y(t,x) := \frac{1}{2}\int_{x-t}^{x+t} g(r)dr$$

(*Jean Baptiste Le Rond d'Alembert,* 16.11.1717 – 29.10.1783).
We have

$$\begin{aligned} y_t(t,x) &= \frac{1}{2}(g(x+t) + g(x-t)), \\ y_x(t,x) &= \frac{1}{2}(g(x+t) - g(x-t)) \end{aligned}$$

whence it is obvious that y solves the initial value problem (2.1), (2.2). Moreover

$$\forall t \geq 0: \quad \|Dw(t)g\|_\infty \leq \|g\|_\infty \leq c\|g\|_{1,1}$$

by Sobolev's imbedding theorem. This proves (ii) for the case $n = 1$.
Now let $n = 3$: Kirchhoff's formula says that y defined by

$$y(t,x) := \frac{t}{4\pi}\int_{S^2} g(x+tz)dz, \tag{2.3}$$

is the solution, where $S^2 = \partial B(0,1)$ denotes the unit sphere in $\mathbb{R}^3$ *(Gustav Robert Kirchhoff,* 12.3.1824 – 17.10.1887). This is easily checked. From (2.3) we obtain

$$\begin{aligned} y(t=0) &= 0, \\ 4\pi y_t(t,x) &= \int_{S^2} g(x+tz)dz + t\int_{S^2} (\nabla g)(x+tz)z\,dz, \\ y_t(t=0) &= g. \end{aligned}$$

Moreover

$$4\pi\nabla y(t,x) = t\int_{S^2}(\nabla g)(x+tz)dz,$$

hence

$$\begin{aligned} 4\pi y_{tt}(t,x) &= 2\int_{S^2}(\nabla g)(x+tz)zdz + t\int_{S^2}\nabla\{(\nabla g)(x+tz)z\}zdz \\ &= 3t\int_{B(0,1)}(\Delta g)(x+tz)dz + t^2\int_{B(0,1)}(\nabla\Delta g)(x+tz)zdz, \end{aligned}$$

$$\begin{aligned} 4\pi\Delta y(t,x) &= t\int_{S^2}(\Delta g)(x+tz)dz = t\int_{S^2}\{(\Delta g)(x+tz)z\}zdz \\ &= t^2\int_{B(0,1)}(\nabla\Delta g)(x+tz)zdz + 3t\int_{B(0,1)}(\Delta g)(x+tz)dz. \end{aligned}$$

This implies

$$y_{tt} - \Delta y = 0.$$

Now we shall prove (ii).
First let $t \geq 1$:
1.

$$\begin{aligned} -\int_{S^2} g(x+tz)dz &= \int_{S^2}\int_t^\infty \frac{d}{ds}g(x+sz)dsdz = \int_{S^2}\int_t^\infty(\nabla g)(x+sz)zdsdz \\ &= \int_{S^2}\int_t^\infty \frac{s^2}{s^3}(\nabla g)(x+sz)szdsdz \\ &= \int_{|z|>t}|z|^{-3}(\nabla g)(x+z)zdz. \end{aligned}$$

This implies

$$|\int_{S^2} g(x+tz)dz| \leq t^{-2}\int_{|z|>t}|(\nabla g)(x+z)|\,dz \leq t^{-2}\|g\|_{1,1}.$$

2. Analogously one obtains

$$|t\int_{S^2}(\nabla g)(x+tz)zdz| \leq t^{-1}\|g\|_{2,1}$$

and

$$|t\int_{S^2}\nabla g(x+tz)dz| \leq t^{-1}\|g\|_{2,1}.$$

Hence we get for $t \geq 1$:

$$\|Dw(t)g\|_\infty \leq (4\pi\,t)^{-1}\|g\|_{2,1}. \tag{2.4}$$

3. Now let $0 \leq t < 1$:

$$\begin{aligned} -\int_{S^2} g(x+tz)dz &= \int_{S^2}\int_t^\infty \frac{d}{ds}g(x+sz)dsdz \\ &= -\int_{S^2}\int_t^\infty (s-t)\frac{d^2}{ds^2}g(x+sz)dsdz \\ &= \int_{S^2}\int_t^\infty \frac{(s-t)^2}{2}\frac{d^3}{ds^3}g(x+sz)dsdz \\ &= \int_{|z|>t}\frac{(|z|-t)^2}{2|z|^5}\sum_{i,j,k=1}^3 z_iz_jz_k(\partial_i\partial_j\partial_k g)(x+z)dz. \end{aligned}$$

This implies

$$|\int_{S^2} g(x+tz)dz| \leq \sum_{i,j,k=1} \int_{|z|>t} |\partial_i\partial_j\partial_k g(x+z)|dz \leq \|g\|_{3,1}.$$

Analogously for the terms discussed in 2. Thus we have obtained for $0 \leq t < 1$:

$$\|Dw(t)g\|_\infty \leq c\|g\|_{3,1}. \tag{2.5}$$

(2.4) and (2.5) prove (ii).

Q.E.D.

Remarks: For $g \in W^{n,1}$ there is still a distributional solution y to the initial value problem (2.1), (2.2). Since $W^{n,1}$ is continuously imbedded into L^2 we have

$$y \in C^0([0,\infty), W^{1,2}) \cap C^1([0,\infty), L^2)$$

(see e.g. [98]).
Moreover one can define a trace on $\partial\Omega$ for $g \in W^{1,p}(\Omega), 1 \leq p < \infty$, where Ω is a smoothly bounded domain in $\mathbb{R}^n$ (Lipschitz boundary is sufficient); namely, there is a continuous map B,

$$B : W^{1,p}(\Omega) \longrightarrow L^p(\partial\Omega)$$

with

$$Bg = g_{/\partial\Omega} \quad \text{if} \quad g \in W^{1,p}(\Omega) \cap C^0(\overline{\Omega})$$

(see e.g. the book of H.W. Alt [6]), (*Rudolf Otto Sigismund Lipschitz*, 14.5.1832 – 7.10.1903). Therefore Kirchhoff's formula (2.3) makes sense for $g \in W^{3,1} \hookrightarrow W^{1,2}$ ($\hookrightarrow$ denotes the continuous imbedding).
Thus we obtain the corresponding results for $g \in W^{n,1}$ by approximation with $(g_k)_k \subset C_0^\infty$, expressed in the following theorem.

Theorem 2.2 $\exists c = c(n) > 0 \quad \forall g \in W^{n,1} \quad \forall t \geq 0$:

$$\begin{aligned} (i) \quad \|Dw(t)g\|_2 &= \|g\|_2, \\ (ii) \quad \|Dw(t)g\|_\infty &\leq c(1+t)^{-\frac{n-1}{2}} \|g\|_{n,1}. \end{aligned}$$

In other words, the operator T_t, defined by

$$T_t g := Dw(t)g$$

maps as follows:

$$T_t : W^{n,1} \longrightarrow L^\infty \quad \text{with norm} \quad M_0 \leq c(1+t)^{-\frac{n-1}{2}},$$

$$T_t : L^2 \longrightarrow L^2 \quad \text{with norm} \quad M_1 = 1.$$

By interpolation we obtain

$$T_t : [W^{n,1}, L^2]_\theta \longrightarrow [L^\infty, L^2]_\theta, \quad 0 \le \theta \le 1,$$

$$\text{with norm} \quad M_\theta \le c M_0^{1-\theta} M_1^\theta, \quad c = c(\theta, n).$$

The interpolation spaces $[\cdot,\cdot]_\theta$ are described in Appendix A. We have

$$1 \le q_0, q_1 \le \infty \quad \Rightarrow \quad [L^{q_0}, L^{q_1}]_\theta = L^{q_\theta}, \tag{2.6}$$

where q_θ is defined by the relation

$$\frac{1}{q_\theta} = \frac{1-\theta}{q_0} + \frac{\theta}{q_1}.$$

In particular we get

$$[L^\infty, L^2]_\theta = L^{q_\theta} \quad \text{with} \quad q_\theta = \frac{2}{\theta}.$$

The proof of (2.6) is not very difficult after having given an appropriate meaning to $[\cdot,\cdot]_\theta$. This is possible for example in general Banach spaces. The proof uses the Three-Line-Theorem of J. Hadamard (*Jacques Hadamard,* 8.12.1865 – 17.10.1963). The interpolation of $W^{n,1}$ and L^2 is much more difficult. For this purpose Besov spaces and Bessel potential spaces are used (*Friedrich Wilhelm Bessel,* 22.7.1784 – 17.3.1846). We refer the reader to Appendix A for a survey and to the books of Bergh & Löfström [11] and H. Triebel [181] for details. One special result suitable for our purposes is:

$$W^{N,p_\theta} \hookrightarrow [W^{n,1}, L^2]_\theta \quad \text{if} \quad N > (1-\theta)n,$$

where

$$\frac{1}{p_\theta} + \frac{1}{q_\theta} = 1$$

defines p_θ (see Theorem A.10 in Appendix A).

Remark: For $\theta \in \{0,1\}$ we may allow $N = (1-\theta)n$.

Thus we obtain the following theorem on the L^p–L^q-decay of solutions to the linear wave equation.

Theorem 2.3 *Let* $2 \le q \le \infty$, $1/p + 1/q = 1$, $N_p > n(1-2/q)$. *Then*

$$\exists c = c(q,n) > 0 \quad \forall g \in W^{N_p,p} \quad \forall t \ge 0: \quad \|Dw(t)g\|_q \le c(1+t)^{-\frac{n-1}{2}(1-\frac{2}{q})}\|g\|_{N_p,p}.$$

Remarks: $N_p = n(1-2/q)$ is possible if $q \in \{2,\infty\}$.

Since

$$N_p \cdot p > n(1-2/q)p = n(2-p)$$

we have

$$W^{N_p,p} \hookrightarrow L^2$$

and hence

$$Dw(\cdot)g \in C^0([0,\infty), L^2).$$

If $(g_m)_m \subset C_0^\infty$ converges to g in $W^{N_p,p}$, then $(Dw(t)g_m)_m$ converges in L^2 to $Dw(t)g$. Several sharper results for solutions to linear wave equations are contained in Section 11.5 and in the paper of W. v. Wahl [187] respectively. Another method of proving L^p–L^q-decay estimates (at least for $q < \infty$) is to use the Fourier representation of the solution (*Jean-Baptiste-Joseph Fourier*, 21.3.1768 – 16.5.1830). This has been carried out by H. Pecher in [138] and the result is essentially expressed in Lemma 11.16 in Section 11.7.

3 Linear symmetric hyperbolic systems

Let $u = u(t,x) = (u_1, \ldots, u_N)(t,x)$, $t \geq 0, x \in \mathbb{R}^n, N \in \mathbb{N}$, and let the formal linear differential operator L be defined by

$$Lu := A^0(t,x)\partial_t u + \sum_{j=1}^{n} A^j(t,x)\partial_j u + B(t,x)u. \tag{3.1}$$

Here $A^0, A^1, \ldots, A^n$ and B are complex $N \times N$-matrices depending on t and x. A^j, for $0 \leq j \leq n$, is assumed to be hermitian and A^0 is assumed to be positive definite, uniformly with respect to t and to x (*Charles Hermite,* 24.12.1822 – 14.1.1901).
With these assumptions L is a symmetric hyperbolic differential operator and the (formal) system of equations

$$Lu = f, \tag{3.2}$$

$$u(t=0) = u_0 \tag{3.3}$$

is a symmetric hyperbolic system with data

$$f = f(t,x) \quad \text{and} \quad u_0 = u_0(x).$$

Every scalar hyperbolic equation of second order can be transformed into a symmetric hyperbolic system. Let

$$\partial_t^2 v = \sum_{i,j=1}^{n} a_{ij}(t,x)\partial_i\partial_j v + \sum_{i=1}^{n} b_i(t,x)\partial_i v + c(t,x)\partial_t v + d(t,x)v,$$

where all functions are real-valued and $(a_{ij}(t,x))_{ij}$ is a symmetric positive definite $n \times n$-matrix, uniformly with respect to t and x. (We do not care about differentiability questions for the moment.)
Let

$$u_1 := \partial_1 v, \ldots, u_n := \partial_n v, u_{n+1} := \partial_t v, u_{n+2} := v.$$

Then we obtain the following system of differential equations for the $N := n+2$ functions $u_1, \ldots, u_{n+2}$:

$$\sum_{j=1}^{n} a_{ij}(t,x)\partial_t u_j - \sum_{j=1}^{n} a_{ij}(t,x)\partial_j u_{n+1} = 0, \quad i = 1, \ldots, n, \tag{3.4}$$

$$\partial_t u_{n+1} - \sum_{i,j=1}^{n} a_{ij}(t,x)\partial_j u_i - \sum_{i=1}^{n} b_i(t,x)u_i - c(t,x)u_{n+1} - d(t,x)u_{n+2} = 0, \tag{3.5}$$

$$\partial_t u_{n+2} - u_{n+1} = 0. \tag{3.6}$$

(3.4) – (3.6) are equivalent to a symmetric hyperbolic system

$$Lu = 0$$

of the type (3.1) with

$$A^0 := \begin{pmatrix} a_{11} & \cdots & a_{1n} & 0 & 0 \\ \vdots & & \vdots & \vdots & \vdots \\ a_{n1} & \cdots & a_{nn} & 0 & 0 \\ 0 & \cdots & 0 & 1 & 0 \\ 0 & \cdots & 0 & 0 & 1 \end{pmatrix}, \quad B := \begin{pmatrix} 0 & \cdots & 0 & 0 & 0 \\ \vdots & & \vdots & \vdots & \vdots \\ 0 & \cdots & 0 & 0 & 0 \\ -b_1 & \cdots & -b_n & -c & -d \\ 0 & \cdots & 0 & -1 & 0 \end{pmatrix},$$

$$A^j := \begin{pmatrix} 0 & \cdots & 0 & -a_{1j} & 0 \\ \vdots & & \vdots & \vdots & \vdots \\ 0 & \cdots & 0 & -a_{nj} & 0 \\ -a_{1j} & \cdots & -a_{nj} & 0 & 0 \\ 0 & \cdots & 0 & 0 & 0 \end{pmatrix}, \quad j = 1, \ldots, n.$$

We shall prove in this section an existence theorem for the system (3.2), (3.3). This will be done first for analytic data, then an approximation will be carried out. For this purpose a priori (energy) estimates are required.

3.1 Energy estimates

We assume

$$A^0, A^1, \ldots, A^n \in C_b^1, B \in C_b^0,$$

where C_b^k denotes the space of k-times continuously differentiable functions with bounded derivatives up to order k, $k \in \mathbb{N}_0 \cup \{\infty\}$.
Let

$$\begin{aligned} a_0 &:= \min_{v,t,x;|v|=1} A^0(t,x)v \cdot \overline{v} > 0, \\ a_1 &:= \max_{v,t,x,j;|v|=1} |A^j(t,x)v \cdot \overline{v}| > 0, \end{aligned}$$

$$\rho := \frac{a_0}{na_1}. \tag{3.7}$$

Let $K^\rho \equiv K^\rho(t_0)$ be the truncated cone

$$K^\rho := \{(t,x) \mid x \in K_t, 0 \leq t \leq t_0\}, \tag{3.8}$$

where

$$K_s := B\left(0, \frac{T_0 - s}{\rho}\right) \subset \mathbb{R}^n, \quad 0 \le s \le T_0,$$

and t_0, T_0 are arbitrary but fixed, satisfying

$$0 < t_0 \le T_0.$$

The boundary ∂K^ρ of K^ρ consists of three parts:
bottom: $\{0\} \times K_0$; top: $\{t_0\} \times K_{t_0}$; lateral surface: M; see Figure 3.1.

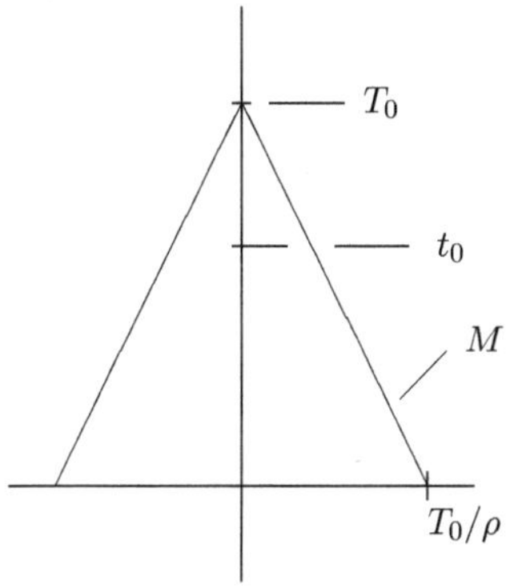

Figure 3.1: Characteristic cone

The cone K^ρ, or more precisely, ρ has been chosen in a way such that for integrals of the type:

$$\int_{K^\rho} Lu \cdot \overline{u}$$

the terms

$$\int_M \cdots,$$

arising through partial integration, have an appropriate sign (cf. (3.11) below). In this situation M is called *space-like* for L.
We introduce the following notation:

$$|u(t)|_{K_t} := \left(\int_{K_t} A^0(t,x)u(t,x) \cdot \overline{u(t,x)}dx\right)^{1/2}.$$

With the assumptions made for A^0 we have

$$\exists a_0' > 0 \quad \forall t \in [0, T_0] \; : \; a_0 \|u(t)\|^2_{L^2(K_t)} \le |u(t)|^2_{K_t} \le a_0' \|u(t)\|^2_{L^2(K_t)}.$$

(We write $u(t)$ short for $u(t, \cdot)$ in most places.) The first basic energy estimate is given in the next theorem.

Theorem 3.1 *Let $K^\rho = K^\rho(t_0)$ and let $u \in C^1(K^\rho(T_0))$ be a solution to*

$$Lu = f \in C^0(K^\rho(T_0)),$$

$$u(t=0) = u_0 \in C^0(K_0).$$

Then

$$\exists c = c(\|(A^0, \partial_t A^0, \partial_1 A^1, \dots, \partial_n A^n, B)\|_{C^0(K^\rho(T_0))}) > 0 \quad \forall t \in [0, T_0] :$$

$$|u(t)|_{K_t} \leq c\{|u_0|_{K_0} + (\int_0^{T_0} |f(r)|^2_{K_r} dr)^{1/2}\}e^{ct}.$$

Proof:

$$\text{Re } Lu \cdot \overline{u} = \text{Re}\,(A^0\partial_t u \cdot \overline{u} + \sum_{j=1}^n A^j\partial_j u \cdot \overline{u} + Bu \cdot \overline{u}) = \text{Re } f \cdot \overline{u}.$$

This implies

$$\begin{aligned} \text{Re}\{\frac{1}{2}\partial_t(A^0 u \cdot \overline{u}) &- \frac{1}{2}(\partial_t A^0)u \cdot \overline{u} + \frac{1}{2}\sum_{j=1}^n \partial_j(A^j u \cdot \overline{u}) - \frac{1}{2}\sum_{j=1}^n (\partial_j A^j)u \cdot \overline{u} \\ &+ Bu \cdot \overline{u}\} = \text{Re}\, f \cdot \overline{u} \end{aligned}$$

or

$$D' \begin{pmatrix} A^0 u \cdot \overline{u} \\ A^1 u \cdot \overline{u} \\ \vdots \\ A^n u \cdot \overline{u} \end{pmatrix} = \text{Re}\,\{\partial_t A^0 + \sum_{j=1}^n \partial_j A^j - 2B\}u \cdot \overline{u} + 2 \text{ Re } f \cdot \overline{u}.$$

Let

$$H := \partial_t A^0 + \sum_{j=1}^n \partial_j A^j - 2B.$$

Then we obtain by integration over K^ρ,

$$\int_{\partial K^\rho} (\nu_t A^0 u \cdot \overline{u} + \sum_{j=1}^n \nu_j A^j u \cdot \overline{u}) = \int_{K^\rho} (\text{Re } Hu \cdot \overline{u} + 2 \text{ Re } f \cdot \overline{u}), \tag{3.9}$$

where

$$(\nu_t, \nu_1, \nu_2, \dots, \nu_n) =: \nu$$

denotes the exterior normal vector on ∂K^ρ.

We have

$$\begin{aligned} \nu &= (-1, 0, 0, \dots, 0) \quad \text{on} \quad \{0\} \times K_0, \\ \nu &= (+1, 0, 0, \dots, 0) \quad \text{on} \quad \{t_0\} \times K_{t_0} \quad (0 < t_0 < T_0). \end{aligned}$$

The lateral surface M of K^ρ can be parametrized in the following way:

$$M = \{(t,x) \quad | \quad t = \gamma(x) := T_0 - \rho|x|,\, x \in K_0,\, 0 \le t \le t_0\}.$$

Then the normal vector ν is given by

$$\nu = \frac{1}{\sqrt{1+|\nabla\gamma|^2}}(1, -\nabla\gamma) = \frac{1}{\sqrt{1+\rho^2}}(1, \frac{\rho x}{|x|}).$$

Thus, (3.9) turns into

$$\int\limits_{K_{t_0}} A^0(t_0,x)u(t_0,x)\cdot\overline{u(t_0,x)}dx - \int\limits_{K_0} A^0(0,x)u_0(x)\cdot\overline{u_0(x)}dx \tag{3.10}$$

$$+\frac{1}{\sqrt{1+\rho^2}}\int\limits_M (A^0 u\cdot\overline{u} - \sum_{j=1}^n(\partial_j\gamma)A^j u\cdot\overline{u}) = \int\limits_{K^\rho}(\text{Re } Hu\cdot\overline{u} + 2\text{ Re } f\cdot\overline{u}).$$

By the definition of γ and ρ (see (3.7)) we obtain

$$\begin{aligned} |\sum_{j=1}^n(\partial_j\gamma)A^j u\cdot\overline{u}| &\le \sum_{j=1}^n|\partial_j\gamma|a_1 u\cdot\overline{u} \le |\nabla\gamma|n\, a_1 u\cdot\overline{u} \\ &= a_0 u\cdot\overline{u} \le A^0 u\cdot\overline{u}. \end{aligned} \tag{3.11}$$

Hence we see that the integrand in $\int\limits_M\cdots$ is pointwise nonnegative.
Using (3.11) we obtain from (3.10)

$$\begin{aligned} |u(t_0)|^2_{K_{t_0}} &\le |u_0|^2_{K_0} + \int_0^{t_0}\int\limits_{K_r}|Hu\cdot\overline{u}|(r,x)dx\,dr + 2\int_0^{t_0}\int\limits_{K_r}(|f||u|)(r,x)dx\,dr \\ &\le |u_0|^2_{K_0} + c\int_0^{t_0}|u(r)|^2_{K_r}dr + c\int_0^{t_0}|f(r)|_{K_r}|u(r)|_{K_r}dr, \end{aligned}$$

where

$$c = c\left(\|(A^0, \partial_t A^0, \partial_1 A^1, \dots, \partial_n A^n, B)\|_{C^0(K^\rho(T_0))}\right).$$

The assertion of Theorem 3.1 now follows from an application of Gronwall's inequality, Lemma 4.1.

Q.E.D.

We shall now prove a corollary with the corresponding estimates for higher derivatives. Let

$$|u(t)|_{s,K_t} := \left(\sum_{|\alpha|\le s}|\nabla^\alpha u(t)|^2_{K_t}\right)^{1/2}, \quad s\in\mathbb{N}_0.$$

Corollary 3.2 *Let $s \in \mathbb{N}, A^0, A^1, \ldots, A^n, B \in C_b^s$ and let $u \in C^{s+1}(K^\rho(T_0))$ be a solution to*

$$Lu = f \in C^s(K^\rho(T_0)),$$

$$u(t=0) = u_0 \in C^s(K_0).$$

Then

$$\exists\, c = c(\|(A^0, A^1, \ldots, A^n, B)\|_{C^s(K^\rho(T_0))}) > 0 \quad \forall t \in [0, T_0]:$$

$$|u(t)|_{s,K_t} \leq c\,\{|u_0|_{s,K_0} + (\int_0^{T_0} |f(r)|^2_{s,K_r} dr)^{1/2}\} e^{ct}.$$

PROOF: Differentiating the equation for u:

$$A^0 \partial_t u + \sum_{j=1}^n A^j \partial_j u + Bu = f, \tag{3.12}$$

with respect to $x_k, k = 1, \ldots, n$, we obtain

$$\begin{aligned} A^0 \partial_t \partial_k u \;&+\; \sum_{j=1}^n A^j \partial_j \partial_k u + B\, \partial_k u \\ &+\; (\partial_k A^0)\partial_t u + \sum_{j=1}^n (\partial_k A^j)\partial_j u \;+\; (\partial_k B)u = \partial_k f, \quad k = 1, \ldots, n. \end{aligned}$$

Using (3.12) we may express $\partial_t u$ in terms of $\partial_1 u, \ldots, \partial_n u$ and u. So we get a differential equation for $V := (u, \partial_1 u, \ldots, \partial_n u)$ of the following type:

$$\mathcal{A}_1^0(t,x)\partial_t V + \sum_{j=1}^n \mathcal{A}_1^j(t,x)\partial_j V + \mathcal{B}_1 V = \mathcal{F}_1, \tag{3.13}$$

with initial value

$$V(t=0) = (u_0, \partial_1 u_0, \ldots, \partial_n u_0), \tag{3.14}$$

where

$$\mathcal{A}_1^0 := \begin{pmatrix} A^0 & & 0 \\ & \ddots & \\ 0 & & A^0 \end{pmatrix}, \quad \mathcal{A}_1^j := \begin{pmatrix} A^j & & 0 \\ & \ddots & \\ 0 & & A^j \end{pmatrix}, \quad j = 1, \ldots, n,$$

$\mathcal{B}_1$ is a matrix composed of $B, \partial_k A^0, \partial_k A^j$ and $\partial_k B, j, k = 1, ..., n$ the detailed structure of which does not matter, and

$$\mathcal{F}_1 = \mathcal{F}_1(f, \partial_1 f, \ldots, \partial_n f).$$

An application of Theorem 3.1 to (3.13), (3.14) yields the assertion of Corollary 3.2 for $s = 1$. Analogously one obtains the assertion for $s > 1$.

Q.E.D.

Remark: Using the differential equation (3.12) we also obtain estimates for

$$|\partial_t^k u(t)|_{K_t},\ 1 \le k \le s.$$

We notice that Theorem 3.1 implies properties of propagation of signals. Let $f = 0$. Then we may conclude the finite propagation speed. Theorem 3.1 says that the solution u at time t only depends on values of the initial datum u_0 in K_0. In particular, if u_0 has compact support then $u(t)$ has compact support (with respect to x) for each $t > 0$ ("finite propagation speed"); see Figure 3.2.

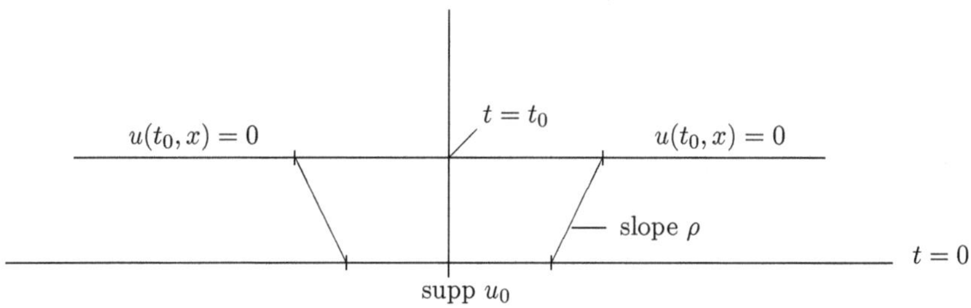

Figure 3.2: Finite propagation speed

This phenomenon is typical for hyperbolic problems. In contrast to this we see that the parabolic initial value problem

$$\begin{aligned} w_t - \Delta w &= 0, \\ w(t=0) &= w_0, \end{aligned}$$

for a real-valued function $w = w(t,x), t \ge 0, x \in \mathbb{R}^n$, is solved for $w_0 \in C_0^\infty$ (for simplicity) by

$$w(t,x) := (4\pi t)^{-n/2} \int\limits_{\mathbb{R}^n} \mathrm{e}^{-\frac{|x-y|^2}{4t}} w_0(y)dy.$$

This shows that we have infinite propagation speed (cf. Section 11.2).

3.2 A global existence theorem

Let L be given as in the introduction to this chapter.

Theorem 3.3 *Let $s \in \mathbb{N}, s > n/2+1$, let $A^0, A^1, \ldots, A^n, B \in C_b^{s+1}$ $[C_b^\infty]$ and let $u_0 \in W^{s,2}$ $[\cap C^\infty]$. Then there exists a unique solution $u \in C^1([0,\infty) \times \mathbb{R}^n)$ $[C^\infty([0,\infty) \times \mathbb{R}^n)]$ to the initial value problem $Lu = 0, u(t=0) = u_0$.*
Moreover

$$u \in C^0([0,\infty), W^{s,2}) \cap C^1([0,\infty), W^{s-1,2}).$$

Proof: The uniqueness immediately follows from the energy estimate in Theorem 3.1 applied to the difference of two solutions.
The existence will be proved in four steps. For this purpose let ρ be defined by (3.7), and let $0 < \beta < \rho$. If M^β denotes the lateral surface of K^β, K^β defined via (3.8), then M^β is space-like for all L with coefficients $\tilde{A}^0, \tilde{A}^1, \ldots, \tilde{A}^n$ which are close to $A^0, A^1, \ldots, A^n$ respectively (in K^β).
Step 1: We assume that the coefficients of L, denoted by $\tilde{A}^0, \tilde{A}^1, \ldots, \tilde{A}^n, \tilde{B}$, are analytic. Correspondingly we shall write $\tilde{L}$ instead of L. For a fixed but arbitrary $T_0 > 0$ we approximate u_0 in K_0 by a sequence of polynomials $(u_0^k)_k$ in $W^{s,2}(K_0)$. M^β is assumed to be space-like for $\tilde{L}$. By the theorem of Cauchy–Kowalevsky (Theorem B.1 in Appendix B) there is a local solution u^m to the initial value problem $\tilde{L}u^m = 0$, $u^m(t=0) = u_0^m$ in a truncated cone $K(\delta)$ with

$$K(\delta) := K^\beta \cap \{(t,x) \,|\, t \le \delta\}$$

for some δ with $0 < \delta \le T_0$. (*Sophie von Kowalevsky*, 15.1.1850 – 10.2.1891; *Augustin Louis Cauchy*, 21.8.1789 – 23.5.1857).
We have

$$\begin{aligned} \delta &= \delta((\tilde{A}^0, \tilde{A}^1, \ldots, \tilde{A}^n, \tilde{B})_{/K^\beta}, T_0/\beta) \\ &= \delta((A^0, A^1, \ldots, A^n, B)_{/K^\beta}, T_0/\beta) \end{aligned}$$

if $\tilde{A}^0, \tilde{A}^1, \ldots, \tilde{A}^n, \tilde{B}$ are approximations of $A^0, A^1, \ldots, A^n, B$ respectively.
Using Corollary 3.2 we obtain for $k, m \in \mathbb{N}$:

$$\|u^k(t) - u^m(t)\|_{W^{s,2}(K_t)} \le c\|u_0^k - u_0^m\|_{W^{s,2}(K_0)},$$

where c only depends on T_0. Hence $(u^m(t))_m$ converges in $W^{s,2}(K_t), 0 \le t \le \delta$. Let $u(t)$ be the limit in $W^{s,2}(K_t)$.
The convergence of $(u^m(t))_m$ is uniform with respect to t. This implies

$$u \in C^0([0,\delta], W^{s,2}(K_\delta))$$

and

$$u \in C^0([0,\delta] \times K_\delta)$$

because $s > n/2$.
The following identity holds in $W^{s-1,2}(K_\delta)$:

$$u^m(t) = u_0^m + \int_0^t \partial_t u^m(r)dr, \quad 0 \le t \le \delta. \tag{3.15}$$

Since

$$\partial_t u^m(t) = \tilde{A}_0^{-1}(-\sum_{j=1}^n \tilde{A}^j \partial_j u^m(t) - \tilde{B}u^m(t))$$

we see that $(\partial_t u^m(t))_m$ converges to some $v(t) \in W^{s-1,2}(K_\delta)$, again uniformly with respect to t, i.e.

$$v \in C^0([0,\delta], W^{s-1,2}(K_\delta)).$$

Using this information in (3.15) we get

$$u(t) = u_0 + \int_0^t v(r)dr$$

which implies

$$u \in C^0([0,\delta], W^{s,2}(K_\delta)) \cap C^1([0,\delta], W^{s-1,2}(K_\delta)),$$

in particular

$$u \in C^1([0,\delta] \times K_\delta)$$

and u satisfies the system of differential equations $\tilde{L}u = 0$ with initial value $u(t=0) = u_0$ (in $[0,\delta] \times K_\delta$). Considering δ' with $0 < \delta' < \delta$ instead of δ we finally obtain

$$\tilde{L}u = 0, u(t=0) = u_0 \quad \text{in} \quad K(\delta).$$

Now we may consider a new initial value problem in $t = \delta$ and we obtain an extension of u into $K(\delta + \delta_1)$ where

$$\delta_1 = \delta_1((A^0, A^1, \dots, A^n, B)_{/K^\beta}, T_0/\beta) = \delta.$$

($\delta = \delta_1$ does not depend on the specific polynomial approximation, cf. Appendix B.) Thus we obtain a solution in K^β successively.

Step 2: Let $A^0, A^1, \dots, A^n, B \in C_b^{s+1}$ as in the assumption above, but let $u_0 \in W^{s+1,2}$. We approximate $A^0, A^1, \dots, A^n, B$ by analytic functions $(A_k^0)_k, (A_k^1)_k, \dots, (A_k^n)_k, (B_k)_k$ respectively, uniformly with respect to all the derivatives up to order $s+1$ in K^β and such that M^β is space-like for each operator L_k, where

$$L_k := A_k^0 \partial_t + \sum_{j=1}^n A_k^j \partial_j + B_k.$$

The problem

$$L_k u^k = 0, \quad u^k(t=0) = u_0,$$

can be solved according to Step 1. We have

$$u^k \in C^0([0,T], W^{s+1,2}(K_T)) \cap C^1([0,T], W^{s,2}(K_T)),$$

where

$$0 < T \le t_0 < T_0.$$

Corollary 3.2 implies

$$\|u^k(t)\|_{W^{s+1,2}(K_t)} \le c\|u_0\|_{s+1,2}, \quad 0 \le t \le T_0, \tag{3.16}$$

where c depends only on T_0.
A difference $u^k - u^j$ satisfies

$$L_k(u^k - u^j) = -L_k u^j = L_j u^j - L_k u^j = (L_j - L_k)u^j =: f_{kj},$$

$$(u^k - u^j)(t=0) = 0.$$

Applying again Corollary 3.2 we obtain

$$\begin{aligned} \|u^k(t) - u^j(t)\|^2_{W^{s,2}(K_t)} &\le c\int_0^t \|f_{kj}(r)\|^2_{W^{s,2}(K_r)} dr \\ &\le c\,\varepsilon_{kj}\int_0^t \|u^j(r)\|^2_{W^{s+1,2}(K_r)} dr, \end{aligned}$$

where c depends only on T_0 and $0 \le \varepsilon_{kj} \to 0$ as $k, j \to \infty$.
Observing (3.16) we conclude that $(u^k(t))_k$ converges in $W^{s,2}(K_t)$ to some $u(t) \in W^{s,2}(K_t)$.
With the same arguments as in Step 1 we obtain

$$Lu = 0, \quad u(t=0) = u_0, \quad \text{in} \quad K^\beta,$$

$$u \in C^0([0,T], W^{s,2}(K_T)) \cap C^1([0,T], W^{s-1,2}(K_T)), \quad 0 < T \le t_0. \tag{3.17}$$

Moreover

$$\|u(t)\|_{W^{s,2}(K_t)} \le c\|u_0\|_{W^{s,2}(K_0)} \le c\|u_0\|_{s,2}, \quad 0 \le t \le t_0, \tag{3.18}$$

where c depends only on T_0.
Step 3: Let $A^0, A^1, \ldots, A^n, B \in C_b^{s+1}$ and $u_0 \in W^{s,2}$. There is a sequence $(u_0^k)_k \subset W^{s+1,2}$ approximating u_0 in $W^{s,2}$.
According to Step 2 there are solutions u^k of $Lu^k = 0, u^k(t=0) = u_0^k, k \in \mathbb{N}$, in K^β and they satisfy (cf. (3.18))

$$\sup_{0 \le t \le t_0} \|u^k(t) - u^j(t)\|_{W^{s,2}(K_t)} \le c\|u_0^k - u_0^j\|_{W^{s,2}}.$$

This implies the existence of a solution u satisfying (3.17) in K^β.
[Under the additional assumption $A^0, A^1, \ldots, A^n, B \in C_b^\infty, u_0 \in C^\infty$, we obtain $u \in C^\infty(K^\beta)$ (consider e.g. $W^{s',2}(K_T), s' \in \mathbb{N}$ arbitrary).]
Step 4: The coefficients are assumed to be uniformly bounded. Therefore, for every truncated cone which is congruent to K^β and which is obtained by translation at any place in $[0,\infty) \times \mathbb{R}^n$ we can find a local solution. Using the uniqueness properties for overlapping cones we can construct this way a global solution

$$u \in C^1([0,\infty) \times \mathbb{R}^n) \quad [\text{resp. } u \in C^\infty([0,\infty) \times \mathbb{R}^n)].$$

It remains to prove

$$u \in C^0([0,\infty), W^{s,2}) \cap C^1([0,\infty), W^{s-1,2}).$$

This easily follows by a triangle inequality, e.g. let $\varepsilon > 0, t_1, t_2 \in [0,\infty)$ be given, $t_1, t_2 \leq T$ for some $T \geq 0$. Then it follows for $R > 0$:

$$\begin{aligned} \|u(t_1) - u(t_2)\|_{s,2} &\leq \|u(t_1) - u(t_2)\|_{W^{s,2}(B(0,R))} + \|u(t_1) - u(t_2)\|_{W^{s,2}(\mathbb{R}^n \setminus B(0,R))} \\ &\equiv I_1 + I_2. \end{aligned}$$

We have

$$\|u(t_j)\|_{W^{s,2}(\mathbb{R}^n \setminus B(0,R))} \leq c\|u_0\|_{W^{s,2}(\mathbb{R}^n \setminus B(0,R_1))}, \quad j = 1,2, \tag{3.19}$$

where c depends only on T and

$$R_1 = R_1(\beta, R) < R$$

with

$$R_1 \to \infty \quad \text{as} \quad R \to \infty.$$

This is a consequence of Corollary 3.2. Now (3.19) implies that $I_2 < \varepsilon/2$ if $R \geq R_0(\varepsilon)$ independent of t_1, t_2. If R is fixed then $I_1 < \varepsilon/2$ for sufficiently small $|t_1 - t_2|$ according to the local continuity properties of u we already know. This completes the proof of Theorem 3.3.

Q.E.D.

3.3 Remarks on other methods

(References: F. John [71], Courant & Hilbert [23], K.O. Friedrichs [34, 35], T. Kato [79, 80, 81, 82, 83]).
The method used above is called the method of Schauder (*Pawel Juliusz Schauder*, 21.9.1899 – September 1943). (Cf. also [157].)

1. A "weak" solution $u \in L^2([0,T] \times \mathbb{R}^n), T > 0$ arbitrary but fixed, is easily obtained with the Riesz representation theorem in a suitable Hilbert space (*Friedrich Riesz*, 22.1.1880 – 28.6.1956; *David Hilbert*, 23.1.1862 – 14.2.1943).
Without loss of generality we assume u_0 to be zero and we wish to solve the inhomogeneous system

$$Lu = f \quad \text{in} \quad R_T := [0,T] \times \mathbb{R}^n.$$

Let

$$\tilde{C}^1(R_T) := \{v \in C^1(R_T) \mid v(T) = 0, \operatorname{supp} v(t) \subset\subset \mathbb{R}^n, 0 \le t \le T\}.$$

($\subset\subset$ denotes compactly supported in.)
We have

$$\begin{aligned} u \in C^1(R_T) \quad \text{is a solution of} \quad Lu \;&=\; f \in C^0(R_T), u(t=0) = 0 \\ \iff \quad \forall v \in \tilde{C}^1(R_T) \quad &: \quad \langle f, v\rangle_{L^2(R_T)} = \langle u, \tilde{L}v\rangle_{L^2(R_T)}. \end{aligned} \tag{3.20}$$

Here $\tilde{L}$ denotes the formal adjoint operator to L which appears through partial integration. An inner product in $\tilde{C}^1(R_T)$ is defined by

$$\langle v, w\rangle_{\mathcal{H}} := \langle \tilde{L}v, \tilde{L}w\rangle_{L^2(R_T)}.$$

The positive definiteness is a consequence of

$$\exists c = c(T) > 0 \quad \forall v \in \tilde{C}^1(R_T) : \|v\|_{L^2(R_T)} \le c\, \|\tilde{L}v\|_{L^2(R_T)}. \tag{3.21}$$

(3.21) follows from Theorem 3.1 applied to $\tilde{L}$ instead of L.
Let $\mathcal{H}$ denote the completion of $\tilde{C}^1(R_T)$ with respect to the norm induced by $\langle \cdot, \cdot\rangle_{\mathcal{H}}$. The relation (3.21) implies that F defined by

$$F : \mathcal{H} \longrightarrow \mathbb{C},$$

$$v \mapsto Fv := \langle v, f\rangle_{L^2(R_T)},$$

is a continuous, linear function and thus we get by the Riesz representation theorem that there exists a $u_1 \in \mathcal{H}$ with the property

$$\begin{aligned} \forall v \in \mathcal{H} : \langle v, f\rangle_{L^2(R_T)} \;&=\; \langle v, u_1\rangle_{\mathcal{H}} = \langle \tilde{L}v, \tilde{L}u_1\rangle_{L^2(R_T)} \\ &=\; \langle \tilde{L}v, u\rangle_{L^2(R_T)}, \quad \text{where} \quad u := \tilde{L}u_1. \end{aligned}$$

Looking at (3.20) we call u a weak solution to $Lu = f, u(t=0) = 0$.
The difficulty now consists in proving regularity of u (for regular f), catchword "weak" = "strong", see the papers of K.O. Friedrichs, [34, 35] or the related papers of T. Kasuga [77] and Meyers & Serrin [125].

2. For f with compact support in x (for each t) we also mention the difference method. There, derivatives are replaced by quotients of differences, then discrete "energy estimates" are proved and the limit "mesh width $\to 0$" is studied (see for example [71]).
3. Another important approach is that of T. Kato using the theory of semigroups and evolution operators respectively, see [79, 80, 81, 82, 83]. Formally we have

$$L = \partial_t + A(t)$$

and

$$u(t) = \mathrm{e}^{-\int_0^t A(r)dr} u_0$$

represents the solution of $Lu = 0, u(t = 0) = u_0$. In order to make this precise in Banach spaces or Hilbert spaces, an enormous technical set up is required. On the other hand it provides a general abstract theory and — with respect to our application — detailed results on the existence and the regularity of the solution u under weaker assumptions on the coefficients. Of course this would also have consequences for corresponding existence theorems to nonlinear hyperbolic systems, cf. the remarks at the end of Chapter 5. Moreover, this approach works for parabolic and other problems, cf. Appendix C.

4 Some inequalities

We start with Gronwall's inequality (*Thomas Hakon Gronwall* (orig.: *Hakon Tomi Grönwall)*, 16.1.1877 – 9.5.1932):

Lemma 4.1 *Let* $a > 0$, $\varphi, h \in C^0([0,a])$, $h \geq 0$, *and* $g : [0,a] \longrightarrow \mathbb{R}$ *increasing.*
If

$$\forall t \in [0,a]: \quad \varphi(t) \leq g(t) + \int_0^t h(r)\varphi(r)dr$$

then

$$\forall t \in [0,a]: \quad \varphi(t) \leq g(t)\exp\{\int_0^t h(r)dr\}.$$

Proof: Let $\varepsilon > 0$ and ψ_ε given by

$$\psi_\varepsilon(t) := \exp\{\int_0^t h(r)dr\}\left(\int_0^t g'(r)\exp\{-\int_0^r h(s)ds\}dr + g(0) + \varepsilon\right).$$

Then ψ_ε solves $\psi_\varepsilon' = g' + h\,\psi_\varepsilon$ (a.e.), $\psi_\varepsilon(0) = g(0) + \varepsilon$, and hence

$$\psi_\varepsilon(t) = \varepsilon + g(t) + \int_0^t h(r)\psi_\varepsilon(r)dr.$$

We have

$$\varphi(0) \leq g(0) < g(0) + \varepsilon = \psi_\varepsilon(0).$$

We prove that $\varphi(t) < \psi_\varepsilon(t)$ for all $t \in [0,a]$. Namely, let $t_0 \in (0,a]$ be the first point with $\varphi(t_0) = \psi_\varepsilon(t_0)$, in particular $\varphi(t) < \psi_\varepsilon(t)$ for $0 \leq t < t_0$.
Then

$$\begin{aligned}\varphi(t_0) &\leq g(t_0) + \int_0^{t_0} h(r)\varphi(r)dr < \varepsilon + g(t_0) + \int_0^{t_0} h(r)\varphi(r)dr \\ &\leq \varepsilon + g(t_0) + \int_0^{t_0} h(r)\psi_\varepsilon(r)dr = \psi_\varepsilon(t_0),\end{aligned}$$

which is a contradiction.
Hence the inequalities

$$\varphi < \psi_\varepsilon$$

and

$$\psi_\varepsilon(t) \leq \exp\{\int_0^t h(r)dr\}(g(t) + \varepsilon)$$

complete the proof by letting ε tend to zero.

Q.E.D.

The proofs of the inequalities for composite functions below require some facts about the Friedrichs mollifiers which we present first (see [1]) (*Kurt Otto Friedrichs*, 28.9.1901 – 31.12.1982).

Let $j : \mathbb{R}^n \longrightarrow \mathbb{R}$ be given by

$$j(x) := k \begin{cases} e^{-1/(1-|x|^2)} & \text{if } |x| < 1, \\ 0 & \text{if } |x| \geq 1, \end{cases}$$

where k is chosen in a way that

$$\int_{\mathbb{R}^n} j(x)dx = 1$$

holds.

For a given $\varepsilon > 0$ the Friedrichs mollifier j_ε is defined by

$$j_\varepsilon(x) := \varepsilon^{-n} j(x/\varepsilon), \quad x \in \mathbb{R}^n$$

and J_ε denotes the corresponding convolution operator

$$(J_\varepsilon u)(x) := \int_{\mathbb{R}^n} j_\varepsilon(x-y)u(y)dy \equiv (j_\varepsilon * u)(x)$$

for $u \in L^1_{loc}(\mathbb{R}^n)$.

Remark: Instead of the special j from above one may take any $j \in C_0^\infty$ with $j(x) = 0$ if $|x| \geq 1$, $j \geq 0$ and $\int_{\mathbb{R}^n} j(x)dx = 1$.

Lemma 4.2 *Let $1 \leq p < \infty$ and let $u \in L^p$. Then*

(*i*) $J_\varepsilon u \in C^\infty$,

(*ii*) $||J_\varepsilon u||_p \leq ||u||_p$,

(*iii*) $\lim_{\varepsilon \downarrow 0} ||J_\varepsilon u - u||_p = 0$,

(*iv*) $\forall m \in \mathbb{N}_0 \; \forall q \geq p : \; J_\varepsilon u \in W^{m,q}$.

Proof: Since $j_\varepsilon \in C_0^\infty$ we have

$$\nabla^\alpha (J_\varepsilon u)(x) = \int_{\mathbb{R}^n} \nabla_x^\alpha j_\varepsilon(x-y)u(y)dy \tag{4.1}$$

for every multi-index α. Thus (i) is obvious.

Let $1 < p < \infty$, $p' := p/(p-1)$. Then by Hölder's inequality (*Ludwig Otto Hölder*, 22.12.1859 – 29.8.1937)

$$\begin{aligned} |J_\varepsilon u(x)| &= |\int_{\mathbf{R}^n} j_\varepsilon(x-y)u(y)dy| \\ &\leq \{\int_{\mathbf{R}^n} j_\varepsilon(x-y)dy\}^{1/p'}\{\int_{\mathbf{R}^n} j_\varepsilon(x-y)|u(y)|^p dy\}^{1/p} \\ &= \{\int_{\mathbf{R}^n} j_\varepsilon(x-y)|u(y)|^p dy\}^{1/p}. \end{aligned}$$

Hence

$$\begin{aligned} \|J_\varepsilon u\|_p^p &\leq \int_{\mathbf{R}^n}\int_{\mathbf{R}^n} j_\varepsilon(x-y)|u(y)|^p dy dx \\ &= \int_{\mathbf{R}^n} |u(y)|^p dy \int_{\mathbf{R}^n} j_\varepsilon(x-y)dx = \|u\|_p^p. \end{aligned}$$

This proves (ii) for $1 < p < \infty$; the case $p = 1$ follows right from the definition of $J_\varepsilon u$. Let $\eta > 0$ be given. Since C_0^∞ is dense in L^p, there is a $\varphi \in C_0^\infty$ with

$$\|u - \varphi\|_p < \eta/3 \tag{4.2}$$

which implies

$$\|J_\varepsilon u - J_\varepsilon \varphi\|_p < \eta/3 \tag{4.3}$$

by (ii).

$$\begin{aligned} |J_\varepsilon\varphi(x) - \varphi(x)| &= |\int_{\mathbf{R}^n} j_\varepsilon(x-y)(\varphi(y) - \varphi(x))dy| \\ &\leq \sup_{|y-x|<\varepsilon} |\varphi(y) - \varphi(x)|. \end{aligned}$$

Since φ is uniformly continuous, the last term tends to zero as $\varepsilon \downarrow 0$. Hence we have

$$\|J_\varepsilon \varphi - \varphi\|_p < \eta/3 \tag{4.4}$$

for sufficiently small ε, and for these ε we obtain from (4.2) – (4.4)

$$\|J_\varepsilon u - u\|_p < \eta.$$

This proves (iii).
(iv) follows from (4.1) and the convolution inequality

$$\|\nabla^\alpha j_\varepsilon * u\|_q \leq \|\nabla^\alpha j_\varepsilon\|_r \|u\|_p \tag{4.5}$$

where $r \in [1, \infty]$ is given by
$$\frac{1}{r} = 1 + \frac{1}{q} - \frac{1}{p}$$
(see [46, p. 117]).

Q.E.D.

A possible rate of convergence in (iii) from Lemma 4.2 is obtained by making a stronger assumption on u.

Lemma 4.3

$$\forall \varepsilon_0 > 0\ \exists c > 0\ \forall \varepsilon \in (0, \varepsilon_0]\ \forall u \in W^{1,2} : \|J_\varepsilon u - u\|_2 \le c\varepsilon \|u\|_{1,2}.$$

PROOF: First let $\varphi \in C_0^\infty$. Then

$$\begin{aligned} |\varphi(x) - J_\varepsilon \varphi(x)| &= \left| \int_{\mathbf{R}^n} j_\varepsilon(x-y)(\varphi(x) - \varphi(y))dy \right| \\ &\le \int_{\mathbf{R}^n} j_\varepsilon(x-y) \int_0^1 |\nabla\varphi|(x + s(y-x))ds|x-y|dy \\ &\le \varepsilon \int_0^1 \int_{\mathbf{R}^n} j_\varepsilon(x-y)|\nabla\varphi|(x+s(y-x))dyds \\ &= \varepsilon \int_0^1 \int_{\mathbf{R}^n} j_{s\varepsilon}(z)|\nabla\varphi|(x-z)dzds \\ &= \varepsilon \int_0^1 (J_{s\varepsilon}|\nabla\varphi|)(x)ds. \end{aligned}$$

This implies
$$\|J_\varepsilon\varphi - \varphi\|_2^2 \le \varepsilon^2 \int_0^1 \|J_{s\varepsilon}|\nabla\varphi|\,\|_2^2\, ds \le \varepsilon^2 \|\nabla\varphi\|_2^2, \tag{4.6}$$
where we have used Lemma 4.2, (ii).

Now let $u \in W^{1,2}$. Without loss of generality let $\|u\|_{1,2} = 1$. For a given $\varepsilon > 0$ there is a $\varphi \in C_0^\infty$ with
$$\|u - \varphi\|_{1,2} < \varepsilon,$$
hence
$$\|\varphi\|_{1,2} < 1 + \varepsilon.$$
Thus, we obtain from (4.6), using Lemma 4.2, (ii), that
$$\|J_\varepsilon u - u\|_2 \le \|J_\varepsilon(u-\varphi)\|_2 + \|u - \varphi\|_2 + \|J_\varepsilon\varphi - \varphi\|_2 \le \varepsilon(3+\varepsilon) \le c\varepsilon$$
with $c := 3 + \varepsilon_0$.

Q.E.D.

In the sequel we shall prove some inequalities (of Sobolev type) for composite functions. First we present an interpolation inequality due to E. Gagliardo and L. Nirenberg. This inequality holds under more general assumptions, namely in domains $\Omega \neq \mathbb{R}^n$, see [36, 136] or the book of A. Friedman [33] to which we also refer for a proof for bounded domains.

Notation: $\|\nabla^i w\|_\tau := (\sum\limits_{|\alpha|=i} \|\nabla^\alpha w\|_\tau^\tau)^{1/\tau}$, $i \in \mathbb{N}_0$, $1 \le \tau < \infty$ ($\tau = \infty$ as usual).

Theorem 4.4 *Let $1 \le r,\, p \le \infty$, $m \in \mathbb{N}$. Then there is a constant $c > 0$ such that for all $w \in W^{m,p} \cap L^r$ the inequality*

$$\|\nabla^j w\|_q \le c\|\nabla^m w\|_p^{j/m} \|w\|_r^{1-j/m} \tag{4.7}$$

holds, where $j \in \{0, 1, \ldots, m\}$ and

$$\frac{1}{q} = \frac{j}{m}\frac{1}{p} + (1 - \frac{j}{m})\frac{1}{r}.$$

The theorem will be proved in detail in a series of estimates first for the mollified function $J_\varepsilon w$. This is done succesively for the case $n = 1$ and $m = 2$, then for $m = 2$ and n arbitrary and finally for n and m arbitrary by induction on m. The last step will be to let ε tend to zero in the estimate for $J_\varepsilon w$.

Case A: Let $\varepsilon > 0$ be fixed and let $v := J_\varepsilon w$. We shall prove (4.7) for v.

Lemma 4.5 *Let $-\infty < a < b < \infty$, $f \in C^2([a,b])$, $1 \le p,\, r \le \infty$, $2/q = 1/r + 1/p$, and $q \neq \infty$. Then*

$$\|f'\|_{L^q((a,b))}^q \le 18^q\{(b-a)^{1+q-q/p}\|f''\|_{L^p((a,b))}^q + (b-a)^{-(1+q-q/p)}\|f\|_{L^r((a,b))}^q\}.$$

PROOF:

(i) Assume $p, r \neq \infty$.

By the mean value theorem we have

$$\forall \psi, \eta \in (0, \frac{b-a}{3})\ \exists \lambda \in (a+\psi, b-\eta) : |f'(\lambda)| \le \frac{3}{b-a}\{|f(b-\eta)| + |f(a+\psi)|\}.$$

This implies for $x \in [a,b]$ that

$$|f'(x)| \le \frac{3}{b-a}\{|f(b-\eta)| + |f(a+\psi)|\} + \int_a^b |f''(t)|dt.$$

Integration over $[0, \frac{b-a}{3}]^2$ with respect to η and ψ yields

$$\begin{aligned} \frac{(b-a)^2}{9}|f'(x)| &\le \int_a^b |f(t)|dt + \frac{(b-a)^2}{9}\int_a^b |f''(t)|dt \\ &\le \|f\|_{L^r((a,b))}(b-a)^{\frac{r-1}{r}} + \frac{(b-a)^{2+\frac{p-1}{p}}}{9}\|f''\|_{L^p((a,b))}. \end{aligned}$$

Hence

$$|f'(x)|^q \leq 2^{q-1}\{9^q(b-a)^{q(\frac{r-1}{r}-2)}\,\|f\|^q_{L^r((a,b))} + (b-a)^{q\frac{p-1}{p}}\,\|f''\|^q_{L^p((a,b))}\}$$

and because of $1-q-q/r = -(1+q-q/p)$:

$$\|f'\|^q_{L^q((a,b))} \leq 18^q\{(b-a)^{-1-q+q/p}\|f\|^q_{L^r((a,b))} + (b-a)^{1+q-q/p}\|f''\|^q_{L^p((a,b))}\}.$$

(ii) $p=\infty$ or $r=\infty$. The proof is analogous to the above.

Q.E.D.

Lemma 4.6 *Let* $1 \leq p, r \leq \infty$, $f \in C^2(\mathbb{R}) \cap L^r(\mathbb{R})$, $f'' \in L^p(\mathbb{R})$, $2/q = 1/p + 1/r$. *Then we have* $f' \in L^q(\mathbb{R})$ *and there is a constant* $c = c(q) > 0$ *such that*

$$\|f'\|_q \leq c\|f''\|_p^{1/2}\|f\|_r^{1/2}.$$

PROOF: Without loss of generality we assume $\|f''\|_p = 1$. The proof shall be divided into two cases, $q=\infty$ and $q \neq \infty$.

(1) $\underline{q \neq \infty}$.

Lemma 4.5 implies for $-\infty < a < b < \infty$:

$$\|f'\|^q_{L^q((a,b))} \leq 18^q\{T_1(a,b) + T_2(a,b)\} \tag{4.8}$$

where

$$\begin{aligned} T_1(a,b) &:= (b-a)^{1+q-q/p}\|f''\|^q_{L^p((a,b))}, \\ T_2(a,b) &:= (b-a)^{-(1+q-q/p)}\|f\|^q_{L^r((a,b))}. \end{aligned}$$

$T_1 = T_2$ would imply $T_1 + T_2 = 2\sqrt{T_1}\sqrt{T_2}$ and the assertion would follow immediately. Now, for any given interval $[-L, L]$, $L > 0$, we can find a covering such that on each subinterval $[a_{i-1}, a_i]$ the inequality

$$T_1(a_{i-1}, a_i) \geq T_2(a_{i-1}, a_i)$$

holds. This is done as follows. Let $a_0 := -L$, $a_1' := -L + \frac{2L}{k}$, $k \in \mathbb{N}$ arbitrary but fixed.

If

$$T_1(a_0, a_1') > T_2(a_0, a_1')$$

then set

$$a_1 := a_1'.$$

Then (4.8) implies

$$\|f'\|^q_{L^q((a_0,a_1))} \le 2 \cdot 18^q \left(\frac{2L}{k}\right)^{1+q-q/p} \|f''\|^q_{L^p((a_0,a_1))}. \tag{4.9}$$

If

$$T_1(a_0, a_1') \le T_2(a_0, a_1')$$

then we choose $a_1 \ge a_1'$ sufficiently large such that

$$T_1(a_0, a_1) = T_2(a_0, a_1).$$

This implies

$$\|f'\|^q_{L^q((a_0,a_1))} \le 2 \cdot 18^q \|f''\|^{q/2}_{L^p((a_0,a_1))} \|f\|^{q/2}_{L^r((a_0,a_1))}. \tag{4.10}$$

Proceeding in the same way for $i = 2, \ldots, k'$, $k' \le k$, with a_{i-1} replacing a_0 and $a_i' := a_{i-1} + \frac{2L}{k}$ replacing a_1, as long as $a_{i-1} < L$ holds, we obtain from (4.9), (4.10)

$$\|f'\|^q_{L^q((a_0,a_{k'}))} \le 2 \cdot 18^q \sum_{i=1}^{k'} \|f''\|^{q/2}_{L^p((a_{i-1},a_i))} \|f\|^{q/2}_{L^r((a_{i-1},a_i))} + R, \tag{4.11}$$

where $a_{k'} \ge L$ and

$$R := 2 \cdot 18^q \left(\frac{2L}{k}\right)^{1+q-q/p} \sum_{i=1}^{k'} \|f''\|^q_{L^p((a_{i-1},a_i))}.$$

(i) <u>Claim:</u> $\sum_{i=1}^{k'} \|f''\|^{q/2}_{L^p((a_{i-1},a_i))} \|f\|^{q/2}_{L^r((a_{i-1},a_i))} \le \|f''\|_p^{q/2} \|f\|_r^{q/2}$.

PROOF:

(α) <u>$r \ne \infty, p \ne \infty$:</u>

The relation

$$2p/q = 1 + p/r > 1$$

implies

$$\sum_{i=1}^{k'} \|f''\|^{q/2}_{L^p((a_{i-1},a_i))} \|f\|^{q/2}_{L^r((a_{i-1},a_i))}$$
$$\le \{\sum_{i=1}^{k'} \int_{a_{i-1}}^{a_i} |f''(x)|^p dx\}^{q/(2p)} \{\sum_{i=1}^{k'} \int_{a_{i-1}}^{a_i} |f(x)|^r dx\}^{q/(2r)}$$
$$\le \|f''\|_p^{q/2} \|f\|_r^{q/2}.$$

(β) <u>$r = \infty$ or $p = \infty$.</u> Analogously.

(Q.E.D.)

(ii) <u>Claim:</u> $R \to 0$ as $k \to \infty$.

PROOF: Let $c_0 := 2 \cdot 18^q (2L)^{1+q-q/p}$.

(α) $\underline{p = \infty}$:

$$R = c_0 k^{-(1+q)} \sum_{i=1}^{k'} \|f''\|^q_{L^\infty((a_{i-1},a_i))} \leq c_0 \|f''\|^q_\infty k' k^{-(1+q)}$$

$$\to 0 \quad \text{as} \quad k \to \infty.$$

(β) $\underline{p \neq \infty}$.

(a) $\underline{p/q > 1}$:

Hölder's inequality yields

$$\begin{aligned} R &\leq c_0 k^{-(1+q-q/p)} \|f''\|^q_p \cdot (k')^{1-q/p} \\ &\leq c_0 k^{-q} \|f''\|^q_p \\ &\to 0 \quad \text{as} \quad k \to \infty. \end{aligned}$$

(b) $\underline{p/q \leq 1}$:

Since

$$\int_{a_{i-1}}^{a_i} |f''(t)|^p dt \leq \|f''\|^p_p = 1,$$

we get

$$\left(\int_{a_{i-1}}^{a_i} |f''(t)|^p dt\right)^{q/p} \leq \int_{a_{i-1}}^{a_i} |f''(t)|^p dt \leq 1$$

which implies

$$R \leq c_0 k^{-(1+q-q/p)}$$

$$\to 0 \quad \text{as} \quad k \to \infty.$$

(Q.E.D.)

(4.11), (i) and (ii) imply

$$\|f'\|^q_{L^q((-L,L))} \leq 2 \cdot 18^q \|f''\|^{q/2}_p \|f\|^{q/2}_r,$$

which yields the assertion of Lemma 4.6 (for $q \neq \infty$) by letting L tend to infinity.

(2) $\underline{q = \infty}$ (i.e. $p = r = \infty$).

Applying Lemma 4.5 with $r = \infty$, $p' \neq \infty$, $-\infty < a < b < \infty$, we obtain

$$\begin{aligned} \|f'\|^{q'}_{L^{q'}((a,b))} &\leq 18^{q'}\{(b-a)^{1+q'-q'/p'}\|f''\|^{q'}_{L^{p'}((a,b))} \\ &\quad +(b-a)^{-(1+q'-q'/p')}\|f\|^{q'}_{L^\infty((a,b))}\}, \end{aligned}$$

where

$$q' := 2p' > 1.$$

This implies

$$\begin{aligned} \|f'\|_{L^{q'}((a,b))} &\leq 18\{(b-a)^{1/q'+1-1/p'}\|f''\|_{L^{p'}((a,b))} \\ &\quad +(b-a)^{-(1/q'+1-1/p')}\|f\|_{L^\infty((a,b))}\}. \end{aligned}$$

In the limit $q' \to \infty, p' = q'/2 \to \infty$ we get

$$\|f'\|_{L^\infty((a,b))} \leq 18\{(b-a)\|f''\|_{L^\infty((a,b))} + (b-a)^{-1}\|f\|_{L^\infty((a,b))}\}.$$

Since

$$\forall x \in \mathbb{R} \quad \exists a, b \in \mathbb{R} : x \in (a,b),\ b-a = (\|f\|_\infty/\|f''\|_\infty)^{1/2}$$

(without loss of generality: $\|f''\|_\infty \neq 0$)
we obtain

$$\|f'\|_\infty \leq 36\|f\|^{1/2}_\infty\|f''\|^{1/2}_\infty,$$

which completes the proof of Lemma 4.6.

Q.E.D.

Now we shall prove Theorem 4.4 for $v = J_\varepsilon w$ by induction on m. Without loss of generality we assume $m \geq 2$ and $j \in \{1, \ldots, m-1\}$.

(1) Basis of the induction ($m = 2$):

(i) $r, p \neq \infty$:

$$\|\nabla v\|^q_q = \sum_{i=1}^n \int_{\mathbb{R}^{n-1}} \int_{\mathbb{R}} |\partial_i v(x)|^q \, dx_i \, d\hat{x}_i$$

with

$$\hat{x}_i := (x_1, \ldots, x_{i-1}, x_{i+1} \ldots, x_n).$$

This implies by Lemma 4.6 (with constants $c, c_i > 0$)

$$\begin{aligned} \|\nabla v\|^q_q &\leq \sum_{i=1}^n \int_{\mathbb{R}^{n-1}} c_i\{\int_{\mathbb{R}} |\partial_i^2 v(x)|^p dx_i\}^{q/(2p)}\{\int_{\mathbb{R}} |v(x)|^r dx_i\}^{q/(2r)} d\hat{x}_i \\ &\leq c\sum_{i=1}^n \|\partial_i^2 v\|^{q/2}_p \|v\|^{q/2}_r \\ &\leq c\|\nabla^2 v\|^{q/2}_p \|v\|^{q/2}_r. \end{aligned}$$

(ii) $\underline{r = \infty, p \neq \infty}$ or $\underline{r \neq \infty, p = \infty}$:

Analogously to (i), observe $q = 2p$ and $q = 2r$ respectively.

(iii) $\underline{r = p = \infty}$: Analogously.

(2) Induction step $(m \to m+1)$:

(i) Claim:

$$\|\nabla^m v\|_{q'} \leq c\|\nabla^{m+1}v\|_p^{m/(m+1)}\|v\|_r^{1-m/(m+1)}, \tag{4.12}$$

where

$$\frac{1}{q'} = \frac{m}{m+1}\frac{1}{p} + \frac{1}{m+1}\frac{1}{r}.$$

PROOF: We notice that either

$$(\alpha) \quad p \leq q' \leq r \quad \text{or} \quad (\beta) \quad r \leq q' \leq p \tag{4.13}$$

holds.

Let r' be given by

$$\frac{2}{q'} = \frac{1}{p} + \frac{1}{r'}.$$

The induction hypothesis yields

$$\|\nabla^{m-1}v\|_{r'} \leq c\|\nabla^m v\|_{p'}^{(m-1)/m}\|v\|_r^{1/m}, \tag{4.14}$$

where

$$\frac{1}{r'} = \frac{m-1}{m}\frac{1}{p'} + \frac{1}{m}\frac{1}{r}.$$

Notice that we may apply the induction hypothesis because $q' = p'$ by the definition of q' and p', and because $v \in W^{m,q'}$. The latter follows from (4.13) which yields

$$(\alpha) \quad p \leq q' \,(\leq r' \leq r) \quad \text{or} \quad (\beta) \quad r \leq r' \leq q' \,(\leq p).$$

Since

$$\forall k \in \mathbb{N}_0 \quad \forall \tau \geq \min(p, r): \quad v \in W^{k,\tau}$$

this implies

$$v \in W^{m,q'}$$

(by Lemma 4.2, (iv)).

Remark: This is the part of the proof in which we cannot argue directly with w instead of $v = J_\varepsilon w$.

Applying part (1) ("case $m = 2$") to $\nabla^\alpha w, |\alpha| = m-1$, and using (4.14) yields

$$\begin{aligned}\|\nabla^m v\|_{q'} &\leq c\|\nabla^{m+1}v\|_p^{1/2}\|\nabla^{m-1}v\|_{r'}^{1/2} \\ &\leq c\|\nabla^{m+1}v\|_p^{1/2}\|\nabla^m v\|_{p'}^{(m-1)/(2m)}\|v\|_r^{1/(2m)}.\end{aligned}$$

Since $q' = p'$ we arrive at

$$\|\nabla^m v\|_{q'}^{1-(m-1)/(2m)} \leq c\|\nabla^{m+1} v\|_p^{1/2}\|v\|_r^{1/(2m)}$$

which gives the assertion (4.12).

(Q.E.D.)

(ii) <u>Claim</u>:

$$\|\nabla^j v\|_q \leq c\|\nabla^{m+1} v\|_p^{j/(m+1)}\|v\|_r^{1-j/(m+1)}$$

for any $j \in \{1, \ldots, m\}$,

$$\frac{1}{q} = \frac{j}{m+1}\frac{1}{p} + (1 - \frac{j}{m+1})\frac{1}{r}.$$

PROOF: The induction hypothesis yields

$$\|\nabla^j v\|_q \leq c\|\nabla^m v\|_{p'}^{j/m}\|v\|_r^{1-j/m},$$

where

$$\frac{1}{q} = \frac{j}{m}\frac{1}{p'} + (1 - \frac{j}{m})\frac{1}{r}. \tag{4.15}$$

The inequality (4.12) with $q' = p'$ yields

$$\begin{aligned}\|\nabla^j v\|_{p'} &\leq c\|\nabla^{m+1} v\|_p^{j/(m+1)}\|v\|_r^{j/m-j/(m+1)+1-j/m} \\ &= c\|\nabla^{m+1} v\|_p^{j/(m+1)}\|v\|_r^{1-j/(m+1)},\end{aligned}$$

where

$$\frac{1}{p'} = \frac{m}{m+1}\frac{1}{p} + \frac{1}{m+1}\frac{1}{r}. \tag{4.16}$$

The relations (4.15), (4.16) imply

$$\begin{aligned}\frac{1}{q} &= \frac{j}{m}(\frac{m}{m+1}\frac{1}{p} + \frac{1}{m+1}\frac{1}{r}) + (1 - \frac{j}{m})\frac{1}{r} \\ &= \frac{j}{m+1}\frac{1}{p} + (1 - \frac{j}{m+1})\frac{1}{r}\end{aligned}$$

as desired in Theorem 4.4. This completes the induction step and the proof of Theorem 4.4 for $v = J_\varepsilon w$.

(Q.E.D.)

In order to complete the proof of Theorem 4.4 we finally have to consider

<u>Case B</u>: Assumptions as in Theorem 4.4.

For $\varepsilon > 0$ let $w_\varepsilon := J_\varepsilon w$. According to the previous discussion in case A we know

$$\|\nabla^j w_\varepsilon\|_q \leq c\|\nabla^m w_\varepsilon\|_p^{j/m}\|w_\varepsilon\|_r^{1-j/m},$$

where c is independent of ε. Without loss of generality: $m \geq 2, j \in \{1, \ldots, m-1\}$.
Notice: $q = \infty \iff p = r = \infty$.

(1) $\underline{p, r \neq \infty:}$

By Lemma 4.2, (iii) we have

$$w_\varepsilon \to w \quad \text{in} \quad L^r,$$

$$\nabla^m w_\varepsilon \to \nabla^m w \quad \text{in} \quad L^p$$

and hence $\{\nabla^j w_\varepsilon\}_\varepsilon$ converges in L^q, necessarily

$$\nabla^j w_\varepsilon \to \nabla^j w \quad \text{in} \quad L^q.$$

Letting $\varepsilon \downarrow 0$ we obtain the desired inequality (4.7) for w.

(2) $\underline{p \neq \infty, r = \infty \quad \text{or} \quad p = \infty, r \neq \infty:}$

Since

$$\|w_\varepsilon\|_\infty \leq \|w\|_\infty, \quad \|\nabla^m w_\varepsilon\|_\infty \leq \|\nabla^m w\|_\infty \tag{4.17}$$

and

$$\nabla^m w_\varepsilon \to \nabla^m w \quad \text{in} \quad L^p \quad (\text{resp. } w_\varepsilon \to w \text{ in } L^r)$$

we may argue in the same manner as in (1); the inequality still holds in the limit as $\varepsilon \downarrow 0$.

(3) $\underline{q = p = r = \infty:}$

We have by (4.17)

$$\|\nabla^j w_\varepsilon\|_\infty \leq c \|\nabla^m w\|_\infty^{j/m} \|w\|_\infty^{1-j/m}.$$

This means that $(\nabla^\alpha w_{\varepsilon_l})_{\varepsilon_l = \frac{1}{l}}$, $|\alpha| = j$, is a bounded sequence in L^∞. Since balls in L^∞ are weak* sequentially compact (see e.g. [6, p. 140]), there is a subsequence $(\nabla^\alpha w_{\varepsilon_l'})_{\varepsilon_l'}$ and a $\tilde{w} \in L^\infty$ such that

$$w_{\varepsilon_l'} \to \tilde{w} \quad \text{weak-}* \quad \text{in } L^\infty \quad \text{as} \quad l \to \infty.$$

Moreover

$$\|\tilde{w}\|_\infty \leq \liminf_{l \to \infty} \|\nabla^j w_{\varepsilon_l'}\|_\infty \leq c \|\nabla^m w\|_\infty^{j/m} \|w\|_\infty^{1-j/m}.$$

(see e.g. [6, p. 139] for the first inequality.) It remains to prove

$$\tilde{w} = \nabla^\alpha w$$

which is an easy consequence of the following identities:

$$\forall \varphi \in C_0^\infty : \int_{\mathbf{R}^n} \nabla^\alpha w \varphi \quad = \quad (-1)^j \int_{\mathbf{R}^n} w \nabla^\alpha \varphi \tag{4.18}$$

$$= (-1)^j \lim_{l\to\infty} \int_{\mathbb{R}^n} w_{\varepsilon_l'} \nabla^\alpha \varphi = \lim_{l\to\infty} \int \nabla^\alpha w_{\varepsilon_l'} \varphi$$

$$= \int_{\mathbb{R}^n} \tilde{w}\varphi.$$

(The second identity holds because $\varphi \in C_0^\infty$ and $w_{\varepsilon_l'} \to w$ in $L^2(\operatorname{supp}\varphi)$.) This completes the proof of the Gagliardo–Nirenberg inequality, Theorem 4.4.

Q.E.D.

Remark: The proof of Theorem 4.4 shows that the assumption "$w \in W^{m,p}$" can be replaced by "$w \in L^p$ and $\nabla^m w \in L^p$".

The following first inequality on composite functions is given for smooth functions.

Lemma 4.7 *Let $r, m, n \in \mathbb{N}, 1 \le p \le \infty, h \in C^r(\mathbb{R}^m), B := \|h\|_{C^r(\overline{B(0,1)})}$. Then there is a constant $c = c(r,m,n,p) > 0$ such that for all $w = (w_1, \ldots, w_m) \in W^{r,p}(\mathbb{R}^n) \cap C^r(\mathbb{R}^n)$ with $\|w\|_\infty \le 1$ the inequality*

$$\|\nabla^r h(w)\|_p \le cB\|\nabla^r w\|_p \tag{4.19}$$

holds.

Remark: This cannot be extended to the case $r = 0$ as the example $h \equiv 1$ shows.

PROOF: Without loss of generality we assume $m = 1$. Let $\beta \in \mathbb{N}_0^n$, $|\beta| = r$. Then

$$\nabla^\beta(h(w)) = \sum_{k=1}^r \frac{\partial^k h(w)}{\partial w^k} \sum_{\alpha(k,r)} C_{k\alpha} \prod_{i=1}^r \{ \sum_{\gamma\in\mathbb{N}_0^n, |\gamma|=i} (\nabla^\gamma w)^{\alpha_i} \},$$

where $\sum_{\alpha(k,r)}$ means summation over all $\alpha \in \mathbb{N}_0^r$ with $|\alpha| = k$ and $\sum_{i=1}^r i\alpha_i = r$.

{Example: $r = 4$, $\beta = (4, 0, \ldots, 0)$:

$$\begin{aligned} \partial_1^4(h(w)) &= \frac{\partial^4 h(w)}{\partial w^4} (\partial_1 w)^4 + \frac{\partial^3 h(w)}{\partial w^3} 6\,(\partial_1 w)^2 \partial_1^2 w \\ &+ \frac{\partial^2 h(w)}{\partial w^2} \{3\,(\partial_1^2 w)^2 + 4\,(\partial_1 w)\partial_1^3 w\} + \frac{\partial h(w)}{\partial w}\,\partial_1^4 w. \end{aligned}$$

The only coefficients $C_{k\alpha}$ which are different from zero are the following:

$$\begin{aligned} C_{4(4,0,0,0)} &= 1,\ C_{3(2,1,0,0)} = 6,\ C_{2(0,2,0,0)} = 3, \\ C_{2(1,0,1,0)} &= 4,\ C_{1(0,0,0,1)} = 1.\} \end{aligned}$$

Abbreviation: $(\nabla_*^i w)^{\alpha^i} := \sum_{\gamma\in\mathbb{N}_0^n, |\gamma|=i} (\nabla^\gamma w)^{\alpha_i}$.

Using Hölder's inequality we obtain

$$\begin{aligned} \|\nabla^\beta h(w)\|_p &\leq cB\sum_{k=1}^{r}\sum_{\alpha(k,r)}\|\prod_{i=1}^{r}(\nabla_*^i w)^{\alpha_i}\|_p \\ &\leq cB\sum_{k=1}^{r}\sum_{\alpha(k,r)}\prod_{i=1}^{r}\|(\nabla_*^i w)^{\alpha_i}\|_{\frac{pr}{i\alpha_i}} \\ &\leq cB\sum_{k=1}^{r}\sum_{\alpha(k,r)}\prod_{i=1}^{r}\|\nabla_*^i w\|_{\frac{pr}{i}}^{\alpha_i}. \end{aligned} \tag{4.20}$$

The inequality of Gagliardo–Nirenberg (4.7) yields

$$\|\nabla_*^i w\|_{\frac{pr}{i}}^{\alpha_i} \leq c\,\|\nabla^r w\|_p^{\frac{i\alpha_i}{r}}\,\|w\|_\infty^{\alpha_i(1-i/r)} \leq c\,\|\nabla^r w\|_p^{\frac{i\alpha_i}{r}}. \tag{4.21}$$

((r,p,m,j,q) in Theorem 4.4 corresponds to $(\infty,p,r,i,pr/i)$ here.) This implies

$$\|\nabla^\beta h(w)\|_p \leq c\,B\sum_{k=1}^{r}\sum_{\alpha(k,r)}\prod_{i=1}^{r}\|\nabla^r w\|_p^{\frac{i\alpha_i}{r}} \leq c\,B\,\|\nabla^r w\|_p.$$

Q.E.D.

Remark: If the assumption "$\|w\|_\infty \leq 1$" is replaced by "$\|w\|_\infty \leq \Gamma$ for some $\Gamma \geq 1$" and "B" is replaced by "$B_\Gamma := \|h\|_{C^r(\overline{B(0,\Gamma)})}$", then the corresponding estimate (4.19) follows with "$c\,B$" replaced by "$c\,B_\Gamma\Gamma^{r-1}$" which can easily be seen from the inequality (4.21).

The C^r-assumption on w shall be replaced by an L^∞-assumption, which we show to be true if $1 < p \leq \infty$.

Lemma 4.8 *Let $r,m,n \in \mathbb{N}$, $1 < p \leq \infty$, $h \in C^r(\mathbb{R}^m)$, $B := \|h\|_{C^r(\overline{B(0,1)})}$. Then there is a constant $c = c(r,m,n,p) > 0$ such that for all $w = (w_1,\ldots,w_m) \in W^{r,p}(\mathbb{R}^n) \cap L^\infty(\mathbb{R}^n)$ with $\|w\|_\infty \leq 1$ the inequality*

$$\|\nabla^r h(w)\|_p \leq c\,B\,\|\nabla^r w\|_p$$

holds.

PROOF: Let $w_k := J_{1/k}w$. Then $w_k \in W^{r,p}(\mathbb{R}^n) \cap C^\infty(\mathbb{R}^n)$ and

$$w_k \to w \quad \text{in} \quad W^{r,p}(\mathbb{R}^n) \quad \text{as} \quad k \to \infty,$$

$$\|w_k\|_\infty \leq \|w\|_\infty \leq 1 \quad \text{for all} \quad k \in \mathbb{N}.$$

Without loss of generality we may assume

$$w_k \to w \quad a.e.$$

(Take a suitable subsequence and denote it by w_k again.) Lemma 4.7 yields

$$\|\nabla^r h(w_k)\|_p \leq c\,B\,\|\nabla^r w_k\|_p.$$

Thus, $(\nabla^r h(w_k))_k$ is a uniformly bounded sequence in $L^p(\mathbb{R}^n)$.
(1) $1 < p < \infty$:
$L^p(\mathbb{R}^n)$ is reflexive. This implies

$$\forall \beta \in \mathbb{N}_0^n,\ |\beta| = r \quad \exists g \in L^p(\mathbb{R}^n) \,:\, \nabla^\beta h(w_k) \rightharpoonup g$$

(weak convergence in $L^p(\mathbb{R}^n)$ for a subsequence which is denoted with the same symbol.) Analogously to (4.18) it follows that

$$\nabla^\beta h(w_k) \rightharpoonup g = \nabla^\beta h(w)$$

whence

$$\begin{aligned} \|\nabla^\beta h(w)\|_p &\leq \liminf_{k\to\infty} \|\nabla^\beta h(w_k)\|_p \leq c\,B \lim_{k\to\infty} \|\nabla^r w_k\|_p \\ &= c\,B\,\|\nabla^r w\|_p. \end{aligned}$$

(2) $p = \infty$:
The result follows with the arguments of the proof of Theorem 4.4, case B, (2) ($q = p = r = \infty$ there) using the weak* sequential compactness of balls in $L^\infty(\mathbb{R}^n)$.

Q.E.D.

The last two inequalities and the following inequalities in Lemma 4.9 often are quoted as "Moser inequalities" or also "Moser-type inequalities".

Lemma 4.9 *Let $m \in \mathbb{N}$. Then there is a constant $c = c(m,n) > 0$ such that for all $f, g \in W^{m,2} \cap L^\infty$ and $\alpha \in \mathbb{N}_0^n, |\alpha| \leq m$, the following inequalities hold:*

$$(i) \qquad \|\nabla^\alpha(fg)\|_2 \leq c(\|f\|_\infty \|\nabla^m g\|_2 + \|\nabla^m f\|_2 \|g\|_\infty),$$

$$(ii) \quad \|\nabla^\alpha(fg) - f\nabla^\alpha g\|_2 \leq c(\|\nabla f\|_\infty \|\nabla^{m-1} g\|_2 + \|\nabla^m f\|_2 \|g\|_\infty).$$

PROOF: Part (i) follows from Lemma 4.8 with

$$h(x,y) := xy,\ w := (f/\|f\|_\infty, g/\|g\|_\infty).$$

We shall give an alternative proof which can be carried over to part (ii) directly.
Proof of (i): Without loss of generality we assume $\alpha \neq 0$.
First let $F := J_\varepsilon f$, $G := J_\varepsilon g$. Then

$$\begin{aligned}\|\nabla^\alpha(FG)\|_2 &\leq c \sum_{\beta+\gamma=\alpha} \|\nabla^\gamma F \nabla^\beta G\|_2 \\ &\leq c \sum_{\beta+\gamma=\alpha} \|\nabla^\gamma F\|_{\frac{2|\alpha|}{|\gamma|}} \|\nabla^\beta G\|_{\frac{2|\alpha|}{|\beta|}} \\ &\leq c \sum_{\beta+\gamma=\alpha} \|F\|_\infty^{1-\frac{|\gamma|}{|\alpha|}} \|\nabla^{|\alpha|}F\|_2^{\frac{|\gamma|}{|\alpha|}} \|G\|_\infty^{1-\frac{|\beta|}{|\alpha|}} \|\nabla^{|\alpha|}G\|_2^{\frac{|\beta|}{|\alpha|}},\end{aligned}$$

where we have used the inequality of Gagliardo–Nirenberg, Theorem 4.4, (with (r,m,j,p,q) from Theorem 4.4 replaced by $(\infty, |\alpha|, |\gamma|(\text{resp. }|\beta|), 2, \frac{2|\alpha|}{|\gamma|}\ (\text{resp. } \frac{2|\alpha|}{|\beta|}))$ here). Since $|\beta| + |\gamma| = |\alpha|$ we obtain

$$\begin{aligned}\|\nabla^\alpha(FG)\|_2 &\leq \sum_{\beta+\gamma=\alpha} (\|F\|_\infty \|\nabla^{|\alpha|}G\|_2)^{\frac{|\beta|}{|\alpha|}} (\|\nabla^{|\alpha|}F\|_2 \|G\|_\infty)^{\frac{|\gamma|}{|\alpha|}} \\ &\leq c \sum_{\beta+\gamma=\alpha} \left(\frac{|\beta|}{|\alpha|} \|F\|_\infty \|\nabla^{|\alpha|}G\|_2 + \frac{|\gamma|}{|\alpha|} \|\nabla^{|\alpha|}F\|_2 \|G\|_\infty \right),\end{aligned}$$

where we used Young's inequality (*William Henry Young*, 20.10.1862 – 7.7.1942). This implies

$$\|\nabla^\alpha(FG)\|_2 \leq c(\|F\|_\infty \|\nabla^m G\|_2 + \|\nabla^m F\|_2 \|G\|_\infty)$$

which is the desired result for

$$F = J_\varepsilon f,\ G = J_\varepsilon g.$$

Now, let $\varepsilon \downarrow 0$ and the result follows for f, g. (Notice again that $\|J_\varepsilon f\|_\infty \leq \|f\|_\infty$).
Proof of (ii): The proof is analogous to that of part (i) observing

$$\|\nabla^\alpha(FG) - F\nabla^\alpha G\|_2 \leq \sum_{\gamma+\beta=\alpha-\sigma;|\sigma|=1} \|\nabla^\beta(\nabla^\sigma F)\nabla^\gamma G\|_2.$$

Q.E.D.

The inequalities given in the previous Lemmata are perfectly suitable for the proof of Theorem 1.1 (the first global existence theorem for the wave equation) and for the proof of the corresponding theorems for the evolution equations to be studied in Chapter 11. For the proof of Theorem 1.2, which is an optimal theorem with respect to the relation between the space dimension and the degree of vanishing of the nonlinearity near zero, we have to exploit special invariance properties of the d'Alembert operator $\Box = \partial_t^2 - \Delta$. This is reflected in the following Lemmata, which are also of interest in themselves.

We consider the Minkowski space $\mathbb{R} \times \mathbb{R}^n$ with co-ordinates $x_0 = t, x = (x_1, \ldots, x_n)$ and the metric

$$\eta = (\eta_{ab})_{a,b=0,1,\ldots,n} = \begin{pmatrix} -1 & 0 & 0 & \cdots & 0 \\ 0 & 1 & 0 & \cdots & 0 \\ 0 & 0 & \ddots & & 0 \\ \vdots & \vdots & & \ddots & \vdots \\ 0 & 0 & \cdots & 0 & 1 \end{pmatrix}$$

(*Hermann Minkowski*, 22.6.1864 – 12.1.1909).
The d'Alembert operator $\Box$ is given by

$$\Box \equiv \Box_n := -\sum_{a,b=0}^{n} \eta_{ab}\partial_a\partial_b = \partial_0^2 - \partial_1^2 - \ldots - \partial_n^2$$

with $\partial_0 = -\partial_t$.
This operator is invariant under the action of the generators of the (inhomogeneous) Lorentz group which are given by

$$\partial_0, \partial_1, \ldots, \partial_n$$

and

$$\Omega_{ab} := x_a\partial_b - x_b\partial_a, \quad a, b = 0, 1, \ldots, n.$$

(*Hendrik Antoon Lorentz*, 18.7.1853 – 4.2.1928).
In particular we use

$$\begin{aligned} \Omega_{ij} &:= x_i\partial_j - x_j\partial_i, \quad i, j = 1, \ldots, n, \\ L_i := \Omega_{0i} &= t\partial_i + x_i\partial_t, \quad i = 1, \ldots, n. \end{aligned}$$

Moreover, let

$$L_0 := \sum_{a,b=0}^{n} \eta_{ab}x_a\partial_b = t\partial_t + x_1\partial_1 + \ldots + x_n\partial_n. \tag{4.22}$$

The (inhomogeneous) Lorentz group consists of isometries of flat space-time. In $\mathbb{R} \times \mathbb{R}^3$ these are the four translations $\partial_a, a = 0, 1, 2, 3$, and the rotations in space-time Ω_{ab}, $0 \le a < b \le 3$, which generate the ten-parameter Lorentz group.
Remark: We have called ∂_a a translation adopting the usual notation. Actually, ∂_j is the infinitesimal generator of the group of translations $\{T_j(h)\}_{h\in\mathbb{R}}$

$$(T_j(h)f)(x) := f(x + he_j),$$

where e_j is the unit vector in the direction of the co-ordinate x_j, (f taken in suitable function spaces). In the same sense Ω_{ab} is called a rotation, e.g. for $n = 2, \Omega_{21}$ is the infinitesimal generator of the group of rotations $\{T_{21}(h)\}_{h\in\mathbb{R}}$ given by

$$(T_{21}(h)f)(t, x) := f\left(\begin{pmatrix} 1 & 0 & 0 \\ 0 & \cos h & \sin h \\ 0 & -\sin h & \cos h \end{pmatrix}\begin{pmatrix} t \\ x \end{pmatrix}\right).$$

The $\Omega_{ij}, i, j = 1, \ldots, n$, can be expressed in terms of angular co-ordinates only, hence they represent tangential vectors on the unit sphere $S^{n-1} \subset \mathbb{R}^n$. We shall prove this for $n = 3$.

Let polar co-ordinates be given by

$$\begin{aligned} x_1 &= r \cos\varphi \sin\theta, \\ x_2 &= r \sin\varphi \sin\theta, \\ x_3 &= r \cos\theta, \end{aligned}$$

where

$$0 \le \theta < \pi \quad \text{and} \quad 0 \le \varphi < \pi,$$

respectively

$$\begin{aligned} r &= |x|, \\ \varphi &= \arctan \frac{x_2}{x_1}, \\ \theta &= \arccos \frac{x_3}{r} \end{aligned}$$

(on appropriate branches of arctan, arccos).

Lemma 4.10 *Let $n = 3$. Then the operators $\Omega_{ij}, 1 \le i < j \le 3$, are given in polar co-ordinates by*

$$\begin{aligned} \Omega_{12} &= \partial_\varphi, \\ \Omega_{13} &= -\cos\varphi\, \partial_\theta + \cot\theta \sin\varphi\, \partial_\varphi, \\ \Omega_{23} &= -\sin\varphi\, \partial_\theta - \cot\theta \cos\varphi\, \partial_\varphi. \end{aligned}$$

Moreover

$$\partial_\theta = -\sin\varphi\, \Omega_{23} - \cos\varphi\, \Omega_{13},$$

where $\partial_\varphi = \frac{\partial}{\partial\varphi}$, $\partial_\theta = \frac{\partial}{\partial\theta}$.

Proof: Expressing the x-derivatives in polar co-ordinates we have

$$\partial_j = \frac{\partial r}{\partial x_j}\partial_r + \frac{\partial \theta}{\partial x_j}\partial_\theta + \frac{\partial \varphi}{\partial x_j}\partial_\varphi, \quad j = 1, 2, 3,$$

with

$$\begin{aligned} \frac{\partial r}{\partial x_j} &= \frac{x_j}{r}, \quad j = 1, 2, 3, \\ \frac{\partial \varphi}{\partial x_1} &= \frac{-x_2}{x_1^2 + x_2^2}, \ \frac{\partial \varphi}{\partial x_2} = \frac{x_1}{x_1^2 + x_2^2}, \ \frac{\partial \varphi}{\partial x_3} = 0, \end{aligned}$$

$$\begin{aligned}
\frac{\partial\theta}{\partial x_j} &= \frac{x_j x_3}{r^2}\frac{1}{\sqrt{r^2-x_3^2}} = \frac{x_j x_3}{r^3}\frac{1}{\sin\theta}, \quad j=1,2,\\
\frac{\partial\theta}{\partial x_3} &= -\frac{\sqrt{r^2-x_3^2}}{r^2} = \frac{-\sin\theta}{r}.
\end{aligned}$$

Hence we obtain

$$\begin{aligned}
\partial_1 &= \frac{x_1}{r}\partial_r + \frac{x_1 x_3}{r^3}\frac{1}{\sin\theta}\partial_\theta - \frac{x_2}{x_1^2+x_2^2}\partial_\varphi,\\
\partial_2 &= \frac{x_2}{r}\partial_r + \frac{x_2 x_3}{r^3}\frac{1}{\sin\theta}\partial_\theta + \frac{x_1}{x_1^2+x_2^2}\partial_\varphi,\\
\partial_3 &= \frac{x_3}{r}\partial_r - \frac{\sin\theta}{r}\partial_\theta,
\end{aligned}$$

and finally

$$\begin{aligned}
\Omega_{12} &= (\frac{x_1^2}{x_1^2+x_2^2} + \frac{x_2^2}{x_1^2+x_2^2})\,\partial_\varphi = \partial_\varphi,\\
\Omega_{23} &= \frac{x_2 x_3}{r}\partial_r - \frac{x_2\sin\theta}{r}\partial_\theta - \frac{x_2 x_3}{r}\partial_r - \frac{x_3^2 x_2}{r^3\sin\theta}\partial_\theta - \frac{x_1 x_3}{x_1^2+x_2^2}\partial_\varphi\\
&= \{\frac{-x_2(x_3^2+r^2\sin^2\theta)}{r^3\sin\theta}\}\partial_\theta - \frac{x_1 x_3}{x_1^2+x_2^2}\partial_\varphi\\
&= -\frac{x_2}{r\sin\theta}\partial_\theta - \frac{x_1 x_3}{x_1^2+x_2^2}\partial_\varphi\\
&= -\sin\varphi\,\partial_\theta - \cot\theta\cos\varphi\,\partial_\varphi.
\end{aligned}$$

Analogously:

$$\Omega_{13} = -\cos\varphi\,\partial_\theta + \cot\theta\sin\varphi\,\partial_\varphi.$$

From the representations of Ω_{23} and Ω_{13} we immediately get

$$\partial_\theta = -\sin\varphi\,\Omega_{23} - \cos\varphi\,\Omega_{13}.$$

Q.E.D.

The invariance of the d'Alembert operator under the Lorentz group is described by the commutator $[\Omega_{ab},\Box]$ where the bracket $[\cdot,\cdot]$ is defined by

$$[A,B]\psi := AB\psi - BA\psi$$

for two operators A, B and a function ψ.

Lemma 4.11

$$[\Omega_{ab},\Box] = 0, \quad a,b = 0,1,\dots,n.$$

PROOF:

$$\begin{aligned}\Box\Omega_{ab} &= \partial_t^2 x_a\partial_b - \partial_t^2 x_b\partial_a - \Delta x_a\partial_b + \Delta x_b\partial_a \\ &= x_a\partial_b\partial_t^2 - x_b\partial_a\partial_t^2 - x_a\partial_b\Delta + x_b\partial_a\Delta + R \\ &= \Omega_{ab}\Box + R,\end{aligned}$$

where

$$R := 2(\partial_t x_a)\partial_t\partial_b - 2(\partial_t x_b)\partial_t\partial_a - \sum_{k=1}^{n} 2\{(\partial_k x_a)\partial_k\partial_b - (\partial_k x_b)\partial_k\partial_a\}.$$

(i) $a, b \neq 0 \quad \Rightarrow \quad R = \sum_{k=1}^{n} 2\{\delta_{ka}\partial_k\partial_b - \delta_{kb}\partial_k\partial_a\} = 0,$

(ii) $a = 0$ or $b = 0$. The proof is similar.

Q.E.D.

We have the following commutator relations, the proof of which is as easy as that of the previous Lemma and we omit it.

Lemma 4.12 *Let* $a, b, c, d \in \{0, 1, \ldots, n\}$.

$$\begin{aligned}&(i) & [L_0, \Box] &= -2\Box, \\ &(ii) & [L_0, \Omega_{ab}] &= 0, \\ &(iii) & [L_0, \partial_a] &= -\partial_a, \\ &(iv) & [\Omega_{ab}, \Omega_{cd}] &= \eta_{bc}\Omega_{ad} + \eta_{ad}\Omega_{bc} - \eta_{bd}\Omega_{ac} - \eta_{ac}\Omega_{bd}, \\ &(v) & [\Omega_{ab}, \partial_c] &= \eta_{bc}\partial_a - \eta_{ac}\partial_b.\end{aligned}$$

We introduce three families of first-order operators:

$$\begin{aligned}\Omega &:= (\Omega_{ij})_{1\le i<j\le n}, \\ \overline{\Omega} &:= (\Omega_{ab})_{0\le a<b\le n}, \\ \Gamma &:= (L_0, \overline{\Omega}, \partial_0, \ldots, \partial_n).\end{aligned} \tag{4.23}$$

The commutator relations imply that the $\mathbb{R}$-linear span of each of the families is a Lie-algebra with bracket $[\cdot,\cdot]$ (*Sophus Lie*, 17.12.1842 – 18.2.1899).

In the same way as the family $\partial = (\partial_1, \ldots, \partial_n)$ generates the usual Sobolev norm in $W^{k,p}(\mathbb{R}^n)$ by

$$\|u\|_{\partial,k,p} := \left(\sum_{|\alpha|\le k} \|\partial^\alpha u\|^p_{L^p(\mathbb{R}^n)}\right)^{1/p},$$

where $\alpha = (\alpha_1, \ldots, \alpha_n)$ is a multi-index,

$$\partial^\alpha = \partial_1^{\alpha_1} \cdot \ldots \cdot \partial_n^{\alpha_n},\ k \in \mathbb{N}_0,\ 1 \le p \le \infty,\ (p = \infty \text{ interpreted as usual}),$$

we may define generalized Sobolev norms for functions $u = u(t,x)$ which are smooth and which decay sufficiently rapidly as $|x| \to \infty$ for each fixed t. For this we use the families $\Omega, \overline{\Omega}, \Gamma$.
If $A = (A_i)_{1\le i\le\sigma}$ is one of them, ($\sigma = \frac{n(n-1)}{2}$ if $A = \Omega$, $\sigma = \frac{n(n+1)}{2}$ if $A = \overline{\Omega}$, $\sigma = 1 + \frac{n(n+1)}{2} + n + 1$ if $A = \Gamma$), we define for $k \in \mathbb{N}_0, 1 \le p \le \infty$:

$$\|u(t)\|_{A,k,p} := \left(\sum_{|\alpha|\le k} \|A^\alpha u(t)\|^p_{L^p(\mathbb{R}^n)} \right)^{1/p},$$

where $A^\alpha = \prod_{i=1}^{\sigma} (A_i^{\alpha_i})$, α being a multi-index, $\alpha = (\alpha_1, \ldots, \alpha_\sigma)$.
Because of the commutator relations any two different orderings of the operator A will produce equivalent norms.
The estimates on composite functions which we proved in Lemma 4.7 (resp. Lemma 4.8) and Lemma 4.9 have their counterparts in terms of the family Γ.

Lemma 4.13 *Let $r, m, n \in \mathbb{N}, h \in C^r(\mathbb{R}^m), B := \|h\|_{C^r(\overline{B(0,1)})}$. Then there is a constant $c = c(r,m,n) > 0$ such that for all $w = (w_1, \ldots, w_m) \in C^\infty(\mathbb{R} \times \mathbb{R}^n), w(t,\cdot)$ having compact support in $x \in \mathbb{R}^n$ for any fixed $t \in \mathbb{R}$, with $\|w\|_{\Gamma,[\frac{r}{2}],\infty} \le 1$, the inequality*

$$\|\Gamma^\alpha h(w(t))\|_2 \le c\, B \|w(t)\|_{\Gamma,r,2}$$

holds for any multi-index $\alpha = (\alpha_1, \ldots, \alpha_\sigma), \sigma = 1 + \frac{n(n+1)}{2} + n + 1$, with $|\alpha| = r$.

PROOF: The proof is analogous to that of Lemma 4.7 observing the following:
(i) $\Gamma_i(fg) = (\Gamma_i f)g + f(\Gamma_i g)$ and $\Gamma_i(f(v)) = \frac{\partial f}{\partial v}(v)\Gamma_i v$ holds for any $\Gamma_i \in \Gamma$ and any smooth functions f, g, v.
(ii) We do not have a corresponding sharp Gagliardo–Nirenberg type estimate. Instead, we use the coarser but more elementary estimate

$$\|v_1 \cdot \ldots \cdot v_k\|_2 \le \prod_{i=1}^{k-1} \|v_i\|_\infty \|v_k\|_2.$$

v_j represents a derivative of w appearing in formula (4.20) (cf. the example preceding that formula). At most one factor is a derivative of w of order greater than $[\frac{r}{2}]$ and then all other terms involve derivatives of at most order $[\frac{r}{2}]$. With (i) and (ii) in mind we obtain the proof in analogy to that of Lemma 4.7.

Q.E.D.

With the same arguments we can prove the following analogue to Lemma 4.9.

Lemma 4.14 *Let $m \in \mathbb{N}$. Then there is a constant $c = c(m,n) > 0$ such that for all $u, v \in C^\infty(\mathbb{R} \times \mathbb{R}^n)$ having compact support in $x \in \mathbb{R}^n$ for any fixed $t \in \mathbb{R}$ the following inequalities hold for any multi-index $\alpha = (\alpha_1, \ldots, \alpha_\sigma), \sigma = 1 + \frac{n(n+1)}{2} + n + 1$, with*

$|\alpha| = m:$

(i) $\|\Gamma^\alpha(uv)(t)\|_2 \;\le\; c\left(\|u(t)\|_{\Gamma,[\frac{m}{2}],\infty}\|v(t)\|_{\Gamma,m,2} + \|u(t)\|_{\Gamma,m,2}\|v(t)\|_{\Gamma,[\frac{m}{2}],\infty}\right),$

(ii) $\|\Gamma^\alpha(uv)(t) - u(t)\Gamma^\alpha v(t)\|_2 \le$

$$c\left(\|\Gamma u(t)\|_{\Gamma,[\frac{m}{2}]-1,\infty}\|v(t)\|_{\Gamma,m-1,2} + \|u(t)\|_{\Gamma,m,2}\|v(t)\|_{\Gamma,[\frac{m}{2}],\infty}\right).$$

Finally we present a Sobolev inequality on the unit sphere $S^{n-1} \subset \mathbb{R}^n, n \in \mathbb{N}$. We shall give a proof for $n = 3$.

Lemma 4.15 *There is a constant $C = C(n) > 0$ such that for all smooth functions $u : S^{n-1} \longrightarrow \mathbb{R}$ the inequality*

$$|u(\xi)| \le C\left(\sum_{|\alpha|\le[\frac{n-1}{2}]+1} \|\Omega^\alpha u\|^2_{L^2(S^{n-1})}\right)^{1/2}$$

holds for each $\xi \in S^{n-1}$.

PROOF: $(n = 3)$:
Step 1: We prove the desired inequality for ξ in a fixed neighbourhood U of the great circle $\{x_3 = 0\}$. Then the factor $\sin\theta$ appearing in the volume element in polar co-ordinates can be uniformly bounded from below by, say, c_1 since θ varies in a neighbourhood of $\pi/2$.
Introducing polar co-ordinates we use the notation

$$u(\xi) = \tilde{u}(\varphi,\theta),$$
$$(\varphi,\theta) \in M := [0,2\pi] \times \text{neighbourhood of}\, \pi/2.$$

The classical Sobolev inequality for $\tilde{u}$ in $\mathbb{R}^2$ yields

$$\exists c > 0\ \forall(\varphi,\theta) \in M:\ |\tilde{u}(\varphi,\theta)| \le c\{\|\tilde{u}\|_{L^2(M)} + \|\nabla_{\varphi,\theta}\tilde{u}\|_{L^2(M)} + \|\nabla^2_{\varphi,\theta}\tilde{u}\|_{L^2(M)}\}.$$

Now we have

$$\begin{aligned}\|\tilde{u}\|^2_{L^2(M)} &= \int_M |\tilde{u}(\varphi,\theta)|^2 d(\varphi,\theta) \le \frac{1}{c_1}\int_M |\tilde{u}(\varphi,\theta)|^2 \sin\theta d(\varphi,\theta)\\ &= \frac{1}{c_1}\int_U |u(\xi)|^2 d\xi \le \frac{1}{c_1}\int_{S^2} |u(\xi)|^2 d\xi.\end{aligned}$$

Using Lemma 4.10 we obtain

$$\begin{aligned}\|\partial_\varphi\tilde{u}\|^2_{L^2(M)} &\le \frac{1}{c_1}\int_M |\partial_\varphi\tilde{u}(\varphi,\theta)|^2 \sin\theta d(\varphi,\theta)\\ &= \frac{1}{c_1}\int_U |\Omega_{12}u(\xi)|^2 d\xi \le \frac{1}{c_1}\int_{S^2} |\Omega_{12}u(\xi)|^2 d\xi,\end{aligned}$$

and

$$\|\partial_\theta \tilde{u}\|^2_{L^2(M)} \leq \frac{2}{c_1} \int_U (|\Omega_{23} u(\xi)|^2 + |\Omega_{13} u(\xi)|^2) d\xi.$$

Analogously:

$$\|\nabla^2_{\varphi,\theta} \tilde{u}\|_{L^2(M)} \leq c \sum_{|\alpha| \leq 2} \|\Omega^\alpha f\|_{L^2(S^2)}.$$

This proves the Lemma ($n = 3$) for $\xi \in U$.

Step 2: Let $\xi_0 \in S^2 \setminus U$, $\xi_0 = (\varphi_0, \theta_0)$ in polar co-ordinates.

There is an orthogonal transformation P — a rotation orthogonal to the great circle $\{x_3 = 0\}$ with angle μ — such that

$$P\xi_0 = \xi_1 \in U,$$

in polar co-ordinates

$$(\varphi_0, \theta_0 + \mu) = (\varphi_1, \theta_1) \in M.$$

If u_1 is defined by

$$u_1(\xi) := u(P^{-1}\xi)$$

respectively

$$\tilde{u}_1(\varphi, \theta) := \tilde{u}(\varphi, \theta - \mu)$$

the results from Step 1 imply

$$|u(\xi_0)| = |u_1(\xi_1)| \leq c\{\|\tilde{u}_1\|_{L^2(M)} + \|\nabla_{\varphi,\theta} \tilde{u}_1\|_{L^2(M)} + \|\nabla^2_{\varphi,\theta} \tilde{u}_1\|_{L^2(M)}\}.$$

Since

$$\begin{aligned} (\partial_\varphi \tilde{u}_1)(\varphi, \theta) &= (\partial_\varphi \tilde{u})(\varphi, \theta - \mu), \\ (\partial_\theta \tilde{u}_1)(\varphi, \theta) &= (\partial_\theta \tilde{u})(\varphi, \theta - \mu), \end{aligned}$$

and because P is an orthogonal transformation from S^2 onto itself, we obtain, as before in Step 1,

$$\begin{aligned} |u(\xi)| &\leq c\{\|\tilde{u}_1\|_{L^2(M))} + \|\nabla_{\varphi,\theta} \tilde{u}_1\|_{L^2(M)} + \|\nabla^2_{\varphi,\theta} \tilde{u}_1\|_{L^2(M)}\} \\ &\leq c\{\|u\|_{L^2(S^2)} + \|\Omega u\|_{L^2(S^2)} + \|\Omega^2 u\|_{L^2(S^2)}\}. \end{aligned}$$

Q.E.D.

5 Local existence for quasilinear symmetric hyperbolic systems

Theorem 1.1 and Theorem 1.2 will be proved in detail for the initial value problem

$$y_{tt} - \Delta y = f(Dy, \nabla Dy), \tag{5.1}$$

$$y(t=0) = y_0, \quad y_t(t=0) = y_1, \tag{5.2}$$

with

$$f(Dy, \nabla Dy) = \sum_{i,j=1}^{n} a_{ij}(Dy)\partial_i\partial_j y, \tag{5.3}$$

where

$$a_{ij} = \bar{a}_{ji} \in C^\infty(\mathbb{R}^{n+1}), \quad i,j = 1,\ldots,n, \tag{5.4}$$

$$a_{ij}(0) = 0 \quad , \quad i,j = 1,\ldots,n. \tag{5.5}$$

(The latter assumption corresponds to the case $\alpha = 1$.)
The assumptions (5.3), (5.4) are made without loss of generality, see the remarks in Chapter 8 ("quasilinearization"). The assumption (5.5) implies

$$\exists m > 0 \quad \exists \eta > 0 \qquad \forall \xi \in \mathbb{C}^n \quad \forall v \in \mathbb{C}^{n+1} \quad , \quad |v| < \eta : \tag{5.6}$$
$$|\xi|^2 + \sum_{i,j=1}^{n} a_{ij}(v)\xi_i\bar{\xi}_j \geq m|\xi|^2.$$

In analogy to the procedure in Chapter 3 we may now write the initial value problem (5.1), (5.2) for the function y as a first-order quasilinear symmetric hyperbolic system for the vector $u := Dy = (\partial_t y, \partial_1 y, \ldots, \partial_n y)$:

$$A^0(u)\partial_t u + \sum_{j=1}^{n} A^j(u)\partial_j u = 0, \tag{5.7}$$

$$u(t=0) = u_0, \tag{5.8}$$

where

$$u_0 := (y_1, \nabla y_0),$$

$$A^0 := \begin{pmatrix} 1 & 0 & \cdots & 0 \\ 0 & b_{11} & \cdots & b_{1n} \\ \vdots & \vdots & & \vdots \\ 0 & b_{n1} & \cdots & b_{nn} \end{pmatrix}, \quad A^j := -\begin{pmatrix} 0 & b_{1j} & \cdots & b_{nj} \\ b_{1j} & 0 & \cdots & 0 \\ \vdots & \vdots & & \vdots \\ b_{nj} & 0 & \cdots & 0 \end{pmatrix}, \quad j = 1,\ldots,n,$$

with

$$b_{ij} = b_{ij}(u) := a_{ij}(u) + \delta_{ij}, \quad i,j = 1,\ldots,n,$$

(δ_{ij} : Kronecker delta (*Leopold Kronecker*, 7.12.1823 – 29.12.1891)).

The initial value problems (5.1), (5.2) for y and (5.7), (5.8) for u, respectively, are equivalent for smooth functions. $A^0(u)$ will be positive definite for small u according to (5.6). This will be no restriction of generality since we are looking for global *small* solutions later.

In this chapter we shall prove a local existence theorem in a more general situation, namely for the following initial value problem.

$$A^0(u)\partial_t u + \sum_{j=1}^{n} A^j(u)\partial_j u + B(u)u = 0, \tag{5.9}$$

$$u(t=0) = u_0, \tag{5.10}$$

where $u = (u_1, \ldots, u_N) \in \mathbb{C}^N$, $u = u(t,x)$, $t \in \mathbb{R}$, $x \in \mathbb{R}^n$. The following assumption is made:

$$\left.\begin{array}{l} A^0, A^1, \ldots, A^n, B \text{ are complex } N \times N\text{- matrices and } C^\infty\text{-functions of} \\ \text{their arguments } v \in \mathbb{C}^N.\ A^j(v), j = 0, 1, \ldots, n \text{ is hermitian and } A^0(v) \\ \text{is positive definite, uniformly in each compact set with respect to } v. \end{array}\right\} \tag{5.11}$$

By κ_s we shall denote the Sobolev constant characterizing the continuous imbedding of $W^{s,2}$ into the space of uniformly bounded, continuous functions if $s > n/2$, i.e.

$$|w(x)| \leq \kappa_s \|w\|_{s,2}$$

for $w \in W^{s,2}$ and almost all $x \in \mathbb{R}^n$.

In this chapter we shall use the following additional notation:

$$|u|_{s,T} := \sup_{0 \leq t \leq T} \|u(t)\|_{s,2}$$

if $u \in L^\infty([0,T], W^{s,2})$, $T > 0$, $s \in \mathbb{N}_0$.

The two main theorems of this chapter will be Theorem 5.1 and Theorem 5.8.

Theorem 5.1 *Assume (5.11) and let $u_0 \in W^{s,2}$, $s \in \mathbb{N}$, $s > \frac{n}{2} + 1$. Let $g_1 := \kappa_s \|u_0\|_{s,2}$ and $g_2 > g_1$ arbitrary but fixed.*

Then there is a $T > 0$ such that there exists a unique classical solution $u \in C_b^1([0,T] \times \mathbb{R}^n)$ of the initial value problem (5.9), (5.10) with

$$\sup_{(t,x) \in [0,T] \times \mathbb{R}^n} |u(t,x)| \leq g_2$$

and

$$u \in C^0([0,T], W^{s,2}) \cap C^1([0,T], W^{s-1,2}).$$

T is a function of $\|u_0\|_{s,2}$ and g_2.

In order to prove the existence of a solution we shall state and prove a series of Lemmata. First we show the uniqueness in the following class of functions $\mathcal{U}$,

$$\mathcal{U} := C^0([0,T], W^{1,2}) \cap C^1([0,T], L^2) \cap L^\infty([0,T], W^{s,2}). \tag{5.12}$$

Let $u_1, u_2 \in \mathcal{U}$ be two solutions of (5.9), (5.10) (in the strong sense with respect to the derivatives which appear). Then we have

$$A^0(u_k)\partial_t u_k + \sum_{j=1}^n A^j(u_k)\partial_j u_k + B(u_k)u_k = 0,$$

and

$$u_k(t=0) = u_0, \quad k = 1, 2.$$

This implies

$$\begin{aligned} & A^0(u_2)\partial_t(u_2 - u_1) + \sum_{j=1}^n A^j(u_2)\partial_j(u_2 - u_1) + B(u_2)(u_2 - u_1) \\ = \; & (A^0(u_1) - A^0(u_2))\partial_t u_1 + \sum_{j=1}^n (A^j(u_1) - A^j(u_2))\partial_j u_1 + (B(u_1) - B(u_2))u_1, \end{aligned}$$

that is, v, defined by $v := u_2 - u_1$, satisfies

$$\tilde{A}^0(t,x)\partial_t v + \sum_{j=1}^n \tilde{A}^j(t,x)\partial_j v + \tilde{B}(t,x)v = \tilde{F}(t,x), \tag{5.13}$$

$$v(t=0) = 0,$$

where the notation is obvious: $\tilde{A}^0(t,x) := A^0(u_2(t,x))$ etc.
We take the inner product of both sides of equation (5.13) with v in $L^2(\mathbb{R}^n)$ and we obtain (cf. Chapter 3):

$$\frac{1}{2}\frac{d}{dt}\int_{\mathbf{R}^n} \tilde{A}^0 v \cdot \bar{v} = \frac{1}{2}\int_{\mathbf{R}^n} (\partial_t \tilde{A}^0) v \cdot \bar{v} + \frac{1}{2}\int_{\mathbf{R}^n} \Big(\sum_{j=1}^n \partial_j \tilde{A}^j\Big) v \cdot \bar{v} - \mathrm{Re}\int_{\mathbf{R}^n} \tilde{B} v \cdot \bar{v} + \mathrm{Re}\int_{\mathbf{R}^n} \tilde{F} \cdot v. \tag{5.14}$$

All coefficients in the preceding equation are bounded by Sobolev's imbedding theorem because we have by assumption

$$|u_1|_{s,T} + |u_2|_{s,T} \le c < \infty.$$

(c will denote various constants which do not depend on t, T, u_0, but at most on s or n.) Therefore, we obtain from (5.14), using the positive definiteness of $\tilde{A}^0$,

$$\frac{d}{dt}\int_{\mathbf{R}^n} \tilde{A}^0 v \cdot \bar{v} \le c\int_{\mathbf{R}^n} \tilde{A}^0 v \cdot \bar{v} + c\int_{\mathbf{R}^n} |\tilde{F}|\,|v|.$$

The last term may be estimated as follows:

$$\begin{aligned}|\tilde{F}| &\leq \left|\int_0^1 \frac{d}{dr} A^0(ru_1 + (1-r)u_2)dr\right| |\partial_t u_1| + \ldots \\ &\leq \int_0^1 \left|(\nabla_u A^0)(ru_1 + (1-r)u_2)\right| dr\, |\partial_t u_1|\, |u_1 - u_2| + \ldots \\ &\leq c|v|.\end{aligned}$$

This implies

$$\frac{d}{dt}\int_{\mathbb{R}^n} \tilde{A}^0 v \cdot \bar{v} \leq c \int_{\mathbb{R}^n} \tilde{A}^0 v \cdot \bar{v},$$

whence $v = 0$ follows with Gronwall's inequality, Lemma 4.1, because $v(t = 0) = 0$. This proves the *uniqueness* (in the larger class $\mathcal{U}$).

Q.E.D.

Now we turn to the proof of *existence*. The proof is a slight modification of that outlined by A. Majda in [114].

Let

$$\varepsilon_k := 2^{-k}\varepsilon_0, \quad k \in \mathbb{N}_0, \quad 0 < \varepsilon_0 \leq 1,$$

where ε_0 is arbitrary but fixed.

Let

$$u_0^k := J_{\varepsilon_k} u_0, \quad k \in \mathbb{N}_0,$$

denote the smoothed initial value, where J_{ε_k} denotes the convolution with the Friedrichs mollifier j_{ε_k}, cf. Chapter 4.

Then, according to Lemma 4.2, we have

$$\forall k, m \in \mathbb{N}_0 : \qquad u_0^k \in W^{m,2} \cap C^\infty,$$

$$\|u_0^k\|_{m,2} \leq \|u_0\|_{m,2}.$$

Let

$$u^0(t,x) := u_0^0(x), \quad t \geq 0, \quad x \in \mathbb{R}^n,$$

and let $u^{k+1} = u^{k+1}(t,x)$ be defined by iteration for $k \in \mathbb{N}_0$ as the solution of the linear initial value problem

$$A^0(u^k)\partial_t u^{k+1} + \sum_{j=1}^n A^j(u^k)\partial_j u^{k+1} + B(u^k)u^{k+1} = 0, \tag{5.15}$$

$$u^{k+1}(t=0) = u_0^{k+1},$$

$k \in \mathbb{N}_0$.

By Theorem 3.3 u^{k+1} is well-defined (inductively) and we have

$$\forall k, m \in \mathbb{N}_0: \quad u^{k+1} \in C^0([0,\infty), W^{m,2}) \cap C^1([0,\infty), W^{m-1,2}) \cap C^\infty([0,\infty) \times \mathbb{R}^n).$$

(Observe that all coefficients in the equation (5.15) belong to C_b^{m+1}.)
Our aim will be to show that a subsequence of $(u^k)_k$ converges towards a solution. For this purpose we first prove some boundedness properties of $(u^k)_k$, namely boundedness in high norms.

Lemma 5.2 *There are $R, L, T_* > 0$ such that for all $k \in \mathbb{N}_0$ we have:*

(i) $|u^k|_{s,T_*} \le R$,
(ii) $|\partial_t u^k|_{s-1,T_*} \le L$,
(iii) $\forall (t,x) \in [0,T_*] \times \mathbb{R}^n$: $\quad |u^k(t,x)| \le g_2$.

R, L and T_* are functions of $\|u_0\|_{s,2}$ and g_2.

In the following proof $c_0(g_2)$ will denote a constant for which

$$\forall v, w \in \mathbb{C}^N, \quad |w| \le g_2: \quad c_0^{-1}(g_2)|v|^2 \le A^0(w)v \cdot \bar{v} \le c_0(g_2)|v|^2$$

holds. $c(g_2)$ will denote various constants, which depend only on g_2 and on values of the coefficients $A^0(w), \ldots$ for $|w| \le g_2$, respectively. We shall not write all parameters $t, x, \ldots$ in each place.

Proof of Lemma 5.2 (by induction on k):

For $\underline{k=0}$ we have $u^0 \equiv u_0^0$, hence we obtain

$$\|u_0^0\|_{s,2} \le \|u_0\|_{s,2} \le R, \tag{5.16}$$

which is to be read as a first condition on R.

$$\partial_t u^0 = 0,$$

L is still arbitrary. By Sobolev's imbedding theorem we have

$$|u^0(t,x)| \le \kappa_s \|u_0^0\|_{s,2} \le \kappa_s \|u_0\|_{s,2} = g_1 < g_2,$$

T_* is still arbitrary.
Induction step, $\underline{k \to k+1}$:

Step 1: "(i) $\Longrightarrow$ (ii)":

Using Lemma 4.9 we obtain

$$\begin{aligned} \|\partial_t u^{k+1}\|_{s-1,2} &= \left\| A^0(u^k)^{-1} \left(\sum_{j=1}^n A^j(u^k)\partial_j u^{k+1} + B(u^k)u^{k+1} \right) \right\|_{s-1,2} \\ &\le c(g_2)\|u^{k+1}\|_{s,2} \le c(g_2)(R^4 + R^2) =: L. \end{aligned} \tag{5.17}$$

(5.17) is the condition which defines $L = L(R)$.

Step 2: "(ii) $\Longrightarrow$ (iii)":

$$\begin{aligned} |u^{k+1}(t,x)| &\leq |u^{k+1}(0,x)| + \int_0^t |\partial_t u^{k+1}(r,x)| dr \\ &\leq \kappa_s \|u_0\|_{s,2} + \kappa_{s-1} \int_0^t \|\partial_t u^{k+1}(r)\|_{s-1,2} dr \\ &\leq g_1 + \kappa_{s-1} L T_* \\ &\leq g_2 \end{aligned}$$

if

$$T_* \leq T_1 = T_1(\|u_0\|_{s,2}, g_2, R) := \frac{g_2 - g_1}{\kappa_{s-1} L}. \tag{5.18}$$

(5.18) is the first condition on T_*.

Step 3: Proof of (i), determination of R.
For simplicity we abbreviate as follows:

$$u := u^k, \quad v := u^{k+1}, \quad v_0 := u_0^{k+1}.$$

Let $\alpha \in \mathbb{N}_0^N$, $|\alpha| \leq s$.
The differential equation for $v = u^{k+1}$,

$$A^0(u)\partial_t v + \sum_{j=1}^n A^j(u)\partial_j v + B(u)v = 0,$$

implies

$$\partial_t v = -A^0(u)^{-1} \left(\sum_{j=1}^n A^j(u)\partial_j v + B(u)v \right)$$

and

$$\begin{aligned} \partial_t \nabla^\alpha v = &- A^0(u)^{-1} \left(\sum_{j=1}^n A^j(u)\partial_j \nabla^\alpha v + B(u)\nabla^\alpha v \right) \\ &+ A^0(u)^{-1} \sum_{j=1}^n A^j(u)\partial_j \nabla^\alpha v - \nabla^\alpha \left(A^0(u)^{-1} \sum_{j=1}^n A^j(u)\partial_j v \right) \\ &+ A^0(u)^{-1} B(u)\nabla^\alpha v - \nabla^\alpha (A^0(u)^{-1} B(u)v), \end{aligned}$$

or, equivalently,

$$A^0(u)\partial_t \nabla^\alpha v + \sum_{j=1}^n A^j(u)\partial_j \nabla^\alpha v + B(u)\nabla^\alpha v = F_\alpha, \tag{5.19}$$

where

$$F_\alpha := \sum_{j=1}^{n} A^0(u)\Big\{ \big[A^0(u)^{-1}A^j(u)\partial_j\nabla^\alpha v - \nabla^\alpha\big(A^0(u)^{-1}A^j(u)\partial_j v\big)\big] \qquad (5.20)$$
$$+A^0(u)^{-1}B(u)\nabla^\alpha v - \nabla^\alpha(A^0(u)^{-1}B(u)v)\Big\}.$$

Taking the inner product in $L^2(\mathbb{R}^n)$ of both sides of (5.19) with $\nabla^\alpha v$ and summing up for $0 \le |\alpha| \le s$, we obtain (cf. the proof of uniqueness above and Chapter 3, respectively)

$$\frac{d}{dt}\sum_{|\alpha|\le s}\int_{\mathbf{R}^n} A^0(u)\nabla^\alpha v\cdot\overline{\nabla^\alpha v} = \qquad (5.21)$$
$$\mathrm{Re}\sum_{|\alpha|\le s}\int_{\mathbf{R}^n}(D'A(u)-2B(u))\nabla^\alpha v\cdot\overline{\nabla^\alpha v} + 2\,\mathrm{Re}\int_{\mathbf{R}^n} F_\alpha\cdot\overline{\nabla^\alpha v},$$

where A is short for $(A^0, A^1, \ldots, A^n)$.
Using the induction hypothesis and Step 1 we obtain for $t \in [0, T_*]$:

$$\|u(t,\cdot)\|_\infty \le g_2,$$
$$\|\nabla u(t,\cdot)\|_\infty \le \kappa_s |u|_{s,T_*} \le \kappa_s R,$$
$$\|\partial_t u(t,\cdot)\|_\infty \le c(g_2)R.$$

This implies for all $(t,x) \in [0,T_*]\times\mathbb{R}^n$:

$$|D'A(u(t,x))| \le c(g_2)R, \qquad (5.22)$$

$$|B(u(t,x))| \le c(g_2). \qquad (5.23)$$

Remark: We notice the following relation:

$$B(0) = 0 \quad\Longrightarrow\quad \forall (t,x)\in[0,T_*]\times\mathbb{R}^n: \quad |B(u(t,x))| \le c(g_2)R. \qquad (5.24)$$

In particular, the condition $B(0) = 0$ is fulfilled in our application to nonlinear wave equations.
If $\alpha = 0$ we have $F_\alpha = 0$. Let $s' := |\alpha| > 0$, $\nabla^{s'}$ as in Chapter 4. An application of Lemma 4.9 implies

$$\|F_\alpha\|_2 \le c(g_2)\Big\{\sum_{j=1}^{n}\big[\|\nabla(A^0(u)^{-1}A^j(u))\|_\infty\|\nabla^{s'-1}\partial_j v\|_2 \qquad (5.25)$$
$$+\|\nabla^{s'}(A^0(u)^{-1}A^j(u))\|_2\|\partial_j v\|_\infty\big]$$
$$+\big[\|\nabla(A^0(u)^{-1}B(u))\|_\infty\|\nabla^{s'-1}v\|_2 + \|\nabla^{s'}(A^0(u)^{-1}B(u))\|_2\|v\|_\infty\big]\Big\}$$

$$\leq \; c(g_2)\|v\|_{s,2}\Big\{\sum_{j=1}^{n}\Big[\|\nabla(A^0(u)^{-1}A^j(u))\|_\infty + \|\nabla^{s'}(A^0(u)^{-1}A^j(u))\|_2\Big]$$

$$+\,\|\nabla(A^0(u)^{-1}B(u))\|_\infty + \|\nabla^{s'}(A^0(u)^{-1}B(u))\|_2\Big\}$$

$$\leq \; c(g_2)\|v\|_s\left\{\sum_{j=1}^{n}\Big[c(g_2)R + c(g_2)\|u\|_{s,2}\Big] + c(g_2)R + c(g_2)\|u\|_{s,2}\right\}$$

(according to Lemma 4.7 and the remark following the proof

there (g_2 may be larger than 1))

$$\leq \; c(g_2)\|v\|_{s,2}R.$$

Let rhs denote the right-hand side of equation (5.21). From (5.22), (5.23), (5.25) we conclude

$$|rhs| \leq c(g_2)(R+1)\|v\|_{s,2}^2. \tag{5.26}$$

Remark: (Cf. (5.24))

$$B(0) = 0 \quad \Longrightarrow \quad |rhs| \leq c(g_2)R\|v\|_{s,2}^2. \tag{5.27}$$

Integration on both sides of (5.21) from 0 to t yields

$$\|v(t)\|_{s,2}^2 \leq c_0(g_2)\sum_{|\alpha|\leq s}\int_{\mathbf{R}^n} A^0(u)\nabla^\alpha v\cdot\overline{\nabla^\alpha v}$$

$$\leq c_0(g_2)\Big\{\sum_{|\alpha|\leq s} A^0(u(t=0))\nabla^\alpha v(t=0)\cdot\overline{\nabla^\alpha v(t=0)} + c(g_2)(R+1)\int_0^t \|v(r)\|_{s,2}^2 dr\Big\}$$

$$\leq c_0^2(g_2)\|v_0\|_{s,2}^2 + c_0(g_2)c(g_2)(R+1)\int_0^t \|v(r)\|_{s,2}^2 dr.$$

Gronwall's inequality, Lemma 4.1, implies

$$\begin{aligned}\|v(t)\|_{s,2} \;&\leq\; c_0(g_2)\|u_0^{k+1}\|_{s,2}\,\mathrm{e}^{c(g_2)(R+1)t} \\ &\leq\; c_0(g_2)\|u_0\|_{s,2}\,\mathrm{e}^{c(g_2)(R+1)T_*}.\end{aligned} \tag{5.28}$$

Now we choose R such that

$$R \geq c_0(g_2)\|u_0\|_{s,2}\mathrm{e}^{c(g_2)}$$

holds.

Observing the condition on R postulated in (5.16) we choose

$$R := \|u_0\|_{s,2}\max\left\{1, c_0(g_2)\mathrm{e}^{c(g_2)}\right\},$$

in particular we have

$$R = R(\|u_0\|_{s,2}, g_2).$$

Moreover, we choose T_* such that

$$T_* \leq T_2 = T_2(\|u_0\|_{s,2}, g_2) := \frac{1}{R+1}$$

holds.

Observing the condition on T_* postulated in (5.18) we require

$$T_* := \min(T_1, T_2) > 0$$

which means in particular that

$$T_* = T_*(\|u_0\|_{s,2}, g_2).$$

With these choices for R and T_* we obtain from (5.28)

$$|u^{k+1}|_{s,T_*} = |v|_{s,T_*} \leq R.$$

This proves (i) and completes the proof of Lemma 5.2.

Q.E.D.

Remark: If g_2 is fixed we have

$$\|u_0\|_{s,2} \to 0 \quad \Longrightarrow \quad R + L \to 0, \quad T_1 = \frac{g_2 - g_1}{\kappa_{s-1} L} \to \infty.$$

If, additionally, $B(0) = 0$ we have

$$T_2 = \frac{1}{R}$$

and hence

$$\|u_0\|_{s,2} \to 0 \quad \Longrightarrow \quad T_2 \to \infty \quad , \quad T_* \to \infty.$$

More precisely, it holds

$$B(0) = 0 \quad \Longrightarrow \quad T_* \geq \frac{c(g_2)}{\sqrt{\|u_0\|_{s,2}^2 + \|u_0\|_{s,2}^4}} \tag{5.29}$$

that is to say, if $u_0 = \varepsilon\varphi$, $\varepsilon > 0$, $\varphi \in W^{s,2}$, then

$$T_* \geq c(\varphi, g_2)\varepsilon^{-1} \quad \text{as} \quad \varepsilon \downarrow 0.$$

Having proved the boundedness of $(u^k)_k$ in high norms we turn now to the investigation of convergence of subsequences in appropriate low norms. Note that our estimates are not strong enough to prove convergence in the high norms. The idea to prove convergence only in lower norms goes back to P.D. Lax [97] and T. Kato [81]; compare the remarks in [114].

Lemma 5.3 *There are nonnegative real numbers $T, \alpha, \beta_1, \beta_2, \ldots$ with $0 < \alpha < 1$, $\sum_{k=1}^{\infty} \beta_k < \infty$ and $0 < T \le T_*$ such that*

$$\forall k \in \mathbb{N} \qquad |u^{k+1} - u^k|_{0,T} \le \alpha |u^k - u^{k-1}|_{0,T} + \beta_k.$$

Proof: Let $0 \le t \le T \le T_*$, T still arbitrary with $0 < T \le T_*$. From the differential equations for u^{k+1} and u^k, respectively, we obtain

$$\begin{aligned} & A^0(u^k)\partial_t(u^{k+1} - u^k) + \sum_{j=1}^{n} A^j(u^k)\partial_j(u^{k+1} - u^k) + B(u^k)(u^{k+1} - u^k) \qquad (5.30) \\ = \;& (A^0(u^{k-1}) - A^0(u^k))\partial_t u^k + \sum_{j=1}^{n}(A^j(u^{k-1}) - A^j(u^k))\partial_j u^k + (B(u^{k-1}) - B(u^k))u^k \\ =: \;& F_k. \end{aligned}$$

F_k satisfies

$$\begin{aligned} F_k = \int_0^1 \Big\{ & (\nabla_u A^0)(u^{k-1} + r(u^k - u^{k-1}))(u^{k-1} - u^k)\partial_t u^k \\ & + \sum_{j=1}^{n} (\nabla_u A^j)(u^{k-1} + r(u^k - u^{k-1}))(u^{k-1} - u^k)\partial_j u^k \\ & + (\nabla_u B)(u^{k-1} + r(u^k - u^{k-1}))(u^{k-1} - u^k)u^k \Big\} dr. \end{aligned}$$

This implies

$$\|F_k\|_2 \le c(g_2, R)\|u^k - u^{k-1}\|_2,$$

where

$$c(g_2, R) = c(g_2, \|u_0\|_{s,2}) \to 0 \quad \text{as} \quad R \to 0.$$

Taking the supremum for $t \in [0, T]$ we get

$$|F_k|_{0,T} \le c(g_2, R)|u^k - u^{k-1}|_{0,T}. \qquad (5.31)$$

The technique of proving energy estimates — which we multiply used above — finally leads to

$$|u^{k+1} - u^k|_{0,T} \le \sqrt{c(g_2, R)}\, \mathrm{e}^{c(g_2,R)T} \left(\|u_0^{k+1} - u_0^k\|_2 + \sqrt{T}|u^k - u^{k-1}|_{0,T} \right). \qquad (5.32)$$

Using Lemma 4.3 we get the following inequalities:

$$\begin{aligned} \|u_0^{k+1} - u_0^k\|_2 &\le \|u_0^{k+1} - u_0\|_2 + \|u_0 - u_0^k\|_2 \\ &\le c\varepsilon_0 2^{-(k+1)}\|u_0\|_{1,2} + c\varepsilon_0 2^{-k}\|u_0\|_{1,2} \\ &\le c2^{-k}, \end{aligned}$$

where $c = c(\|u_0\|_{1,2})$ is a positive constant.
Choosing $0 < T \leq T_*$ such that

$$\sqrt{c(g_2, R)T}\, e^{c(g_2,R)T} < 1$$

holds, we see that (5.32) implies

$$|u^{k+1} - u^k|_{0,T} \leq \alpha |u^k - u^{k-1}|_{0,T} + \beta_k,$$

with

$$\beta_k := c2^{-k} \quad \text{and} \quad \alpha := \sqrt{c(g_2, R)T} e^{c(g_2,R)T} < 1.$$

Q.E.D.
The constant $c(g_2, R)$ appearing in the proof of Lemma 5.3 satisfies

$$c(g_2, R) \to 0 \quad \text{as} \quad R \to 0,$$

hence if

$$T_* \to T_\infty \in (0, \infty] \quad \text{as} \quad R \to 0,$$

then

$$T \to T_\infty \quad \text{as} \quad R \to 0.$$

More precisely, we have

$$c(g_2, R) \leq c(g_2)R.$$

Let $M = M(g_2)$ be defined by the equation

$$M e^{c(g_2)M^2} = \frac{1}{2\sqrt{c(g_2)}},$$

and let

$$T = T(g_2, R) := \frac{M^2}{R}.$$

With this choice we have $\alpha < 1$ and

$$T \geq \frac{c(g_2)}{\|u_0\|_{s,2}} \quad \text{(provided} \quad T \leq T_*). \tag{5.33}$$

As an easy consequence of Lemma 5.3 we get

Corollary 5.4 *There exists a $u \in C^0([0,T], L^2)$ such that $(u^k)_k$ converges to u in $C^0([0,T], L^2)$.*

Proof: It is an elementary observation that

$$0 \leq a_{k+1} \leq \alpha a_k + \beta_k\,,\ \beta_k \geq 0\,,\ k \in \mathbb{N}\,,\ 0 < \alpha < 1\,,\ \sum_{k=1}^{\infty} \beta_k < \infty$$

implies

$$\sum_{k=1}^{\infty} a_k < \infty.$$

$\Big($ Prove the boundedness of $(a_k)_k$ and then of $(\sum_{k=1}^{N} a_k)_N$. $\Big)$
With $a_k := |u^{k+1} - u^k|_{0,T}$ we conclude that $(u^k)_k$ is a Cauchy sequence in the Banach space $C^0([0,T], L^2)$.

Q.E.D.

Lemma 5.2 means in particular:

$$\forall k \in \mathbb{N}_0 : \quad |u^k|_{s,T} \leq R.$$

The inequality of Gagliardo–Nirenberg, Theorem 4.4, implies

$$\begin{aligned} |u^k - u^m|_{s',T} &\leq c|u^k - u^m|_{0,T}^{1-s'/s} |u^k - u^m|_{s,T}^{s'/s} \\ &\leq c|u^k - u^m|_{0,T}^{1-s'/s}, \quad k, m \in \mathbb{N}_0, \quad s' \in \{0, 1, \ldots, s\}. \end{aligned}$$

Therefore $(u^k)_k$ is a Cauchy sequence in $C([0,T], W^{s',2})$ for $0 \leq s' < s$, $s' \in \mathbb{N}_0$. This implies

$$u \in C^0([0,T], W^{s',2}), \quad \lim_{k\to\infty} |u^k - u|_{s,',T} = 0, \tag{5.34}$$

for all $s' \in \mathbb{N}_0$ with $0 \leq s' < s$.

Remark: If Theorem 4.4 is used in Sobolev spaces with fractional derivatives, i.e. for $s' \in [0, \infty)$ arbitrary then it would immediately follow by Sobolev's imbedding theorem for s' with $s > s' > n/2 + 1$ that

$$u \in C^0([0,T], C_b^1) \quad \text{and} \quad \partial_t u \in C^0([0,T], C_b^0).$$

In particular,

$$u \in C_b^1([0,T] \times \mathbb{R}^n)$$

and u is a classical solution. Since we do not work with fractional derivatives we shall use different arguments. (See the book of R. A. Adams [1] for a discussion of fractional derivatives and associated Sobolev spaces.)

Taking $s' := s - 1 > n/2$ we obtain, using Sobolev's imbedding theorem,

$$u^k \to u \quad \text{in} \quad C^0([0,T], C_b^0) \quad \text{as} \quad k \to \infty,$$

which implies, together with Lemma 4.3,

$$\sup_{(t,x)\in[0,T]\times\mathbb{R}^n} |u(t,x)| \leq g_2.$$

The equation

$$\partial_t u^{k+1} = -A^0(u^k)^{-1}\left(\sum_{j=1}^n A^j(u^k)\partial_j u^{k+1} + B(u^k)u^{k+1}\right)$$

now implies that $(\partial_t u^{k+1})_k$ converges in $C^0([0,T],W^{s'-1,2})$ which is continuously imbedded in $C^0([0,T],L^2)$. Hence

$$u \in C^1([0,T],L^2)$$

and the differential equation

$$A^0(u)\partial_t u = \sum_{j=1}^n A^j(u)\partial_j u + B(u)u$$

is satisfied (being an equality in $C^0([0,T],L^2)$).

Lemma 5.5 *$u \in L^\infty([0,T],W^{s,2})$, $|u|_{s,T} \le R$.*

Proof: From Lemma 5.2 we obtain for all $t \in [0,T]$, $k \in \mathbb{N}_0$,

$$\|u^k(t)\|_{s,2} \le R.$$

Thus, for $t \in [0,T]$ fixed, there is a subsequence (again denoted by $(u^k(t))_k$) and a $w_t \in W^{s,2}$ such that

$$u^k(t) \rightharpoonup w_t \quad \text{in} \quad W^{s,2} \quad \text{as} \quad k \to \infty,$$

$$\|w_t\|_{s,2} \le R.$$

For $h \in L^2$

$$g \to F_h g := \int_{\mathbb{R}^n} g\bar{h}$$

defines a continuous linear map on $W^{s,2}$, hence, by the Riesz representation theorem, we have

$$\forall h \in L^2\ ,\ \exists \psi = \psi(h) \in W^{s,2} \quad \forall g \in W^{s,2}: \quad F_h g = \langle g, \psi(h)\rangle_s,$$

where $\langle\cdot,\cdot\rangle_s$ denotes the inner product in $W^{s,2}$. Consequently, we get for all $h \in L^2$

$$\begin{aligned}\int_{\mathbb{R}^n} w_t\bar{h} &= F_h(w_t) = \langle w_t, \psi(h)\rangle_s = \lim_{k\to\infty}\langle u^k(t), \psi(h)\rangle_s \\ &= \lim_{k\to\infty} F_h(u^k(t)) = \lim_{k\to\infty}\int_{\mathbb{R}^n} u^k(t)\bar{h} = \int_{\mathbb{R}^n} u(t)\bar{h}.\end{aligned} \tag{5.35}$$

The last equality follows from (5.34).
From (5.35) we obtain

$$u(t) = w_t,$$

which completes the proof of Lemma 5.5.

Q.E.D.

Since each subsequence of $(u^k(t))_k$ in the proof of Lemma 5.5 will converge to the same limit $u(t)$ we conclude

$$\forall t \in [0,T]: \quad u^k(t) \rightharpoonup u(t) \quad \text{in} \quad W^{s,2}. \tag{5.36}$$

Let $C_w([0,T],E)$ {and $Lip([0,T],E)$} denote the space of weakly continuous {resp. Lipschitz continuous} functions from $[0,T]$ into a Banach space E.

Lemma 5.6 $u \in C_w([0,T],W^{s,2}) \cap Lip([0,T],W^{s-1,2})$.

PROOF:

1. Let $(t_n)_n \subset [0,1]$, $\lim_{n\to\infty} t_n = t_0$. Since $u \in C^0([0,T], W^{s-1,2})$ we have

$$\lim_{n\to\infty} \|u(t_n)\|_{s-1,2} = \|u(t_0)\|_{s-1,2}$$

and because

$$\|u(t_n)\|_{s,2} \le R$$

we get (cf. the proof of Lemma 5.5)

$$u(t_n) \rightharpoonup u(t_0) \quad \text{in} \quad W^{s,2} \quad \text{as} \quad n \to \infty.$$

This proves $u \in C_w([0,T], W^{s,2})$.

2.

$$u^k(t_2) - u^k(t_1) = \int_{t_1}^{t_2} \partial_t u^k(r)dr \quad \text{for} \quad 0 \le t_1 \le t_2 \le T.$$

Hence we obtain using Lemma 5.2

$$\|u^k(t_2) - u^k(t_1)\|_{s-1,2} \le L|t_2 - t_1|.$$

By the help of (5.34) this implies in the limit as $k \to \infty$ that $u \in Lip([0,T], W^{s-1,2})$ (with the Lipschitz constant L having been defined in the proof of Lemma 5.2).

Q.E.D.

In order to prove Theorem 5.1 it remains to show that

$$u \in C^0([0,T), W^{s,2}) \cap C^1([0,T), W^{s-1,2})$$

(which implies

$$u \in C_b^1([0,T) \times \mathbb{R}^n)$$

by Sobolev's imbedding theorem).
For this purpose three reduction steps will be made first.

Step 1: It is sufficient to prove that $u \in C^0([0,T], W^{s,2})$. Then the differential equation for u implies $\partial_t u \in C^0([0,T], W^{s-1,2})$, hence $u \in C^1([0,T], W^{s-1,2})$.

Step 2: It is sufficient to prove the continuity on the right of $t \in [0,T)$, because we may consider v where

$$v(t) := u(T-t) \quad \text{for} \quad 0 \le t \le T.$$

v satisfies

$$A^0(v)\partial_t v + \sum_{j=1}^n (-A^j(v))\partial_j v + (-B(v))v = 0,$$

$$v(0) = u(T).$$

Then the continuity of v on the right in $t \in [0,T)$ implies the continuity of u on the left of $t \in (0,T]$.

Step 3: It is sufficient to prove the continuity on the right of $t=0$, because for $t = t_0 \in (0,T)$ we may consider the initial value problem

$$A^0(w)\partial_t w + \sum_{j=1}^{n} A^j(w)\partial_j w + B(w)w = 0, \tag{5.37}$$

$$w(t = t_0) = u(t_0), \tag{5.38}$$

for a vector-valued function $w = w(t,x)$, $t \geq t_0$, $x \in \mathbb{R}^n$. From what we have proved above we know that there is a (unique) solution w of (5.37), (5.38) with

$$w \in C^0(I, W^{s-1,2}) \cap C^1(I, L^2) \cap C_w(I, W^{s,2}) \cap Lip(I, W^{s-1,2}) \cap L^\infty(I, W^{s,2})$$

where

$$\begin{aligned} I &:= [t_0, T_{t_0}], \quad \text{for some } T_{t_0} > t_0, \\ T_{t_0} &= T_{t_0}(\|u(t_0)\|_{s,2}, \tilde{g}_2), \\ \tilde{g}_1 &:= \kappa_s \|u(t_0)\|_{s,2} < \tilde{g}_2. \end{aligned}$$

$w(t)$ coincides with $u(t+t_0)$ on $[0, \min(T-t_0, T_{t_0})]$ because of the uniqueness of solutions in the class $\mathcal{U}$ defined in (5.12). Therefore the continuity on the right of w at the initial time $t = t_0$ implies the continuity on the right of u in t_0.
After these three reductions we turn to the proof of the continuity on the right of u in $t=0$. For this purpose we introduce the norm $\|\cdot\|_{s,A^0(t)}$, $0 \leq t \leq T$, which will be equivalent to the norm $\|\cdot\|_{s,2}$ on $W^{s,2}$:

$$\|v\|_{s,A^0(t)} := \left(\sum_{|\alpha| \leq s} \int_{\mathbb{R}^n} A^0(t) \nabla^\alpha v \cdot \overline{\nabla^\alpha v} \right)^{1/2},$$

where

$$A^0(t) := A^0(u(t,\cdot)) \quad \text{for} \quad t \in [0,T].$$

Since

$$c_0^{-1} w \cdot \bar{w} \leq A^0(t) w \cdot \bar{w} \leq c_0 w \cdot \bar{w},$$

we obtain

$$c_0^{-1} \|v\|_{s,2}^2 \leq \|v\|_{s,A_0(t)}^2 \leq c_0 \|v\|_{s,2}^2$$

for all $t \in [0,T]$.

The inequalities

$$
\begin{aligned}
&\left| \|u(t)\|^2_{s,A^0(t)} - \|u(t)\|^2_{s,A^0(0)} \right| \\
\leq\ & \sum_{|\alpha|\leq s} \sup_{x\in\mathbf{R}^n} \left| A^0(u(t,x)) - A^0(u_0(x)) \right| \|\nabla^\alpha u(t)\|_2^2 \\
\leq\ & \sup_{x\in\mathbf{R}^n} \int_0^1 \left| (\nabla_u A^0)(u_0(x) + r(u(t,x) - u_0(x))) \right| dr \, |u(t,x) - u_0(x)| \, \|u(t)\|^2_{s,2} \\
\leq\ & c(g_2) R^2 \|u(t) - u_0\|_{s-1,2}
\end{aligned}
$$

imply

$$\limsup_{t\downarrow 0} \|u(t)\|^2_{s,A^0(t)} = \limsup_{t\downarrow 0} \|u(t)\|^2_{s,A^0(0)}.$$

Since $u \in C_w([0,T], W^{s,2})$ (by Lemma 5.6) it now suffices to prove

$$\|u_0\|^2_{s,A^0(0)} \geq \limsup_{t\downarrow 0} \|u(t)\|^2_{s,A^0(t)}. \tag{5.39}$$

$\Big($ Observe that in a Hilbert space $\mathcal{H}$ with norm $\|\cdot\|_{\mathcal{H}}$ one has:

$$
\begin{aligned}
& w_n \rightharpoonup w \text{ in } \mathcal{H} \quad \text{as } n\to\infty \\
\Longrightarrow\ & \left(w_n \to w \text{ in } \mathcal{H} \text{ as } n\to\infty \Leftrightarrow \|w\|_{\mathcal{H}} \geq \limsup_{n\to\infty} \|w_n\|_{\mathcal{H}} \right). \Big)
\end{aligned}
$$

The inequality (5.39) immediately follows from the following Lemma which remains to be proved.

Lemma 5.7 *There is a constant $c = c(g_2, R) > 0$ such that for all $t \in [0,T]$ the inequality*

$$\|u(t)\|^2_{s,A^0(t)} \leq \|u_0\|^2_{s,A^0(0)} + ct$$

holds.

PROOF: From the proof of Lemma 5.2 we know that

$$
\begin{aligned}
&\sum_{|\alpha|\leq s} \int_{\mathbf{R}^n} A^0(u^k)\nabla^\alpha u^{k+1} \cdot \overline{\nabla^\alpha u^{k+1}} \leq \\
&\qquad \sum_{|\alpha|\leq s} \int_{\mathbf{R}^n} A^0(u_0^k)\nabla^\alpha u_0^{k+1} \cdot \overline{\nabla^\alpha u_0^{k+1}} + \int_0^t cR
\end{aligned}
\tag{5.40}
$$

holds, where $c = c(g_2, R)$.

We have

$$\begin{aligned}
&\left| \int_{\mathbf{R}^n} A^0(u_0^k)\nabla^\alpha u_0^{k+1} \cdot \overline{\nabla^\alpha u_0^{k+1}} - \int_{\mathbf{R}^n} A^0(u_0)\nabla^\alpha u_0 \cdot \overline{\nabla^\alpha u_0} \right| \\
&\quad \le \left| \int_{\mathbf{R}^n} \left(A^0(u_0^k) - A^0(u_0) \right) \nabla^\alpha u_0 \cdot \overline{\nabla^\alpha u_0} \right| \\
&\quad + \left| \int_{\mathbf{R}^n} A^0(u_0^k) \left(\nabla^\alpha u_0^{k+1} - \nabla^\alpha u_0 \right) \cdot \overline{\nabla^\alpha u_0^{k+1}} \right| \\
&\quad + \left| \int_{\mathbf{R}^n} A^0(u_0^k)\nabla^\alpha u_0 \cdot \left(\overline{\nabla^\alpha u_0^{k+1} - \nabla^\alpha u_0} \right) \right|.
\end{aligned}$$

Since

$$\lim_{k\to\infty} \|u_0^k - u_0\|_{s,2} = 0$$

we now obtain

$$\lim_{k\to\infty} \sum_{|\alpha|\le s} \int_{\mathbf{R}^n} A^0(u_0^k)\nabla^\alpha u_0^{k+1} \cdot \overline{\nabla^\alpha u_0^{k+1}} = \|u_0\|^2_{s,A^0(0)}. \tag{5.41}$$

From (5.36) (in $W^{s,2}$ with respect to $\|\cdot\|_{s,A^0(t)}$) we get

$$\|u(t)\|_{s,A^0(t)} \le \liminf_{k\to\infty} \|u^{k+1}(t)\|_{s,A^0(t)}. \tag{5.42}$$

The relations (5.40) – (5.42) imply

$$\|u(t)\|^2_{s,A^0(t)} \le \|u_0\|^2_{s,A^0(0)} + ct.$$

Q.E.D.

This also completes the proof of Theorem 5.1.

Q.E.D.

In the case $B(0) = 0$ (in particular for the application to nonlinear wave equations) we can choose T such that

$$T \ge \frac{c(g_2)}{\|u_0\|_{s,2}},$$

(according to (5.33), (5.29)). This means that the life span T_∞ of a classical solution can be estimated from below by

$$T_\infty \ge \frac{c(g_2)}{\|u_0\|_{s,2}}, \tag{5.43}$$

in particular if $u_0 = \varepsilon\varphi$, $\varepsilon > 0$, $\varphi \in W^{s,2}$, we get

$$T_\infty = T_\infty(\varepsilon) \ge c \cdot \varepsilon^{-1}, \tag{5.44}$$

where

$$c = c(g_2, \varphi) > 0.$$

This elemantary life span estimate can be sharpened in space dimensions $n \geq 2$, e.g. for $n = 3$ to

$$\lim_{\varepsilon \downarrow 0} \varepsilon \log T_\infty(\varepsilon) > 0.$$

This result is contained in Theorem 1.2, see also the papers by F. John [72] or John & Klainerman [75] and Section 10 for further results.

Theorem 5.1 can be improved with respect to the length T of the interval of existence: The dependence on $\|u_0\|_{s,2}$ and g_2 can be weakened as follows.

Theorem 5.8 *Assume (5.11) and $u_0 \in W^{s,2}$, $s, m \in \mathbb{N}$ with $s \geq m > n/2 + 1$. Let $g_1^m := \kappa_m \|u_0\|_{m,2}$ and $g_2^m > g_1^m$ arbitrary but fixed.*
Then there is a $T > 0$ such that there exists a unique classical solution $u \in C_b^1([0,T] \times \mathbb{R}^n)$ to the initial value problem (5.9), (5.10) with

$$\sup_{(t,x) \in [0,T] \times \mathbb{R}^n} |u(t,x)| \leq g_2^m$$

and

$$u \in C^0([0,T], W^{s,2}) \cap C^1([0,T], W^{s-1,2}).$$

T depends only on $\|u_0\|_{m,2}$ and on g_2^m.

PROOF: First we obtain a solution

$$u \in C^0([0,T_m], W^{m,2}) \cap C^1([0,T_m], W^{m-1,2})$$

according to Theorem 5.1, where

$$T_m = T_m(\|u_0\|_{m,2}, g_2^m)$$

is determined in the proof of Lemma 5.3.

Now we show that the approximating sequence $(u^k)_k$ is still bounded in high norms. We prove:

$$\exists R_s, L_s > 0 \quad \exists T \in (0, T_m] \quad \forall k \in \mathbb{N}_0 : \; |u^k|_{s,T} \leq R_s \,, \; |\partial_t u^k|_{s-1,T} \leq L_s \tag{5.45}$$

and T depends only on $\|u_0\|_{m,2}$ and g_2^m.

For this purpose we only have to look at the proof of Lemma 5.2 once more.

The basis of the induction ($k = 0$) yields the condition

$$R_s \geq \|u_0\|_{s,2}. \tag{5.46}$$

Since

$$\begin{aligned} |D'A(u) - 2B(u)| &\leq c(g_2^m)(R_m + 1), \\ \|\nabla^s B(u)\|_2 &\leq c(g_2^m) R_s, \\ \|v\|_{1,\infty} &\leq \kappa_{m-1} \|v\|_{m,2}, \end{aligned}$$

where $u := u^k$, $v := u^{k+1}$, (cf. (5.19), (5.20)), we obtain

$$\|F_\alpha\|_2 \le c(g_2^m)(R_m\|v\|_{s,2} + R_s\|v\|_{m,2}).$$

Denoting by rhs the right-hand side of (5.21) again we get

$$|rhs| \le c(g_2^m)\left((R_m+1)\|v\|_{s,2}^2 + R_sR_m\|v\|_{s,2}\right).$$

This implies

$$\begin{aligned}\|v(t)\|_{s,2}^2 &\le c(g_2^m)\left\{\|v_0\|_{s,2}^2 + \int_0^t \left((R_m+1)\|v(r)\|_{s,2}^2 + R_sR_m\|v(r)\|_{s,2}\right)dr\right\}\\ &\le c(g_2^m)\left\{\|v_0\|_{s,2}^2 + \int_0^t (R_m+1)\|v(r)\|_{s,2}^2\,dr + tR_s^2R_m + R_m\int_0^t \|v(r)\|_{s,2}^2\,dr\right\}.\end{aligned}$$

Gronwall's inequality, Lemma 4.1, implies

$$\|v(t)\|_{s,2}^2 \le c(g_2^m)\left(\|v_0\|_{s,2}^2 + TR_s^2R_m\right)\mathrm{e}^{c(g_2^m)(1+2R_m)t}$$

or, equivalently,

$$\|v(t)\|_{s,2} \le c(g_2^m)\left(\|v_0\|_{s,2} + \sqrt{TR_m}R_s\right)\mathrm{e}^{c(g_2^m)(1+2R_m)t}$$

which implies

$$|v|_{s,T} \le c(g_2^m)\left(\|v_0\|_{s,2} + \sqrt{TR_m}R_s\right)\mathrm{e}^{c(g_2^m)(1+2R_m)T}.$$

We choose $T \in (0, T_m]$ such that

$$c(g_2^m)\sqrt{TR_m}\,\mathrm{e}^{c(g_2^m)(1+2R_m)T} \le \frac{1}{2} \tag{5.47}$$

and R_s (observing (5.46)) such that

$$R_s \ge 2c(g_2^m)\|v_0\|_s\,\mathrm{e}^{c(g_2^m)(1+2R_m)T}.$$

Then

$$|v|_{s,T} \le R_s$$

and

$$|\partial_t v|_{s-1,T} \le c(g_2^m)R_s =: L_s.$$

This proves (5.45).
T is determined by (5.47) and depends only on $\|u_0\|_{m,2}$ and g_2^m. The remaining considerations are literally the same as those in the proof of Theorem 5.1. This completes the

proof of Theorem 5.8.

Q.E.D.

In the case $B(0) = 0$ we have for the life span T_∞

$$T_\infty \geq \frac{c(g_2^m)}{\|u_0\|_{m,2}}$$

(cf. (5.43), (5.44)).

We conclude this chapter with some remarks on the regularity assumption (C^∞) on the coefficients. This assumption was made for simplicity in order to be able to give a self-contained proof of the local existence theorems above only using the elementary results from Chapter 3. It was already mentioned there that these assumptions can be weakened in remarkable ways, for example using the theory of T. Kato culminating in a local existence theorem for quasilinear hyperbolic systems with less restrictive requirements on the coefficients, see [81, 82] and the paper of Hughes, Kato & Marsden [51].

6 High energy estimates

In this chapter we shall prove an energy estimate for the local solution $u = (\partial_t y, \nabla y)$ of the initial value problem (5.7), (5.8), where y solves (5.1), (5.2) with the assumptions (5.3) – (5.6) on the nonlinearity (case $\alpha = 1$). Without loss of generality for further investigations we assume the constant g_2^m appearing in the local existence Theorem 5.8 to be sufficiently small as it will be needed in the proof of the next theorem.

Theorem 6.1 *There is a constant $c > 0$ which is independent of T and u_0 such that the local solution u satisfies*

$$\forall t \in [0,T]: \quad \|u(t)\|_{s,2} \leq c\|u_0\|_{s,2} \exp\{c\int_0^t \|Du(r)\|_\infty dr\}. \tag{6.1}$$

The proof below literally works for $s > n/2+2$ and only in this case we shall make use of Theorem 6.1. If one uses the Gagliardo–Nirenberg inequality, Theorem 4.4, in Sobolev spaces $W^{\tau,2}$, where $\tau \in [0,\infty)$ is not necessarily an integer, then the proof below also works for arbitrary $s > n/2+1$. Otherwise one can get the result for the remaining case $s = [\frac{n}{2}]+2$ by considering $u^\varepsilon := J_\varepsilon u$ (the convolution with the Friedrichs mollifier j_ε, cf. Chapter 4), then proving the energy estimate for u^ε and finally letting ε tend to zero. For this purpose corresponding commutator estimates for the nonlinear terms are needed. These considerations are carried out e.g. by F. Willems [198], see also S. Kawashima [84].
Now let $s > n/2+2$.
PROOF of Theorem 6.1:
Let $(u_0^k)_k \subset W^{s+1,2}$ be a sequence that approximates u_0 in $W^{s,2}$ as $k \to \infty$ and let u^k be the solution to (5.7) with initial value $u^k(t=0) = u_0^k$.
We have

$$\forall k \in \mathbb{N}: \quad u^k \in C^0([0,T_1], W^{s+1,2}) \cap C^1([0,T_1], W^{s,2}),$$

where

$$T_1 := \inf\{\text{length of the existence interval for } u^k \text{ and } u \text{ according to Theorem 5.8}, k \in \mathbb{N}\}.$$

Without loss of generality we may assume

$$0 < T_1 = T.$$

(Observe that $\|u_0^k\|_{m,2} \to \|u_0\|_{m,2}$ as $k \to \infty$, from Theorem 5.8, $s \geq m > n/2+1$).
Remark: Theorem 5.1 would not have been sufficient because

$$\|u_0^k\|_{s+1,2} \to \infty \quad \text{as} \quad k \to \infty \quad \text{if } u_0 \notin W^{s+1,2}.$$

Since

$$u^k \in C^1([0,T], W^{s,2})$$

we have for y^k with

$$u_1^k = \partial_t y^k, u_2^k = \partial_1 y^k, \ldots, u_{n+1}^k = \partial_n y^k,$$

the property

$$y_{tt}^k \in C^0([0,T], W^{s,2}).$$

We apply $\nabla^\alpha, 0 \leq |\alpha| \leq s$, on both sides of the differential equation for y^k:

$$y_{tt}^k - \Delta y^k = \sum_{i,j=1}^n a_{ij}(Dy^k)\partial_i\partial_j y^k,$$

and then take the inner product in L^2 with $\nabla^\alpha y_t^k$. For simplicity we omit the index k for y^k and we assume without loss of generality that all functions are real-valued. Then we obtain, dropping the parameter t most of the time,

$$\begin{aligned} \frac{1}{2}\frac{d}{dt}\{\|\nabla^\alpha y_t\|_2^2 + \|\nabla^\alpha \nabla y\|_2^2\} &= \sum_{i,j=1}^n \langle a_{ij}(Dy)\partial_i\partial_j\nabla^\alpha y, \nabla^\alpha y_t\rangle \\ &+ \sum_{i,j=1}^n \langle \nabla^\alpha\{a_{ij}(Dy)\partial_i\partial_j y\} - a_{ij}(Dy)\partial_i\partial_j\nabla^\alpha y, \nabla^\alpha y_t\rangle \\ &\equiv \mathrm{I} + \mathrm{II}. \end{aligned} \tag{6.2}$$

Integrating by parts in I we obtain

$$\begin{aligned} \mathrm{I} &= -\sum_{i,j=1}^n \langle(\partial_j a_{ij}(Dy))\partial_i\nabla^\alpha y, \nabla^\alpha y_t\rangle - \sum_{i,j=1}^n \langle a_{ij}(Dy)\nabla^\alpha\partial_i y, \nabla^\alpha\partial_j y_t\rangle \\ &\equiv \mathrm{I.1} + \mathrm{I.2}. \end{aligned} \tag{6.3}$$

The term I.1 can be estimated directly by

$$\begin{aligned} |\mathrm{I.1}| &\leq \sum_{i,j=1}^n \|(\nabla_u a_{ij})(Dy)\partial_i Dy\|_\infty(\|\nabla^\alpha\partial_j y\|_2^2 + \|\nabla^\alpha y_t\|_2^2) \\ &\leq c\|Du^k\|_\infty \|u^k\|_{s,2}^2, \end{aligned} \tag{6.4}$$

where c denotes here and in the sequel (various) positive constants not depending on T or u_0.

The term I.2 is split as follows.

$$\begin{aligned} \mathrm{I.2} &= -\frac{1}{2}\frac{d}{dt}\sum_{i,j=1}^n \langle a_{ij}(Dy)\partial_i\nabla^\alpha y, \partial_j\nabla^\alpha y\rangle \\ &\quad +\frac{1}{2}\sum_{i,j=1}^n \langle(\partial_t a_{ij}(Dy))\partial_i\nabla^\alpha y, \partial_j\nabla^\alpha y\rangle \\ &\equiv \mathrm{I.2.a} + \mathrm{I.2.b}. \end{aligned} \tag{6.5}$$

The term I.2.b is estimated in the same way as I.1, see (6.4):

$$|\text{I.2.b}| \le c\|Du^k\|_\infty \|u^k\|_{s,2}^2. \tag{6.6}$$

For the term I.2.a we obtain

$$\begin{aligned}\sum_{|\alpha|\le s} \int_0^t \text{I.2.a}(r)dr &= -\frac{1}{2}\sum_{|\alpha|\le s}\sum_{i,j=1}^n \langle a_{ij}(Dy)\partial_i\nabla^\alpha y, \partial_j\nabla^\alpha y\rangle(t) \\ &+\frac{1}{2}\sum_{|\alpha|\le s}\sum_{i,j=1}^n \langle a_{ij}(Dy(t=0))\partial_i\nabla^\alpha y(t=0), \partial_j\nabla^\alpha y(t=0)\rangle.\end{aligned} \tag{6.7}$$

The first term on the right-hand side of equation (6.7) can be incorporated into the left-hand side of equation (6.2) (after integration with respect to t there) due to the assumption (5.6) if g_2^m is sufficiently small.
Now we consider the term II from equation (6.2). We have

$$\text{II} = 0 \quad \text{if} \quad \alpha = 0,$$

hence let $\alpha \in \mathbb{N}_0^n$ with $|\alpha| \ge 1$. Then we get

$$\begin{aligned}|\text{II}| &\le \sum_{i,j=1}^n \|\nabla^\alpha\{a_{ij}(Dy)\partial_i\partial_j y\} - a_{ij}(Dy)\nabla^\alpha\partial_i\partial_j y\|_2\|\nabla^\alpha y_t\|_2 \\ &\le c\sum_{i,j=1}^n \{\|\nabla(a_{ij}(Dy))\|_\infty\|\nabla^{|\alpha|-1}\partial_i\partial_j y\|_2 + \|\nabla^{|\alpha|}a_{ij}(Dy)\|_2\|\partial_i\partial_j y\|_\infty\}\|\nabla^\alpha y_t\|_2 \\ &\qquad \text{(according to Lemma 4.9)} \\ &\le c\{\|\nabla Dy\|_\infty\|\nabla y\|_{s,2} + \|Dy\|_{s,2}\|\nabla^2 y\|_\infty\}\|\nabla^\alpha y_t\|_2 \\ &\qquad \text{(according to Lemma 4.8)} \\ &\le c\|Du^k\|_\infty\|u^k\|_{s,2}^2.\end{aligned} \tag{6.8}$$

Combining (6.2) – (6.8) we obtain

$$\|u^k(t)\|_{s,2}^2 \le c\|u_0^k\|_{s,2}^2 + c\int_0^t \|Du^k(r)\|_\infty\|u^k(r)\|_{s,2}^2 dr$$

and Gronwall's inequality, Lemma 4.1, implies the desired energy estimate for u^k:

$$\|u^k(t)\|_{s,2} \le c\|u_0^k\|_{s,2}\exp\{c\int_0^t \|Du^k(r)\|_\infty dr\}, \quad 0 \le t \le T. \tag{6.9}$$

It remains to investigate the limit $k \to \infty$. For this purpose we notice that

$$\begin{aligned}&A^0(u)\partial_t(u-u^k) + \sum_{j=1}^n A^j(u)\partial_j(u-u^k) \\ &= (A^0(u^k) - A^0(u))\partial_t u^k + \sum_{j=1}^n (A^j(u^k) - A^j(u))\partial_j u^k \\ &=: F_k.\end{aligned}$$

(The zero-order coefficient B is zero here.)
With the help of the arguments already used in the proof of Lemma 5.3 this implies:

$$\begin{aligned} \|F_k\|_2 &\leq c\|u^k - u\|_2, \\ \|(u^k - u)(t)\|_2^2 &\leq c\left(\|u_0^k - u_0\|_2^2 + \int_0^t \|(u^k - u)(r)\|_2^2 dr\right) \end{aligned}$$

and then

$$\|(u^k - u)(t)\|_2 \leq c\|u_0^k - u_0\|_2 \, e^{cT}.$$

Hence we obtain:

$$u^k \to u \quad \text{in} \quad C^0([0,T], L^2) \quad \text{as} \quad k \to \infty.$$

The Gagliardo–Nirenberg inequality, Theorem 4.4, implies (cf. Chapter 5 for similar arguments)

$$u^k \to u \quad \text{in} \quad C^0([0,T], W^{s-1,2}) \tag{6.10}$$

and

$$\partial_t u^k \to \partial_t u \quad \text{in} \quad C^0([0,T], W^{s-2,2}), \quad \text{as} \quad k \to \infty. \tag{6.11}$$

Moreover we have for each $t \in [0,T]$:

$$u^k(t) \rightharpoonup u(t) \quad \text{in} \quad W^{s,2} \quad \text{as} \quad k \to \infty. \tag{6.12}$$

(Cf. the proofs of the Lemmata 5.5, 5.6.)
(6.12) implies

$$\|u(t)\|_{s,2} \leq \liminf_{k\to\infty} \|u^k(t)\|_{s,2} \tag{6.13}$$

(cf. [6, p. 139]) and from (6.10), (6.11) and Sobolev's imbedding theorem we conclude

$$Du^k \to Du \quad \text{in} \quad C^0([0,T], L^\infty). \tag{6.14}$$

Here we used the fact that $s > n/2 + 2$ holds. Combining (6.9), (6.13) and (6.14) we obtain

$$\begin{aligned} \|u(t)\|_{s,2} &\leq c \liminf_{k\to\infty} \|u_0^k\|_{s,2} \exp\{c\int_0^t \|Du^k(r)\|_\infty dr\} \\ &= c\|u_0\|_{s,2} \exp\{c\int_0^t \|Du(r)\|_\infty dr\}. \end{aligned}$$

Q.E.D.

7 Weighted a priori estimates for small data

Besides the energy estimate which we proved in Chapter 6 the following a priori estimate is essential for the proof of the global existence theorem (Theorem 1.1). As a new ingredient it takes advantage from the decay estimates that were obtained in Chapter 2 for solutions of the linearized equation.

Let u again be the local solution as in Chapter 6 (case $\alpha = 1$). Then we shall prove the following weighted a priori estimate.

Theorem 7.1 *Let $n > 5$ and $s_0, s_1 \in \mathbb{N}$ satisfy*

$$s_1 > \left[\frac{s_1 + N_{4/3}}{2}\right], \; s_0 \geq s_1 + N_{4/3} + 1.$$

Let

$$u_0 \in W^{s_0,2} \cap W^{s_1+N_{4/3},4/3}.$$

Then there are $M_0 > 0$ and $\delta_1 > 0$, both independent of T and u_0, such that the following holds:

If

$$\|u_0\|_{s_0,2} + \|u_0\|_{s_1+N_{4/3},4/3} \leq \delta_1$$

then

$$M_{s_1}(T) := \sup_{0 \leq t \leq T} (1+t)^{\frac{n-1}{4}} \|u(t)\|_{s_1,4} \leq M_0.$$

Remarks: The condition $n > 5$ comes from the condition $\frac{n-1}{2} > \frac{1}{\alpha}(1 + \frac{1}{\alpha})$ in Theorem 1.1 for $\alpha = 1$. $N_{4/3} = N_p$ from Theorem 2.3 for $p = 4/3$. We have

$$N_{4/3} > n(1 - 2/4) = n/2.$$

This implies

$$s_1 \geq N_{4/3} > n/2$$

and hence

$$s_0 \geq s_1 + N_{4/3} + 1 > n + 1 \geq n/2 + 2 \quad \text{since} \quad n > 5.$$

The nonlinearity is given as in Chapters 5 and 6 by

$$f(u, \nabla u) = \sum_{i,j=1}^{n} a_{ij}(u) \partial_j u_{i+1}.$$

c will denote various constants not depending on T or u_0.

Lemma 7.2

$$\|f(u, \nabla u)\|_{s_1+N_{4/3},4/3} \leq c \|u\|_{s_1,4} \|u\|_{s_0,2} \tag{7.1}$$

(Observe that $W^{s_0,2} \hookrightarrow W^{s_1,4}$ since $s_0 > s_1 + n/4$.)

PROOF:

$$\begin{aligned}\|a_{ij}(u)\|_q &= \|a_{ij}(u) - a_{ij}(0)\|_q \\ &= \|\int_0^1 (\nabla_u a_{ij})(ru)udr\|_q \leq c\|u\|_q \leq c\|u\|_{s_0,2}\end{aligned} \tag{7.2}$$

for all $i,j = 1,\cdots,n$ and all $q \in [2,\infty]$.
Using the special form of f we obtain

$$\begin{aligned}\|f(u,\nabla u)\|_{s_1+N_{4/3},4/3} &\leq \sum_{i,j=1}^{n} \sum_{0\leq|\alpha|\leq s_1+N_{4/3}} \|\nabla^\alpha(a_{ij}(u)\partial_j u_{i+1})\|_{4/3} \\ &\leq c\sum_{i,j=1}^{n} \sum_{0\leq|\alpha|+|\beta|\leq s_1+N_{4/3}} \|(\nabla^\alpha a_{ij}(u))\nabla^\beta\partial_j u_{i+1}\|_{4/3}.\end{aligned}$$

First let $|\alpha| > s_1$. Then $|\beta| + 1 < N_{4/3} \leq s_1$.
This implies

$$\begin{aligned}\|(\nabla^\alpha a_{ij}(u))\nabla^\beta\partial_j u_{i+1}\|_{4/3} &\leq \|\nabla^\alpha a_{ij}(u)\|_2\|\nabla^\beta\partial_j u_{i+1}\|_4 \\ &\leq c\|\nabla^{|\alpha|}u\|_2\|u\|_{|\beta|+1,4} \\ &\quad \text{(using Lemma 4.8 and (7.2))} \\ &\leq c\|u\|_{s_0,2}\|u\|_{s_1,4}.\end{aligned} \tag{7.3}$$

Now let $|\beta| + 1 > s_1$. Then $|\alpha| \leq s_1$ and

$$\begin{aligned}\|\nabla^\alpha(a_{ij}(u))\nabla^\beta\partial_j u_{i+1}\|_{4/3} &\leq \|\nabla^\alpha a_{ij}(u)\|_4\|\nabla^\beta\partial_j u_{i+1}\|_2 \\ &\leq c\|\nabla^{|\alpha|}u\|_4\|u\|_{|\beta|+1,2} \\ &\leq c\|u\|_{s_1,4}\|u\|_{s_0,2}.\end{aligned} \tag{7.4}$$

Finally let $|\alpha| \leq s_1$ and $|\beta| + 1 \leq s_1$. Then we obtain

$$\begin{aligned}\|\nabla^\alpha(a_{ij}(u))\nabla^\beta\partial_j u_{i+1}\|_{4/3} &\leq \|\nabla^\alpha a_{ij}(u)\|_2\|\nabla^{|\beta|+1}u\|_4 \\ &\leq c\|u\|_{s_0,2}\|u\|_{s_1,4}.\end{aligned} \tag{7.5}$$

From (7.3) – (7.5) we conclude the claim of Lemma 7.2.

Q.E.D.

Let $w(t)g$ be defined as in Chapter 2 as the solution v of the linear initial value problem

$$v_{tt} - \Delta v = 0,\ v(t=0) = 0,\ v_t(t=0) = g.$$

Remember that

$$u(t=0) = u_0 = (y_1, \nabla y_0).$$

Writing $f(t)$ short for $f(u(t,\cdot), \nabla u(t,\cdot))$ we have the following representation for the local solution u:

Lemma 7.3

$$u(t) = Dw(t)y_1 + D\partial_t w(t)y_0 + \int_0^t Dw(t-r)f(r)dr, \quad 0 \le t \le T.$$

PROOF: Since $u = Dy$ it is sufficient to prove

$$\begin{aligned} y(t) &= w(t)y_1 + \partial_t w(t)y_0 + \int_0^t w(t-r)f(r)dr \\ &\equiv v_1(t) + v_2(t) + v_3(t). \end{aligned} \tag{7.6}$$

We have, using the definition of w(t),

$$\partial_t^2 v_1 - \Delta v_1 = 0,\ v_1(t=0) = 0,\ \partial_t v_1(t=0) = y_1, \tag{7.7}$$

$$\partial_t^2 v_2 - \Delta v_2 = 0,\ v_2(t=0) = y_0,\ \partial_t v_2(t=0) = 0. \tag{7.8}$$

Moreover

$$\partial_t^2 \int_0^t w(t-r)f(r)dr = \partial_t\{w(0)f(t) + \int_0^t \partial_t w(t-r)f(r)dr\}$$

$$= \partial_t \int_0^t \partial_t w(t-r)f(r)dr = (\partial_t w)(t=0)f(t) + \int_0^t \partial_t^2 w(t-r)f(r)dr$$

$$= f(t) + \int_0^t \Delta w(t-r)f(r)dr = f(t) + \Delta \int_0^t w(t-r)f(r)dr.$$

This implies

$$\partial_t^2 v_3 - \Delta v_3 = f,\ v_3(t=0) = 0,\ \partial_t v_3(t=0) = 0. \tag{7.9}$$

(7.7) – (7.9) yield (7.6) (uniqueness of solutions).

Q.E.D.

Now we come to the PROOF of Theorem 7.1.

We write

$$u(t) \equiv \sum_{j=1}^{3} u^j(t)$$

according to Lemma 7.3.

Let

$$d := \frac{n-1}{4}.$$

Then u^1 satisfies

$$\begin{aligned} \|u^1(t)\|_{s_1,4} &\le c(1+t)^{-\frac{n-1}{2}(1-2/4)}\|y_1\|_{s_1+N_{4/3},4/3} \qquad (7.10)\\ &\qquad \text{(according to Theorem 2.3)}\\ &\le c(1+t)^{-d}\delta_1. \end{aligned}$$

Writing u^2 as

$$\begin{aligned} u^2(t) &= D\partial_t w(t)y_0 = (\partial_t^2 w(t)y_0,\, \nabla\partial_t w(t)y_0)\\ &= (\Delta w(t)y_0,\, \nabla\partial_t w(t)y_0) = (\sum_{j=1}^n \partial_j(w(t)\partial_j y_0), \partial_t(w(t)\nabla y_0)) \end{aligned}$$

we obtain by Theorem 2.3

$$\begin{aligned} \|u^2(t)\|_{s_1,4} &\le c\|Dw(t)\nabla y_0\|_{s_1,4} \qquad (7.11)\\ &\le c(1+t)^{-d}\|\nabla y_0\|_{s_1+N_{4/3},4/3}\\ &\le c(1+t)^{-d}\delta_1. \end{aligned}$$

Applying once more Theorem 2.3 and using Lemma 7.2 we get

$$\begin{aligned} \|u^3(t)\|_{s_1,4} &\le c\int_0^t (1+t-r)^{-d}\|f(r)\|_{s_1+N_{4/3},4/3}dr\\ &\le c\int_0^t (1+t-r)^{-d}\|u(r)\|_{s_1,4}\|u(r)\|_{s_0,2}dr. \end{aligned}$$

Using Theorem 6.1 we see that u^3 satisfies

$$\begin{aligned} &\|u^3(t)\|_{s_1,4} \le c\int_0^t (1+t-r)^{-d}\|u(r)\|_{s_1,4}\|u_0\|_{s_0,2}\exp\{c\int_0^t \|Du(\tau)\|_\infty d\tau\}dr \qquad (7.12)\\ &\le c\,\delta_1\exp\{c\int_0^t \|Du(\tau)\|_\infty d\tau\}\int_0^t (1+t-r)^{-d}\|u(r)\|_{s_1,4}dr\\ &\le c\,\delta_1\exp\{c\int_0^t \|Du(\tau)\|_\infty d\tau\}M_{s_1}(t)(1+t)^{-d}\int_0^t (1+t)^d(1+t-r)^{-d}(1+r)^{-d}dr. \end{aligned}$$

Now we make use of the following general estimates which will be proved at the end of this chapter.

Lemma 7.4 *Let $\alpha, \beta, \gamma \geq 0$ satisfy*

$$\alpha + \beta - \gamma \geq 1, \quad \alpha \geq \gamma, \quad \beta \geq \gamma,$$

and

$$\alpha > \gamma \text{ if } \beta = 1, \text{ and } \beta > \gamma \text{ if } \alpha = 1.$$

Then we have

$$(i) \quad \sup_{0 \leq t < \infty} \int_0^t (1+t)^\gamma (1+t-r)^{-\alpha} (1+r)^{-\beta} dr < \infty,$$

$$(ii) \quad \sup_{0 \leq t < \infty} \int_0^\infty (1+t)^\gamma (1+t+r)^{-\alpha} (1+r)^{-\beta} dr < \infty.$$

Applying Lemma 7.4, (i), with $\alpha = \beta = \gamma = d > 1$ we obtain from (7.12)

$$\|u^3(t)\|_{s_1,4} \leq c\delta_1 (1+t)^{-d} M_{s_1}(t) \exp\{c \int_0^t \|Du(\tau)\|_\infty d\tau\}. \tag{7.13}$$

Sobolev's imbedding theorem yields

$$\|\nabla u\|_\infty \leq c\|u\|_{s_1,4}$$

and from the differential equation for u we conclude

$$\|\partial_t u\|_\infty \leq c\|\nabla u\|_\infty.$$

This implies

$$\begin{aligned} \int_0^t \|Du(\tau)\|_\infty d\tau &\leq c \int_0^t (1+\tau)^{-d} (1+\tau)^d \|u(\tau)\|_{s_1,4} d\tau \\ &\leq c\, M_{s_1}(t) \int_0^t (1+\tau)^{-d} d\tau \\ &\leq c\, M_{s_1}(t). \end{aligned} \tag{7.14}$$

From (7.13) and (7.14) we conclude

$$\|u^3(t)\|_{s_1,4} \leq c\delta_1 (1+t)^{-d} M_{s_1}(t) \exp\{c M_{s_1}(t)\}. \tag{7.15}$$

Combining (7.10), (7.11) and (7.15) we get the following estimate for u:

$$\|u(t)\|_{s_1,4} \leq c\delta_1 (1+t)^{-d} + c\delta_1 (1+t)^{-d} M_{s_1}(t) \exp\{c M_{s_1}(t)\}.$$

This implies

$$M_{s_1}(t) \leq c\delta_1 \left(1 + M_{s_1}(t) \exp\{c M_{s_1}(t)\}\right), \quad 0 \leq t \leq T. \tag{7.16}$$

Without loss of generality we assume that c is larger than the imbedding constant $\tilde{\kappa}$ in

$$W^{s_0,2} \quad \hookrightarrow \quad W^{s_1,4}.$$

Let

$$\varphi : [0,\infty) \longrightarrow \mathbb{R}$$

be defined by

$$\varphi(x) := c\delta_1(1 + x\,\mathrm{e}^{cx}) - x.$$

We have

$$\varphi(0) = c\delta_1,\ \varphi'(0) = c\delta_1 - 1.$$

φ has a first positive zero at x_0 with $\varphi'(x_0) < 0$ if δ_1 is sufficiently small ($\delta_1 = \delta_1(c)$).

$$0 = \varphi(x_0) = c\delta_1(1 + x_0\mathrm{e}^{cx_0}) - x_0$$

implies

$$\delta_1 = \frac{x_0}{c(1 + x_0\mathrm{e}^{cx_0})} < \frac{x_0}{c}$$

whence

$$M_{s_1}(0) = \|u_0\|_{s_1,4} \leq \tilde{\kappa}\|u_0\|_{s_0,2} \leq \tilde{\kappa}\delta_1 < x_0 \tag{7.17}$$

follows. The relation (7.16) implies

$$\varphi(M_{s_1}(t)) \geq 0, \quad 0 \leq t \leq T$$

which together with (7.17) and a continuous dependence argument leads to

$$M_{s_1}(t) \leq x_0, \quad 0 \leq t \leq T,$$

which yields the claim of Theorem 7.1 with

$$M_0 := x_0 = x_0(\delta_1).$$

Q.E.D.

It remains to prove Lemma 7.4.

PROOF: (ii) is an easy consequence of (i). Let c denote various constants not depending on t. Then

$$\int_0^t (1+t)^\gamma(1+t-r)^{-\alpha}(1+r)^{-\beta}dr = \int_0^{t/2} \cdots + \int_{t/2}^t \cdots$$

and

$$\begin{aligned}
\int_0^{t/2} \cdots &\leq c \int_0^{t/2} (1+t)^{\gamma}(1+t)^{-\alpha}(1+r)^{-\beta} dr \\
&= \frac{c}{(1+t)^{\alpha-\gamma}} \int_0^{t/2} (1+r)^{-\beta} dr \\
&\leq \frac{c}{(1+t)^{\alpha-\gamma}} \begin{cases} \log(1+t/2) & \text{if } \beta = 1, \\ (1+t/2)^{-\beta+1} + 1 & \text{if } \beta \neq 1 \end{cases} \\
&\leq c
\end{aligned}$$

because $\alpha > \gamma$ if $\beta = 1$ and $\alpha \geq \gamma$ and $\alpha + \beta - \gamma \geq 1$.
Analogously we conclude

$$\begin{aligned}
\int_{t/2}^{t} \cdots &\leq \frac{c}{(1+t)^{\beta-\gamma}} \int_{t/2}^{t} (1+t-r)^{-\alpha} dr \\
&= \frac{c}{(1+t)^{\beta-\gamma}} \int_0^{t/2} (1+r)^{-\alpha} dr \\
&\leq c
\end{aligned}$$

because $\beta > \gamma$ if $\alpha = 1$ and $\beta \geq \gamma$ and $\alpha + \beta - \gamma \geq 1$.

Q.E.D.

8 Global solutions to wave equations — proofs

8.1 Proof of Theorem 1.1

We are now able to give a proof of Theorem 1.1, first again for the case

$$\alpha = 1, \quad f(Dy, \nabla Dy) = \sum_{i,j=1}^{n} a_{ij}(Dy)\partial_i\partial_j y \tag{8.1}$$

as discussed in the previous chapters.
Let δ_1, T, s_0, s_1, u be given as in Theorem 7.1, and let $s \geq s_0$.
By successively using Theorem 6.1, formula (7.14) and Theorem 7.1 we obtain the following sequence of inequalities, where $u(t) = u(t, \cdot)$:

$$\begin{aligned} \|u(t)\|_{s,2} &\leq c\|u_0\|_{s,2} \exp\{c\int_0^t \|Du(\tau)\|_\infty d\tau\} \\ &\leq c\|u_0\|_{s,2} \exp\{cM_{s_1}(t)\} \\ &\leq c\|u_0\|_{s,2} \exp\{cM_0\} \\ &\leq K\|u_0\|_{s,2}, \end{aligned} \tag{8.2}$$

for $0 \leq t \leq T$ and with

$$K := c\exp\{cM_0\}$$

being independent of T and of u_0.
If we choose δ such that

$$0 < \delta < \frac{\delta_1}{K},$$

we obtain

$$\|u(T)\|_{s,2} \leq K\|u_0\|_{s,2} \leq K\delta < \delta_1.$$

Applying the local existence theorem, Theorem 5.8 (at initial time T), we conclude that there exists a continuation of u onto $[T, T+T_1(\delta_1)]$ for some positive number T_1 only depending on δ_1. The inequality (8.2) applied for

$$0 \leq t \leq T + T_1(\delta_1)$$

implies

$$\|u(T+T_1(\delta_1))\|_{s,2} \leq K\|u_0\|_{s,2} < \delta_1.$$

Hence we may apply the same argument once more to conclude that we can continue u onto $[T+T_1(\delta_1), T+2T_1(\delta_1)]$. Proceeding in this way we prove the existence of a global solution

$$u \in C^0([0,\infty), W^{s,2}) \cap C^1([0,\infty), W^{s-1,2}).$$

In particular, we obtain

$$\forall 0 \le t < \infty: \quad \|u(t)\|_{s,2} \le K\delta < \delta_1$$

and, with Sobolev's imbedding theorem and Theorem 7.1,

$$\begin{aligned}\forall 0 \le t < \infty: \quad \|u(t)\|_\infty &\le c\|u(t)\|_{s_1,4} \le cM_{s_1}(t)(1+t)^{-(n-1)/4} \\ &\le cM_0(1+t)^{-(n-1)/4}.\end{aligned}$$

This proves Theorem 1.1 (for the case given in (8.1)).

Q.E.D.

The general case (f, α as given in Theorem 1.1) is proved in the same way. The following remarks point out the differences and show that the restriction to f, α as in (8.1) in the previous chapters were made without loss of generality.

First let f again have the form given in (8.1) but let $\alpha \in \mathbb{N}$ be arbitrary, i.e.

$$a_{ij}(u) = \mathcal{O}(|u|^\alpha) \quad \text{as} \quad |u| \to 0.$$

Then the calculations in the proof of Theorem 6.1 show that the inequality (6.1) claimed in this theorem can be replaced by the following inequality.

$$\forall t \in [0,T]: \quad \|u(t)\|_{s,2} \le c\|u_0\|_{s,2} \exp\{c \int_0^t \|\overline{D}u(\tau)\|_\infty^\alpha d\tau\}. \tag{8.3}$$

The estimate (7.1) for the nonlinearity f now reads as

$$\|f(u,\nabla u)\|_{s_1+N_p,p} \le c\|u\|_{s_1,q}^\alpha \|u\|_{s_0,2}, \tag{8.4}$$

where s_1 is sufficiently large and q, p satisfy the relations

$$\frac{1}{q} + \frac{1}{p} = 1 \quad \text{and} \quad \frac{\alpha}{q} + \frac{1}{2} = \frac{1}{p} \tag{8.5}$$

(the latter simply arising from an application of Hölder's inequality in the proof of (8.4)). The relations (8.5) for p and q are equivalent to

$$p = \frac{2\alpha+2}{2\alpha+1} \quad \text{and} \quad q = 2(\alpha+1), \tag{8.6}$$

which are just the conditions on p and q given in Theorem 1.1.

With the help of the inequality (8.4) the analogue of Theorem 7.1 is proved in the same way by estimating the corresponding term $M_{s_1}(T)$, namely

$$M_{s_1}(T) := \sup_{0 \le t \le T} (1+t)^{\frac{n-1}{2}(1-\frac{2}{q})} \|u(t)\|_{s_1,q} \tag{8.7}$$

(q is given in (8.6)), $s_1 \in \mathbb{N}$ is chosen appropriately large enough).
To carry over the proof of Theorem 7.1 we only mention that the integration exponent d appearing in the formulae (7.13) and (7.15), respectively, having to be larger than 1, now is the following:

$$d = \alpha\frac{n-1}{2}\left(1-\frac{2}{q}\right) = \alpha\frac{\alpha}{\alpha+1}\frac{n-1}{2}.$$

This is an immediate consequence of the relations in the formulae (8.3) and (8.4).
The necessary condition $d > 1$ explains the condition

$$\frac{1}{\alpha}\left(1+\frac{1}{\alpha}\right) < \frac{n-1}{2}$$

in Theorem 1.1.
With the definition $M_{s_1}(T)$ in (8.7) it is obvious how the decay rate claimed in Theorem 1.1, namely $\frac{n-1}{2}\frac{\alpha}{\alpha+1}$, arises.
The considerations up to now demonstrate that higher-order nonlinear terms can be handled even easier. Moreover, the method of differentiating the original differential equation (1.1) leads to a quasilinear differential equation for y and its derivatives. Hence the special form

$$f(Dy, \nabla Dy) = \sum_{i,j=1}^{n} a_{ij}(Dy)\partial_i\partial_j y$$

can be assumed without loss of generality (see also the corresponding remarks in Section 11.2 between the formulae (11.58) and (11.59)). This completes the proof of Theorem 1.1.

Q.E.D.

We remark that the regularity assumptions on f ("C^∞") can be weakened using the theory of evolution operators by Kato, see [51, 79, 80, 81, 82, 83]. Another idea would be to approximate a less regular f by C^∞-functions f_m, $m = 1, 2, \ldots$, and to prove the convergence of the associated solutions u_m, for which energy estimates like those proved in Chapter 5 are needed.
The only reason for having studied C^∞-nonlinearities f was to be able to present a local existence theorem, the proof of which is as simple as possible. See the remarks at the end of Chapter 5.
Theorem 1.1 can be regarded as a kind of stability result which expresses that the solution u behaves asymptotically for large t like a solution of the linearized equation. This means that there should be a function u_+ which is a solution of a linear wave equation (more precisely: $u_+ = Dy_+$ for some function y_+, and y_+ solves the linear wave equation), and $u(t)$ behaves like $u_+(t)$ as t tends to infinity.
We define u_+ by

$$u_+(t) := u(t) + \int_t^\infty W(t-r)F(u(r), \nabla u(r))dr, \tag{8.8}$$

where $F(u(r), \nabla u(r)) = (f(u(r), \nabla u(r)), 0)$ and where $V(t) := W(t)H$ solves the linear initial value problem

$$V_t + AV = 0, \quad V(t=0) = H$$

for all $t \in (-\infty, \infty)$ — not only for $t \geq 0$. Here A denotes the linear differential operator with symbol

$$\begin{pmatrix} 0 & -\nabla' \\ -\nabla & 0 \end{pmatrix}$$

which naturally arises by a change from y with

$$y_{tt} - \Delta y = 0$$

to $V = (\partial_t y, \nabla y)$ with

$$V_t + AV = 0.$$

u_+ is well-defined in $W^{s_1,q}$ with s_1, q as in Theorem 1.1 since

$$\int_t^\infty \|W(t-r)F(u(r), \nabla u(r))\|_{s_1,q} dr \leq c(1+t)^{-\frac{n-1}{2}(1-\frac{2}{q})}.$$

This follows, using (8.4) and Theorem 1.1, from the inequalities:

$$\begin{aligned}
&\int_t^\infty \|W(t-r)F(u(r), \nabla u(r))\|_{s_1,q}\, dr \qquad (8.9)\\
\leq\ & c\int_t^\infty (1+|t-r|)^{-\frac{n-1}{2}(1-\frac{2}{q})} \|F(u(r), \nabla u(r))\|_{s_1+N_p,p}\, dr \\
\leq\ & c\int_t^\infty (1+|t-r|)^{-\frac{n-1}{2}(1-\frac{2}{q})} \|u(r)\|_{s_1,q}^\alpha \|u(r)\|_{s_0,2}\, dr \\
\leq\ & c\int_t^\infty (1+|t-r|)^{-\frac{n-1}{2}(1-\frac{2}{q})} (1+r)^{-\alpha\frac{n-1}{2}(1-\frac{2}{q})} dr \\
=\ & c\int_0^\infty (1+r)^{-\frac{n-1}{2}(1-\frac{2}{q})} (1+r+t)^{-\alpha\frac{n-1}{2}(1-\frac{2}{q})} dr \\
\leq\ & c(1+t)^{-\frac{n-1}{2}(1-\frac{2}{q})}
\end{aligned}$$

(using Lemma 7.4).

By the definition of u_+ we obtain

$$\begin{aligned}
\partial_t u_+(t) &= \partial_t u(t) - F(u(t), \nabla u(t)) + \int_t^\infty \partial_t (W(t-s)F(u(s), \nabla u(s)))ds \\
&= -Au(t) - \int_t^\infty AW(t-s)F(u(s), \nabla u(s))ds \\
&= -Au_+(t),
\end{aligned}$$

hence u_+ solves the linearized equation and we obtain the following Corollary to Theorem 1.1.

Corollary 8.1 *Using the notation of Theorem 1.1 and the definition of u_+ in (8.8), one has that u behaves in L^2, asymptotically as $t \to \infty$, like u_+, where u_+ is the solution of the linearized equation to the initial value*

$$u_+(t=0) = u_0 + \int_0^\infty W(-s)F(u(s), \nabla u(s))ds.$$

More precisely,

$$\lim_{t\to\infty} \|u(t) - u_+(t)\|_2 = 0.$$

PROOF:

$$\begin{aligned} \|u(t) - u_+(t)\|_2 &\leq c\int_t^\infty \|F(u(r), \nabla u(r))\|_{s_1,q}\, dr \\ &\leq c(1+t)^{1-\alpha\frac{n-1}{2}(1-\frac{2}{q})} \longrightarrow 0 \quad \text{as} \quad t \to \infty. \end{aligned}$$

Q.E.D.

We observe that the formulation of the result in scattering theory in Corollary 8.1 requires the solvability of the linearized problem for all real t, not only on the positive time axis. This solvability will not be given for example for the heat conduction problems in Chapter 11.

Up to now we have only considered wave equations. However, we shall see in Chapter 11 in the discussion of other evolution equations that there is a common structure underlying the proofs leading to similar theorems as Theorem 1.1. In particular we mention the paper of J. Shatah [158] where similar results are obtained in a more general functional analytic setup. We have taken the presentation given by Klainerman & Ponce [94].

Now we turn to the proof of Theorem 1.2. This improvement of Theorem 1.1 strongly relies on special properties of the wave equation and does not have counterparts (up to now) for each of the other systems which shall be studied in Chapter 11; however, for Klein–Gordon equations see [89], for Schrödinger equations see [22], for the equations of elasticity see [73]. Hence this underlines the necessity of studying each system in detail to obtain specific optimal results; see also the remarks at the end of this chapter.

8.2 Proof of Theorem 1.2

The proof of Theorem 1.2 will follow from a sequence of Lemmata. Let

$$\partial_r := \sum_{i=1}^n \frac{x_i}{|x|}\partial_i. \tag{8.10}$$

Lemma 8.2 *There is a constant $C > 0$ such that for any smooth function $y = y(x)$ in $\mathbb{R}^n$, being compactly supported (or vanishing sufficiently rapidly at infinity), the following inequality holds for all $x \in \mathbb{R}^n \setminus \{0\}$:*

$$|y(x)| \leq C\left(\frac{1}{|x|}\right)^{\frac{n-1}{2}} \|y\|^{\frac{1}{2}}_{\Omega,[\frac{n-1}{2}]+1,2}\|\partial_r y\|^{\frac{1}{2}}_{\Omega,[\frac{n-1}{2}]+1,2}$$

(where Ω was defined in (4.23)).

Proof: We introduce polar co-ordinates:

$$x = r\xi, \quad r = |x|, \quad \xi \in S^{n-1}.$$

Then we have

$$y^2(r\xi) = -\int_r^\infty 2y(\lambda\xi)\partial_r y(\lambda\xi)d\lambda.$$

This implies

$$y^2(r\xi) \leq \frac{2}{r^{n-1}}\int_r^\infty |y(\lambda\xi)|\,|\partial_r y(\lambda\xi)|\lambda^{n-1}d\lambda,$$

hence

$$\int_{S^{n-1}} |y^2(r\xi)|d\xi \leq \frac{2}{r^{n-1}}\|y\|_2\|\partial_r y\|_2.$$

Observing that Ω — containing only angular derivatives — commutes with ∂_r, we obtain the same inequality for $\Omega^\alpha y$. Summing up for $0 \leq |\alpha| \leq \left[\frac{n-1}{2}\right] + 1$, we get

$$\left(\sum_{|\alpha|\leq[\frac{n-1}{2}]+1}\int_{S^{n-1}} |\Omega^\alpha y(r\xi)|^2 d\xi\right)^{\frac{1}{2}} \leq \frac{C}{r^{\frac{n-1}{2}}}\|y\|^{\frac{1}{2}}_{\Omega,[\frac{n-1}{2}]+1,2}\|\partial_r y\|^{\frac{1}{2}}_{\Omega,[\frac{n-1}{2}]+1,2},$$

which yields the proposition with the help of Lemma 4.15.

Q.E.D.

Remark: All constants in this section naturally depend on the space dimension n. The estimate in Lemma 8.2, applied to a function $y = y(t,x)$ in $\mathbb{R} \times \mathbb{R}^n$, already gives us the desired kind of L^2–L^∞-estimate in a part of the exterior of the *light cone* $(= \{(t,x) \mid t = |x|\})$, namely for (t,x) with $2|x| > t$, see below. To get an estimate for (t,x) with $t \geq 2|x|$, we use the following representation for ∂_r:

$$\partial_r = \frac{1}{t^2 - r^2}(tL_r - rL_0), \tag{8.11}$$

where

$$L_r := \sum_{i=1}^n \frac{x_i}{|x|}L_i. \tag{8.12}$$

It follows from the definitions of ∂_r, L_r, L_0, in (8.10), (8.12), and (4.22) respectively, that

$$L_r = r\partial_t + t\partial_r, \quad L_0 = t\partial_t + r\partial_r,$$

which immediately implies (8.11).

Lemma 8.3 *For all $k \in \mathbb{N}$ there is a constant $C = C(k) > 0$ such that for all smooth functions y in $\mathbb{R} \times \mathbb{R}^n$ the following inequality holds for all (t,x) with $t \neq r = |x| \neq 0$, $t > 0$:*

$$|\partial_r^k y(t,x)| \leq C \frac{1}{|t-r|^k} \sum_{|\alpha| \leq k} |L^\alpha y(t,x)|,$$

where Ly denotes the vector $(L_0 y, L_1 y, \ldots, L_n y)$.

Proof: The proof follows by induction on k.

$k = 1$:

$$\partial_r y(t,x) = \frac{1}{t^2 - r^2} \left(t \sum_{i=1}^n \frac{x_i}{|x|} L_i y(t,x) - r L_0 y(t,x) \right).$$

This implies

$$\begin{aligned} |\partial_r y(t,x)| &\leq \frac{1}{|t^2 - r^2|} \left(t \sum_{i=1}^n |L_i y(t,x)| + r |L_0 y(t,x)| \right) \\ &\leq \frac{1}{|t-r|} |Ly(t,x)|. \end{aligned}$$

$1 \leq k \to k+1$:

$$\begin{aligned} \partial_r^{k+1} y &= \partial_r^k \left\{ \frac{1}{t^2 - r^2} (tL_r - rL_0) y \right\} \\ &= \sum_{j=0}^k \binom{k}{j} \partial_r^j \left\{ \frac{1}{t^2 - r^2} \right\} \partial_r^{k-j} \left\{ (tL_r - rL_0) y \right\}. \end{aligned} \tag{8.13}$$

We have

$$\left| \partial_r^j \left\{ \frac{1}{t^2 - r^2} \right\} \right| \leq C \cdot \frac{1}{|t-r|^{j+1}(t+r)}, \quad 0 \leq j \leq k, \tag{8.14}$$

(which may be easily proved by induction again).
By the induction hypothesis we obtain

$$\begin{aligned} |\partial_r^{k-j} \left\{ (tL_r - rL_0) y(t,x) \right\}| &\leq \frac{C}{|t-r|^{k-j}} \sum_{|\alpha| \leq k-j} |L^\alpha (tL_r - rL_0) y(t,x)| \\ &\leq \frac{C}{|t-r|^{k-j}} \sum_{|\alpha| \leq k} |L^\alpha (tL_r - rL_0) y(t,x)| \\ &\leq \frac{C(t+r)}{|t-r|^{k-j}} \sum_{|\alpha| \leq k+1} |L^\alpha y(t,x)|. \end{aligned} \tag{8.15}$$

Combining (8.13) – (8.15) we get

$$\begin{aligned} |\partial_r^{k+1} y(t,x)| &\leq C \frac{t+r}{|t-r|^{j+1}(t+r)|t-r|^{k-j}} \sum_{|\alpha|\leq k+1} |L^\alpha y(t,x)| \\ &= C \frac{1}{|t-r|^{k+1}} \sum_{|\alpha|\leq k+1} |L^\alpha y(t,x)|. \end{aligned}$$

Q.E.D.

The estimate in Lemma 8.3 will lead to the desired L^2–L^∞-estimate for an arbitrary smooth function $y = y(t,x)$ as we shall show now.

Let $v = v(t,r,\xi)$ be defined by

$$v(t,r,\xi) := (t-r)^k y(t,r\xi),$$

$k \in \mathbb{N}$ fixed, $0 \leq r = |x| < t$, $\xi \in S^{n-1}$, $x = r\xi$. Since by the definition of v we have

$$\frac{\partial^j}{\partial r^j} v(t,r,\xi)_{|r=t} = 0, \quad j = 0,\ldots,k-1,$$

we obtain the representation

$$v(t,r,\xi) = \frac{(-1)^k}{(k-1)!} \int_r^t (\lambda - r)^{k-1} \frac{\partial^k}{\partial \lambda^k} v(t,\lambda,\xi) d\lambda.$$

Applying Lemma 8.3 we get

$$\left| \frac{\partial^k}{\partial \lambda^k} v(t,\lambda,\xi) \right| \leq C M_k(t,\lambda,\xi),$$

where

$$M_k(t,\lambda,\xi) := \sum_{|\alpha|\leq k} |L^\alpha y(t,x)|, \quad x = \lambda\xi.$$

This implies

$$|v(t,r,\xi)| \leq C I_k(t,r) \left(\int_r^t \lambda^{n-1} M_k^2(t,\lambda,\xi) d\lambda \right)^{\frac{1}{2}},$$

where

$$I_k(t,r) := \left(\int_r^t (\lambda - r)^{2k-2} \lambda^{1-n} d\lambda \right)^{\frac{1}{2}}.$$

Integrating $\xi \mapsto |v(t,r,\xi)|$ with respect to $\xi \in S^{n-1}$ yields

$$\left(\int_{S^{n-1}} |y(t,r\xi)|^2 d\xi \right)^{\frac{1}{2}} \leq C \cdot I_k(t,r) \frac{1}{(t-r)^k} \|M_k(t,\cdot)\|_2.$$

Applying Lemma 4.15 and observing that $L_0, L_1, \ldots, L_n$ are contained in the family Γ defined in Chapter 4 we obtain

$$\begin{aligned} |y(t,x)|^2 &= |y(t,r\xi)|^2 \le C \sum_{|\alpha|\le\left[\frac{n-1}{2}\right]+1} \|\Omega^\alpha y\|^2_{L^2(S^{n-1})} \\ &\le C \sum_{|\alpha|\le\left[\frac{n-1}{2}\right]+1} I_k^2(t,r)\frac{1}{(t-r)^{2k}} \sum_{|\beta|\le k} \|L^\beta \Omega^\alpha y(t)\|_2^2 \\ &\le C I_k^2(t,r)\frac{1}{(t-r)^{2k}} \|y(t)\|^2_{\Gamma,k+\left[\frac{n-1}{2}\right]+1,2}. \end{aligned}$$

If n is odd we take $k := \frac{n+1}{2}$ for which

$$I_k(t,r) = \left(\int_r^t \left(1-\frac{r}{\lambda}\right)^{n-1} d\lambda\right)^{\frac{1}{2}} \le (t-r)^{\frac{1}{2}}.$$

If n is even we take $k := \frac{n+2}{2}$ for which

$$I_k(t,r) = \left(\int_r^t (\lambda - r)\left(1-\frac{r}{\lambda}\right)^{n-1} d\lambda\right)^{\frac{1}{2}} \le \left(\int_r^t \lambda d\lambda\right)^{\frac{1}{2}} = \frac{1}{\sqrt{2}}(t^2-r^2)^{\frac{1}{2}}.$$

In both cases we conclude for $t \ge 2|x|$:

$$|y(t,x)| \le C t^{-n/2}\|y(t)\|_{\Gamma,n+1,2}, \tag{8.16}$$

because we have

$$\text{for odd } n: \quad \frac{n+1}{2} + \left[\frac{n-1}{2}\right] + 1 = n+1,$$

and

$$\frac{1}{(t-r)^{\frac{n+1}{2}}}(t-r)^{1/2} = (t-r)^{-n/2} \le c t^{-n/2},$$

while

$$\text{for even } n: \quad \frac{n+2}{2} + \left[\frac{n-1}{2}\right] + 1 = n+1$$

and

$$\frac{1}{(t-r)^{\frac{n+2}{2}}}\frac{1}{\sqrt{2}}(t^2-r^2)^{1/2} \le \sqrt{2}(t-r)^{-n/2} \le c t^{-n/2}$$

(observe $t \ge 2r$).

Lemma 8.4 *There is a constant $c > 0$ such that for any smooth function $y = y(t,x)$ in $\mathbb{R} \times \mathbb{R}^n$, y being compactly supported with respect to $x \in \mathbb{R}^n$ (or vanishing sufficiently rapidly as $|x| \to \infty$) for each fixed $t \ge 0$, the following inequality holds for all $t \ge 0$ and all $x \in \mathbb{R}^n$:*

$$|y(t,x)| \le c(1+t)^{-(n-1)/2}\|y(t)\|_{\Gamma,n+1,2}.$$

PROOF: First let $\underline{t \geq 1}$.
By (8.16) we have for $t \geq 2|x|$

$$|y(t,x)| \leq ct^{-n/2}\|y(t)\|_{\Gamma,n+1,2} \leq c(1+t)^{-(n-1)/2}\|y(t)\|_{\Gamma,n+1,2}. \tag{8.17}$$

By Lemma 8.2 we get for $t < 2|x|$

$$\begin{aligned} |y(t,x)| &\leq c|x|^{-(n-1)/2}\|y(t)\|^{\frac{1}{2}}_{\Omega,[\frac{n-1}{2}]+1,2}\|\partial_r y(t)\|^{\frac{1}{2}}_{\Omega,[\frac{n-1}{2}]+1,2} \\ &\leq ct^{-(n-1)/2}\left(\|y(t)\|_{\Omega,[\frac{n-1}{2}]+1,2} + \|\partial_r y(t)\|_{\Omega,[\frac{n-1}{2}]+1,2}\right) \\ &\leq ct^{-(n-1)/2}\|y(t)\|_{\Gamma,n+1,2} \leq c(1+t)^{-(n-1)/2}\|y(t)\|_{\Gamma,n+1,2}. \end{aligned} \tag{8.18}$$

Now let $\underline{0 \leq t < 1}$.
By Sobolev's imbedding theorem we conclude

$$\begin{aligned} |y(t,x)| &\leq c\|y(t)\|_{[\frac{n}{2}]+1,2} \leq c\|y(t)\|_{n+1,2} \\ &\leq c\|y(t)\|_{\Gamma,n+1,2} \leq c(1+t)^{-(n-1)/2}\|y(t)\|_{\Gamma,n+1,2}. \end{aligned} \tag{8.19}$$

Combining (8.17) – (8.19) we obtain the proof of Lemma 8.4.

Q.E.D.

Remark: The number $n+1$ appearing on the right-hand side in the term $\|y(t)\|_{\Gamma,n+1,2}$ can be replaced by the optimal value $[n/2]+1$ — in analogy to the classical Sobolev inequalities. This is shown by L. Hörmander [47] and S. Klainerman [91].

Now we shall prove Theorem 1.2 with the help of the last lemma. One remarkable fact is that in Lemma 8.4 y is not necessarily a solution of the linear wave equation. Nevertheless a kind of decay rate ($t^{-(n-1)/2}$) is obtained in a special L^2–L^∞-estimate. The price for this, namely the occurence of the $\|\cdot\|_{\Gamma,n+1,2}$-norm of $y(t)$ on the right-hand side, is still good enough — better to say, it is perfectly suitable — for solutions of nonlinear wave equations.

To prove Theorem 1.2 we shall again assume for simplicity that the nonlinearity has the form

$$f(Dy,\nabla Dy) = \sum_{i,j=1}^{n} a_{ij}(Dy)\partial_i\partial_j y,$$

where $a_{ij} = a_{ji}$ is smooth, $a_{ij}(0) = 0$, $1 \leq i,j \leq n$, and also without loss of generality that $\sum\limits_{i,j=1}^{n} |a_{ij}(u)| \leq \frac{1}{2}$ for all u with $|u| \leq 1$ (cf. the proof of Theorem 1.1 for these assumptions).

In analogy to Theorem 6.1 it is proved that there is a constant $C = C_s > 0$ depending only on (at most s derivatives of) f and on $s \in \mathbb{N}$, $s > \frac{n}{2}+1$, such that we have for the local C^∞-solution y on $[0,T]$ (cf. Theorem 5.8)

$$\forall t \in [0,T]: \quad \|Dy(t)\|_{\Gamma,s,2} \le C_s \|Dy(0)\|_{\Gamma,s,2} \exp\left\{ C_s \int_0^t \|Dy(\tau)\|_{\Gamma,[\frac{s}{2}]+1,\infty} d\tau \right\} \tag{8.20}$$

where we have assumed the following relation without loss of generality:

$$\forall t \in [0,T]: \quad \|Dy(t)\|_{\Gamma,[\frac{s}{2}]+1,\infty} \le 1.$$

(Observe that $Dy(0) = \varepsilon(\psi, \nabla\varphi)$ and ε will be small.) The analogy to the proof of Theorem 6.1 consists in multiplying the differential equation for y with $\Gamma^\alpha y_t(t)$ in $L^2(\mathbb{R}^n)$ (instead of multiplying with $\nabla^\alpha y_t(t)$) and using the Lemmata 4.13, 4.14 as well as the commutator relations for the operators of the family Γ given in the Lemmata 4.11, 4.12. This requires more calculations than the proof of Theorem 6.1 but it is still straightforward, (compare John & Klainerman [75]). In analogy to the considerations in Chapter 7 we define for the local solution y and for $t \in [0,T]$:

$$M_t(y) := \sup_{0\le\tau\le t} (1+\tau)^{(n-1)/2} \|Dy(\tau)\|_{\Gamma,N_0,\infty},$$

where $N_0 \in \mathbb{N}$ will be fixed below. (That is, we are interested in proving an a priori L^∞-bound.)
By Lemma 8.4 we conclude that

$$M_t(y) \le c \sup_{0\le\tau\le t} \|Dy(\tau)\|_{\Gamma,N_0+n+1,2}. \tag{8.21}$$

Combining (8.21) and (8.20) we obtain

$$M_t(y) \le C_{N_0+n+1} \|Dy(0)\|_{\Gamma,N_0+n+1,2} \exp\left\{ C_{N_0+n+1} \int_0^t \|Dy(\tau)\|_{\Gamma,[\frac{N_0+n+1}{2}]+1,\infty} d\tau \right\}. \tag{8.22}$$

If

$$N_0 \ge n+2$$

we have

$$\left[\frac{N_0+n+1}{2}\right] + 1 \le N_0.$$

Hence we define

$$N_0 := n+2$$

(and C_{N_0+n+1} becomes C_{2n+3}).
With this choice of N_0, the form of the initial data, the definition of $M_t(y)$ and (8.22) we obtain

$$M_t(y) \le C\varepsilon \exp\left\{ C \int_0^t M_t(y)(1+\tau)^{-(n-1)/2} d\tau \right\}, \tag{8.23}$$

where C is a constant depending on at most $2n+3$ derivatives of F, $\nabla\varphi$ and ψ.
The inequality (8.23) implies
for $\underline{n > 3}$

$$M_t(y) \le C\varepsilon \exp\{CM_t(y)\}, \quad 0 \le t \le T,$$

and for $\underline{n = 3}$

$$M_t(y) \le C\varepsilon \exp\{CM_t(y)\log(1+t)\}, \quad 0 \le t \le T. \tag{8.24}$$

As in Chapter 7 we conclude now that for $\underline{n > 3}$:

$$\forall 0 \le t \le T: \quad M_t(y) \le M_0 < \infty,$$

with a constant M_0 being independent of T, which implies via (8.20)

$$\forall 0 \le t \le T: \quad \|Dy(t)\|_{N_0,2} \le C_{N_0}\|Dy(0)\|_{N_0,2}$$

and hence allows a continuation of the local solution for all $t \in [0,\infty)$ with the same arguments as in the proof of Theorem 1.1, (cf. (8.2) and the arguments following there). This proves part (i) of Theorem 1.2.
To prove part (ii) we define

$$T_0(\varepsilon) := \sup\left\{0 \le t < \infty \,\middle|\, \text{There is a smooth solution in } [0,t) \text{ and } M_t(y) < \frac{\log 2}{C\log(1+t)}\right\}$$

(C equals that C which appears in (8.24).)
By the definitions of $T_0(\varepsilon)$ and $T_\infty(\varepsilon)$ (cf. Chapter 1) it follows that

$$T_\infty(\varepsilon) \ge T_0(\varepsilon),$$

hence, if $T_0(\varepsilon) = \infty$ then

$$T_\infty(\varepsilon) = \infty > \mathrm{e}^{A/\varepsilon} \text{ for all } A, \varepsilon > 0.$$

If $T_0(\varepsilon) < \infty$ we have

either

(a) The solution does not exist for $t \ge T_0(\varepsilon)$

or

(b) $M_{T_0}(y) = \dfrac{\log 2}{C\log(1+T_0(\varepsilon))}$.

In case (a) we conclude that

$$M_t(y) < \frac{\log 2}{C\log(1+t)} \quad \text{for} \quad 0 \le t < T_0$$

which implies

$$M_t(y) \leq C\varepsilon \exp\{CM_t(y)\log(1+t)\} \leq 2C\varepsilon$$

by (8.24) and thus allows a continuation beyond $T_0(\varepsilon)$ with the arguments from above (case $n > 3$), which is a contradiction.
In case (b) we obtain from (8.24)

$$\begin{aligned}\frac{\log 2}{C\log(1+T_0(\varepsilon))} &\leq C\varepsilon \exp\left\{C\frac{\log 2}{C\log(1+T_0(\varepsilon))}\log(1+T_0(\varepsilon))\right\}\\ &= 2C\varepsilon.\end{aligned}$$

This implies

$$T_0(\varepsilon) \geq \exp\left\{\frac{\log 2}{2C^2\varepsilon}\right\} - 1 \geq \mathrm{e}^{A/\varepsilon}$$

with $A := \log 2/(4C^2)$ if $\varepsilon \leq \varepsilon_0 := 1/(2C^2)$. This completes the proof of Theorem 1.2.

Q.E.D.

The proof of Theorem 1.2 followed the paper [88] by S. Klainerman.
In the proof of Theorem 1.1 and of Theorem 1.2 it was important that the nonlinearity did not depend on y explicitly, but only on derivatives of y. This is connected to the fact that one might get L^∞–L^1-estimates for y similar to those for Dy in Chapter 2 by using again the given representation formulae (Kirchhoff's formula, ...), but that it is not easy to find appropriate estimates for the L^2-norm $\|y(t)\|_2$. To overcome this difficulty Li & Chen [104] use a global iteration scheme (global with respect to time t) instead of using a continuation argument for a local solution. The a priori estimates they need also use the invariance properties of $\partial_t^2 - \Delta$ and corresponding Sobolev type estimates as we did in the proof of Theorem 1.2. The result is a global existence theorem for small data as in the Theorems 1.1, 1.2 for the nonlinearity

$$\begin{aligned} f &= f(y, Dy, \nabla Dy),\\ f(W) &= \mathcal{O}(|W|^{\alpha+1}) \quad \text{as} \quad |W| \to 0,\end{aligned}$$

under the condition that the following relation between α and the space dimension n holds:

$$\frac{1}{\alpha} < \frac{n-1}{2}\left(1 - \frac{2}{\alpha n}\right),$$

see Table 8.1.

(Compare this to Table 1.1 in Chapter 1.) In a recent paper by Li & Zhou it is stated that $\alpha \geq 3$ is sufficient for $n = 2$, see [106].

$\alpha =$	1	2, 3	4, 5, ...
$n \geq$	5	3	2

Table 8.1: f depending on y

Remarks on the optimality of the results:

We shall see in Chapter 11 that the method to prove Theorem 1.1 can be carried over almost literally to many other initial value problems of mathematical physics. This great generality of the approach nourishes the expectation that the results will not be optimal results in each special case — although they are optimal in many cases. Here we do not have in mind optimality with respect to the regularity assumptions on the coefficients and on the data; the results will always be theorems for sufficiently smooth coefficients and data without striving for minimal regularity. What we have in mind addressing the question of optimality is the relation between the space dimension n and the order of vanishing of the nonlinearity characterized by the natural number α.
The general method applied to nonlinear wave equations (1.1) leads to Theorem 1.1 and the relation between n and α is expressed in Table 1.1 in Chapter 1. It is determined through the condition

$$\frac{1}{\alpha}\left(1+\frac{1}{\alpha}\right) < \frac{n-1}{2}. \tag{8.25}$$

Theorem 1.2 shows that the following condition is the optimal one

$$\frac{1}{\alpha} < \frac{n-1}{2}.$$

The condition (8.25) is in general sufficient but not necessary. We have for cubic nonlinearities ($\alpha = 2$) that $n > 5/2$, i.e. $n \geq 3$, is sufficient. It is also known that quadratic nonlinearities ($\alpha = 1$) in three space dimensions may lead to the development of singularities, see John [68] and also Chapter 10. In this sense the result is optimal with respect to α in the case $(n, \alpha) = (3, 2)$.
Remark: A similar situation is given for the equations of elasticity in the initially isotropic case (cf. Section 11.1) following S. Klainerman [87] (existence for cubic nonlinearities in $\mathbb{R}^3$) and F. John [69] (development of singularities in the quadratic case).
On the other hand the condition on n, namely $n \geq 6$, which arises in the quadratic case ($\alpha = 1$) from (8.25), is only sufficient, not necessary, as we have shown in Theorem 1.2. To prove this optimal result special properties of the wave equation were used. These special effects are not obviously given for the systems which will be studied in Chapter 11. The results obtained for nonlinear heat equations by the general method (see Section 11.2) will also give us optimal results in many cases but not in all cases; see Table 11.3 in Section 11.2. In particular, if the nonlinearity does not depend on u, the general method

does not take into consideration the special form of the heat equation, i.e. the dissipative term $-\Delta u$, well enough.

This emphasizes that the general method leads to optimal results in many cases, but for each system it may be necessary to exploit its special structure — besides the evolutionary structure needed for the general method — to get optimal results in some special cases.

In this spirit it is interesting to mention other methods for proving global existence theorems which have been developed for special situations. This will be briefly done in the next Chapter.

9 Other methods

As we have pointed out in the preceding section, the classical method used in the proof of Theorem 1.1, which we shall call *energy method* in the sequel, does not lead to optimal results in each case neither for wave equations nor for all the systems in Chapter 11. Special ansätze have turned out to be more efficient for particular systems. Before giving some ideas of such methods we shall first present a method, which historically precedes the energy method concerning general existence results on nonlinear evolution equations; this is the Nash–Moser–Hörmander scheme.

Remark: We have called the energy method *classical* because the basic idea (to prove a nice a priori estimate in order to be able to continue a local solution) is classical. The ingredients of proving the a priori estimate (see Chapters 2, 4, 6, 7) have been developed to their full strength in the last decade.

1. The Nash–Moser–Hörmander scheme

The first general global existence theorems (small, smooth solutions) for nonlinear wave equations, later on also for other evolution equations, were obtained by S. Klainerman 1980 and 1982 respectively in his papers [86, 87]. He used a global iteration scheme for solutions of the linearized equations in $[0, \infty) \times \mathbb{R}^n$ instead of continuing local solutions of the nonlinear problem.

Roughly speaking, this means for the initial value problem

$$y_{tt} - \Delta y = f(Dy, \nabla Dy),$$

$$y(t=0) = y_0, \quad y_t(t=0) = y_1,$$

that first the function $\tilde{y}^{n+1}$ is computed from a given function y^n with the natural iteration:

$$\tilde{y}^{n+1}_{tt} - \Delta \tilde{y}^{n+1} = f(Dy^n, \nabla Dy^n), \qquad \tilde{y}^{n+1}(t=0) = y_0,\, \tilde{y}^{n+1}_t(t=0) = y_1. \tag{9.1}$$

Here decay properties of solutions to the linearized problem are of importance again. This naive iteration leads to a loss of regularity, in particular because of the occurence of the highest-order derivatives in the nonlinearity. To overcome this difficulty a kind of a Nash–Moser–Hörmander scheme is used; for this purpose a smoothing operator $S = S_n$ is introduced and the iteration is $y^n \longrightarrow y^{n+1}$ where

$$y^{n+1} := S_n \tilde{y}^{n+1}.$$

y^{n+1} does not solve the differential equation in (9.1) exactly, but the error produced by S_n only has a quadratic character (compare the classical Newton-iteration (*Isaac Newton*, 25.12.1642 – 20.3.1727)), the smoothing effect provided by S_n compensates for the loss of regularity in the simple iteration (9.1).

This method is technically complicated (and actually much more sophisticated than outlined in the coarse scheme above), but has also been applied to nonlinear wave equations in exterior domains (by Shibata & Tsutsumi [167]). In general this method provides less sharp regularity and decay results (see [86, 87, 94, 167]).

Remarks: Concerning the origin and the name of this method we remark that the crucial loss of regularity, which occurs in the naive iteration, has its counterparts in the so-called "small divisor problems" in celestial mechanics and in isometric imbedding problems in differential geometry. Mainly there are two ways out of the difficulties. The first one was developed by J. Nash in his paper [135] on isometric imbeddings and further developed by L. Hörmander, for example for problems in physical geodesy, see [45]. The second one is based on a modification of the classical Newton-iteration; see for example the paper of J. Moser [133].

It should be mentioned that the main problems for which this method was used to our knowledge essentially have been dealt with by other, simpler methods: the initial value problems by Klainerman & Ponce [94] (instead of [86, 87]) as discussed in the previous chapters, the geodetic problem of Hörmander by K.-J. Witsch in [199] using the Legendre transform (*Adrien-Marie Legendre,* 18.9.1752 – 10.1.1833), wave equation problems in exterior domains with the energy method by Shibata & Tsutsumi [167], and even the starting result of J. Nash was proved by M. Günther in [41] with different methods. But there are certainly other fields of applications for this original method.

2. The method of invariant norms

This method, developed by S. Klainerman for the study of linear and nonlinear wave equations, was used to prove Theorem 1.2. It exploits the invariance properties of $\partial_t^2 - \Delta$ under the inhomogeneous Lorentz group (also called the Poincaré group, consisting of translations and rotations, the transformations of the homogeneous Lorentz group, cf. Chapter 4) (*Henri Poincaré,* 29.4.1854 – 17.7.1912). As described in Chapter 4 one can define generalized Sobolev norms with the generators of the Poincaré group replacing the usual differential operators $\partial_1, \ldots, \partial_n$, which define the classical Sobolev norms. Generalized Sobolev inequalities as in Lemma 8.4 then allow to obtain optimal results for the wave equation in the quadratic case ($\alpha = 1$), namely the result from Theorem 1.2. Moreover, it is possible to get further information in the case $\alpha = 1, n = 3$, where in general a blow-up, a development of singularities in the function or in one of its derivatives, may occur (cf. Chapter 8 and see Chapter 10 for blow-up results). It turns out, with the help of the method of invariant norms, that a so-called "null condition" imposed on the nonlinearity is sufficient for the existence of a global, small solution also in the quadratic case in $\mathbb{R}^3$; see [90].

The *null condition* for a function

$$F = F(y,w) = Q(w) + \mathcal{O}(|(y,w)|^3),$$

where $w = (Dy, D^2y)$, $Q(w) = \mathcal{O}(|w|^2)$ as $|w| \to 0$, reads

$$\begin{aligned}
\frac{\partial^2 Q(w)}{\partial(\partial_a y)\partial(\partial_b y)} \quad \xi_a\xi_b &= 0,\\
\frac{\partial^2 Q(w)}{\partial(\partial_a y)\partial(\partial_b\partial_c y)} \quad \xi_a\xi_b\xi_c &= 0,\\
\frac{\partial^2 Q(w)}{\partial(\partial_a\partial_b y)\partial(\partial_c\partial_d y)} \quad \xi_a\xi_b\xi_c\xi_d &= 0,
\end{aligned}$$

for all

$$\xi = (\xi_0, \xi_1, \xi_2, \xi_3) \in \mathbb{R} \times \mathbb{R}^3 \quad \text{with} \quad \xi_0^2 = \xi_1^2 + \xi_2^2 + \xi_3^2$$

and arbitrary w, and it has to be summed up over all integer indices from 0 to 3.
A typical example for a nonlinearity F satisfying the null condition is

$$F = |\nabla y|^2 - y_t^2,$$

which was studied in the Example 1 in Chapter 1.
For formulations of the null condition see also F. John [72] and W. Strauss [179].
The null condition has been recognized as being a sufficient condition for quadratic nonlinearities in $\mathbb{R}^3$ by D. Christodoulou in [18] too; see paragraph 3. The method of invariant norms has also been applied with appropriate modifications to Klein–Gordon equations by S. Klainerman in [89] (see Section 11.5), to Schrödinger equations by P. Constantin [22] (cf. Section 11.4), and by F. John to the equations of elasticity in [73] (having less invariances at hand and proving an “almost global existence result”; see Section 11.1).
To have a rough idea of the action of the null condition one should notice that the decay of solutions to nonlinear wave equation in t and x in general is better away from the boundary of the light cone (cf. e.g. Lemma 8.3 and [90]). The null condition assures that it can not become too bad on the boundary $= \{\xi \in \mathbb{R} \times \mathbb{R}^3 \mid \xi_0^2 = \xi_1^2 + \xi_2^2 + \xi_3^2\}$.

3. The method of conformal maps

The wave equation is conformally invariant. D. Christodoulou uses in [18] a special conformal map due to R. Penrose, which maps $\mathbb{R} \times \mathbb{R}^n$ into a bounded set in $\mathbb{R} \times S^n$. In this sense it is called a “conformal compactification” (see the notes in [179]). Hence the problem of global existence is carried over to a local problem which has to be solved up to the possible boundary. This method had been applied before to Yang–Mills equations, see [18].

4. The method of normal forms

In order to deal with a quadratic nonlinearity F in $\mathbb{R}^3$ for the nonlinear Klein–Gordon equation

$$(\partial_t^2 - \Delta + m)\, y = f(y, Dy, \nabla Dy), \quad m > 0,$$

(see Section 11.5), J. Shatah applies in [159] a change of the dependent variable, which essentially transforms the quadratic nonlinearity into a cubic one, which may be dealt with by the energy method after some appropriate easy modifications. The name of the method is connected with Poincaré's theory of normal forms appearing in the theory of ordinary differential equations, see V.I. Arnold [7] or Chow & Hale [17].
The ansatz of Shatah can be described as follows. In order to solve

$$v_t + Av = k(v),$$

where k vanishes up to a certain order near $v = 0$, one applies a change of variables of the type

$$w = v + h(w).$$

Now h has to be determined in a way such that w solves the equation

$$w_t + Aw = g(w),$$

where g vanishes near $w = 0$ of one higher order than k near $v = 0$. This means that h has to solve a special differential equation which actually can be solved by Shatah for the case of the Klein–Gordon equation.
We remark that the last three methods are discussed in more detail by W. Strauss in [179]. Of course there are more methods for special systems with specific difficulties, see for example the discussion of parabolic problems in the survey article [10].

10 Development of singularities

The theorems in Chapter 1 are results for small data. The necessity for dealing with small perturbations of the linearized equations is underlined in the sequel by examples which show that, in general, one has to expect the development of singularities in finite time. In particular neither the smallness of the initial data nor the smoothness of data including the coefficients can prevent a solution from blowing up. We shall not go into the details here but we just present an illustration of the typical hyperbolic phenomenon that the solution and/or derivatives of the solution become singular after some time. This will mean in general that norms like the L^∞-norm of the local regular solution or of its derivatives become infinite. The only way to avoid a blow-up are smallness of the data in connection with a sufficiently strong vanishing of the nonlinearity near zero and a sufficiently high space dimension. This is the message of the Theorems 1.1 and 1.2. Moreover we have learned from Theorem 1.2 that a solution of the nonlinear wave equation with a quadratic nonlinearity in $\mathbb{R}^3$ lives at least exponentially long, although the examples below show that in general a blow-up occurs. Nevertheless this result justifies the notion of "almost global existence" in this case (cf. the paper of John & Klainerman [75]). We mention that for large data a blow-up may occur also in the cases where one has global existence for small data, see Example 1 in Chapter 1 and the remarks below.

In this chapter we are only concerned with wave equations (or rather hyperbolic equations and systems). One should however notice that similar results also hold for heat equations and the other systems which are discussed in the next chapter. A few further examples and hints are given there.

The following survey first recalls the simple case of an equation of first order in one space dimension. Already there it will become clear that smoothness and smallness of the data in general cannot assure the existence of global smooth solutions.

Let u be the solution to the following initial value problem:

$$u_t + a(u)u_x = 0, \quad u(t=0) = u_0, \quad (t,x) \in \mathbb{R}^2, \tag{10.1}$$

where a is a smooth function satisfying

$$a' > 0.$$

If the equation for u is in conservation form, i.e. a is the derivative of some function h, then this assumption means the convexity of h.

We have an implicit representation for u:

$$u(t,x) = u_0(x - ta(u(t,x))), \tag{10.2}$$

and the derivatives u_x and u_t are given by

$$u_x(t,x) = \frac{u_0{}'(x - ta(u(t,x)))}{1 + a'(u(t,x))u_0{}'(x - ta(u(t,x)))\,t},$$

$$u_t(t,x) = -\frac{a(u(t,x))u_0{}'(x - ta(u(t,x)))}{1 + a'(u(t,x))u_0{}'(x - ta(u(t,x)))\,t}.$$

It is now obvious that u_x and u_t become singular in finite time if $u_0{}'$ is negative somewhere. This happens independently of the smoothness or the smallness of the datum u_0. If

$$u_0(x) = \varepsilon\phi(x), \quad \phi \in C_0^\infty(\mathbb{R}), \quad \varepsilon > 0,$$

then we have for the maximal length $T = T(\varepsilon)$ of an interval of existence

$$\lim_{\varepsilon \downarrow 0} \varepsilon T(\varepsilon) > 0 \tag{10.3}$$

in the so-called "genuinely nonlinear" case $a'(0) \neq 0$. If $a'(0) = 0$ but a is not constant in a neigbourhood of $u = 0$, then

$$\lim_{\varepsilon \downarrow 0} \varepsilon^2 T(\varepsilon) > 0$$

holds, see F. John [74].
These are typical nonlinear phenomena. In the linear case, where a is constant, the solution of (10.1), given by (10.2), exists globally, also for large data.
F. John [64] and also T.-P. Liu [112] proved a corresponding blow-up result for systems of first order in one space dimension. This is also useful for the treatment of plane waves in higher dimensions, see for example the discussion of elastic waves in [64, 144].
Remark: There is much less known for general systems in higher dimensions even if they are in conservation form. This concerns both the question of existence and the study of singularities. For *large* data we mention a result of T.C. Sideris [172]. Consider a system of m conservation laws in n space dimensions of the form

$$u_t + f(u)_x = 0, \quad u(t=0) = u_0,$$

where $f : \mathbb{R}^m \longrightarrow \mathbb{R}^n \times \mathbb{R}^m$, $f(0) = 0$, $x \in \mathbb{R}^n$. If $B_i(u)$ denotes the matrix with coefficients $\frac{\partial f_{ij}}{\partial u_k}(u)$, then one of the following two conditions 1. or 2. shall hold, where

1. $n = 1$, $B_1(u)$ has only real eigenvalues with corresponding eigenvectors which span the whole space;

2. $B_i(u) = A_0^{-1}(u)A_i(u)$ with symmetric matrices $A_i(u)$, $i = 1, \ldots, n$, $A_0(u)$ positive definite.

Then it is proved (under certain additional assumptions which correspond to the genuine nonlinearity condition $a'(0) \neq 0$ for the system (10.1)) that there are no global smooth solutions for (too) large data.

Now we turn to the type of wave equations that were studied in the previous chapters, i.e.

$$u_{tt} - \Delta u = f(Du, \nabla Du).$$

F. John studied radially symmetric solutions. He proved in [70] for the equation

$$u_{tt} = c^2(u_t)\Delta u \tag{10.4}$$

in $\mathbb{R}^3$, with $c(0) = 1$ and if $c'(0) \neq 0$ (without loss of generality > 0) that there will always appear singularities in the derivatives of u. If

$$u(t = 0) = \varepsilon\phi, \quad u_t(t = 0) = \varepsilon\psi,$$

then

$$\lim_{\varepsilon \downarrow 0} \varepsilon \log T(\varepsilon) \leq \frac{1}{c'(0)K}, \tag{10.5}$$

($K \geq 0$ is determined from the data). Actually, equality holds in (10.5).

Also the following quadratic case in $\mathbb{R}^3$ was studied by F. John [68]:

$$u_{tt} - \Delta u = 2u_t u_{tt}.$$

Again the development of singularities in finite time is proved. More precisely, there is no global C^2-solution (for smooth data with compact support) if

$$\int_{\mathbb{R}^3} [u_t(0, x) - u_t^2(0, x)]\, dx \; > \; 0.$$

This kind of results was extended to radially symmetric quadratic nonlinearities involving the radial derivatives by T.C. Sideris [171].

The special case of nonlinearities of the type $f = f(u)$ is not subject to considerations here; we refer the reader to F. John [74] and W. Strauss [179]. We only remark that in $\mathbb{R}^3$ the quadratic nonlinearity $f(u) = u^2$ leads to a development of singularities; more generally one can try to characterize the critical exponent p in $f(u) = u^p$ depending on the space dimension n; see for example F. John [67], H. Pecher [139].

Having realized that, in general, singularities will develop in finite time, that for example quadratic nonlinearities lead to singularities in $\mathbb{R}^3$ but allow global small solutions in $\mathbb{R}^n$, $n \geq 4$, it is natural to ask for the life span of smooth solutions in dependence of the initial data. We have already given a first result in this direction for the one-dimensional case in (10.3) and for the special equation (10.4) in (10.5).

For the equations

$$u_{tt} - \Delta u = f(Du, \nabla Du),$$

$$u(t=0)=\varepsilon\phi,\quad u_t(t=0)=\varepsilon\psi,\quad (\varepsilon>0),$$

$$|f(W)|=\mathcal{O}(|W|^{\alpha+1}),\quad \text{for } |W|\to 0 \text{ and some } \alpha\in\mathbb{N},$$

we summarize results on estimates for the length $T=T(\varepsilon)$ of the maximal interval of existence of a smooth solution in Table 10.1, see L. Hörmander [48], F. John [74], Li & Yu [105], H. Lindblad [109, 110]. (For the case $f=f(u)$ we refer to [74, 105] and the references there.)

α	n	Lower bounds for $T(\varepsilon)$
1	1	$A\varepsilon^{-1}$
1	2	$A\varepsilon^{-2}$
1	3	$\exp(A\varepsilon^{-1})$
2	2	$\exp(A\varepsilon^{-2})$
1	≥ 4	Global solutions
2	≥ 3	
3	≥ 2	

Table 10.1: (Almost) global existence results

In Table 10.1 A denotes a constant which only depends on ϕ,ψ and f. The statements are to be understood for sufficiently small ε.

The proofs of the relations in Table 10.1 partially use the generalized Sobolev estimates, arising from the invariance of $\partial_t^2-\Delta$ under the Lorentz group, cf. Chapters 4,8,9. The results in the case of non-global existence are named *almost global existence results* (at least in the exponential case, see Table 10.1).

We conclude this chapter with the (as obvious as important) hint at the fact that the previous results require an investigation of weak solutions which might exist globally. One should be aware of the observation that derivatives of smooth solutions may develop singularities but that still a global continuous solution might exist. Here further studies are necessary, in particular on the propagation of discontinuities — shock waves —, see for example the books of A. Majda [114] and J. Smoller [176].

For extensive surveys on initial value problems for nonlinear wave equations we refer to L. Hörmander [48], F. John [74] and W. Strauss [179].

11 More evolution equations

It has been anticipated in the introduction that the proof of the first global existence theorem, Theorem 1.1, follows a general principle which will lead to similar results for the systems of evolution equations in this chapter. The common structure of the proof of global existence of small, smooth solutions can be described as follows.
We consider a system of the type

$$V_t + AV = F(V, \ldots, \nabla^\beta V),\ V(t = 0) = V^0, \tag{11.1}$$

where $V = V(t, x)$ is a vector-valued function taking values in $\mathbb{R}^k$ (or $\mathbb{C}^k$), $t \geq 0$, $x \in \mathbb{R}^n$. A is a linear differential operator of order $m \in \mathbb{N}$, $k, n, m \in \mathbb{N}$. F is a smooth linear function of V and its derivatives up to order $|\beta|$ which may be equal to m, and V^0 is a given initial value.
It is assumed that the nonlinearity $F = F(W)$ behaves near $W = 0$ as follows:

$$F(W) = \mathcal{O}(|W|^{\alpha+1}) \quad \text{as} \quad |W| \longrightarrow 0 \tag{11.2}$$

for some $\alpha \in \mathbb{N}$.

Remark: F may also depend explicitly on t and x, $F = F(t, x, V, \ldots, \nabla^\beta V)$. In that case the condition (11.2) is to be read uniformly in t and x.
A global existence result is proved along the following steps **A**–**E**.

A: Decay of solutions to the linearized system ($F \equiv 0$):
One proves that

$$\|V(t)\|_q \leq c(1+t)^{-d}\|V^0\|_{N,p}, \tag{11.3}$$

where $2 \leq q \leq \infty$, $1/p + 1/q = 1$, $c, d > 0$ and $N \in \mathbb{N}$ are functions of q and of the space dimension n. (Cf. Chapter 2 for the wave equation: $d = \frac{n-1}{2}(1 - \frac{2}{q})$.)

B: Local existence and uniqueness theorem:
The existence of a unique local solution V to (11.1) has to be shown. V shall satisfy:

$$V \in C^0([0, T], W^{s,2}) \cap C^1([0, T], W^{\tilde{s},2}),$$

$s, \tilde{s} \in \mathbb{N}$, $T > 0$ appropriately chosen. (Cf. Chapter 5 for the wave equation: $\tilde{s} = s - 1$.)

C: High energy estimates of the following type should hold:

$$\|V(t)\|_{s,2} \leq C\|V^0\|_{s,2} \exp\{C \int_0^t \|V(r)\|_{b,\infty}^\alpha \, dr\}, \tag{11.4}$$

$t \in [0, T]$. C depends only on s, not on T or V^0, and b is independent of s (which will allow us to close the circle later in Step **E**). (Cf. Chapter 6 for the wave equation: $b = 1$.)

D: Weighted a priori estimates should hold:
With the help of **A** and the standard representation formula

$$V(t) = e^{-tA}V^0 + \int_0^t e^{-(t-r)A}F(V, \ldots, \nabla^\beta V)(r)dr,$$

where $e^{-tA}W$ is the solution to the linearized problem with initial value W, one proves that the following inequality holds for sufficiently small V^0 (small in a sense to be made precise later)

$$\sup_{0 \le t \le T} (1+t)^{d_1} \|V(t)\|_{s_1, q_1} \le M_0 < \infty,$$

where M_0 is independent of T, s_1 is sufficiently large, $q_1 = q_1(\alpha)$ is chosen appropriately, and $d_1 = d(q_1, n)$ according to **A**. (Cf. Chapter 7 for the wave equation: $q_1 = 2\alpha + 2$, $d_1 = \frac{\alpha}{\alpha+1} \frac{n-1}{2}$.)

E: The results in **C** and **D** will lead to the desired final (classical a priori) energy estimate for the local solution:

$$\|V(t)\|_{s,2} \le K \|V^0\|_{s,2}, \quad 0 \le t \le T, \tag{11.5}$$

s being sufficiently large, V^0 being sufficiently small, K being independent of T (and of V^0). The estimate (11.5) allows the continuation of the local solution to the whole time axis $[0, \infty)$ with standard arguments. (Cf. Chapter 8 for the wave equation.)

The examples in the subsequent sections demonstrate the wide range of applicability of the scheme **A–E**, in particular the discussion of the system of thermoelasticity in Section 11.3. There the scheme **A–E** has to be modified because there are different types of nonlinearities, i.e. the assumption (11.2) has to be replaced by a more complicated one, and also there are different decay rates for different components of V, i.e. (11.3) will split into several different estimates. Thus it is clear that also the Steps **B–E** will have to be modified. But since this will be done in the spirit of the scheme above, the example of thermoelasticity is a good demonstration for the applicability of the ideas that are underlying the general scheme.
We shall go through the steps **A–E** for each of the following examples in the Sections 11.1. – 11.7, but we shall not go into all details as we did it discussing the wave equation in the Chapters 2–8. Instead we shall point out the essential results and if necessary, we give further references.

We also remark that most of the systems are studied in their linearized form in the book of R. Leis [98], hence several of the aspects concerning Step **A** are illustrated there. The proof of Theorem 1.2, in particular the use of certain invariance properties of the d'Alembert operator, does not have counterparts for all the following examples. There exist already similar considerations for the equations of elasticity, for Schrödinger equations and for Klein–Gordon equations, cf. the remarks in the corresponding Sections 11.1, 11.4 and 11.5 respectively.

11.1 Equations of elasticity

In this section we consider first the initial value problem for a homogeneous, initially isotropic hyperelastic medium in $\mathbb{R}^3$ and then the initial value problem for a homogeneous, initially cubic hyperelastic medium in $\mathbb{R}^2$. It is not the different space dimension but the different elastic behavior (isotropic or cubic) that will produce different interesting effects. This illustrates that there are many interesting unknown or even unexpected features in these equations as soon as one leaves the most simple situation. This might hold for other equations as well and underlines the necessity of further research on each of these systems.

11.1.1 Initially isotropic media in $\mathbb{R}^3$

Let $U = (U_1, U_2, U_3) = U(t,x)$ be the displacement vector of a three-dimensional elastic medium filling the whole of $\mathbb{R}^3$, i.e. $t \geq 0$, $x \in \mathbb{R}^3$. The equations of motion for a homogeneous medium in the absence of external forces are

$$\partial_t^2 U_i = \sum_{m,j,k=1}^{3} C_{imjk}(\nabla U)\partial_m \partial_k U_j, \quad i = 1,2,3, \tag{11.6}$$

where the C_{imjk} are smooth nonlinear functions, the so-called elastic moduli, which are given by

$$C_{imjk}(\nabla U) = \frac{\partial^2 \psi(\nabla U)}{\partial(\partial_k U_j)\partial(\partial_m U_i)}$$

with a given smooth potential ψ.

Remark: The assumption of the existence of ψ which is mostly made (cf. F. John [66]) corresponds to the assumption that the underlying medium is "hyperelastic", cf. J.M. Ball [8], P.G. Ciarlet [21]. If one does not neglect heat conduction effects then the existence of a corresponding ψ is guaranteed, compare Section 11.3.

The equations (11.6) arise from the classical law of balance of momentum, see the derivation of the equations of thermoelasticity in Section 11.3, where we have assumed in our homogeneous case that the mass density ϱ satisfies $\varrho \equiv 1$ without loss of generality.

For a derivation of the equations from basic physical principles see Gurtin [42], and for the following transformation to a first-order system see F. John [66]. In addition to the differential equations (11.6), the initial values $U(t=0)$ and $U_t(t=0)$ are prescribed:

$$U(t=0) = U^0,\ U_t(t=0) = U^1. \tag{11.7}$$

The equations (11.6) are rewritten as

$$\begin{aligned}\partial_t^2 U_i - \sum_{m,j,k=1}^{3} C_{imjk}(0)\partial_m\partial_k U_j &= \sum_{m,j,k=1}^{3} (C_{imjk}(\nabla U) - C_{imjk}(0))\partial_m\partial_k U_j \\ &=: f_i(\nabla U, \nabla^2 U), \quad i=1,2,3.\end{aligned} \tag{11.8}$$

We assume that the medium is initially isotropic, meaning

$$\begin{aligned}C_{imjk}(0) &= (c_1^2 - 2c_2^2)\delta_{im}\delta_{jk} + c_2^2(\delta_{ij}\delta_{km} + \delta_{jm}\delta_{ik}), \\ &\quad i,m,j,k = 1,2,3,\end{aligned} \tag{11.9}$$

where the constants c_1, c_2 satisfy

$$c_1 > c_2 > 0. \tag{11.10}$$

With the Lamé constants λ, μ (*Gabriel Lamé*, 22.7.1795 – 1.5.1870) one has

$$c_1^2 = \lambda + 2\mu,\ c_2^2 = \mu. \tag{11.11}$$

Then (11.10) is equivalent to

$$\mu > 0,\ \lambda + \mu > 0. \tag{11.12}$$

With this notation and

$$\tau := \lambda + 2\mu$$

the equations (11.8) become

$$U_{tt} + \mu\nabla \times \nabla \times U - \tau\nabla\nabla' U = f(\nabla U, \nabla^2 U), \tag{11.13}$$

$$(f = (f_1, f_2, f_3)').$$

The transformation of the second-order system (11.13) to a first-order system in t is given through defining the vector V by

$$V := (\partial_r U_i)_{ir}, \quad r = 0,1,2,3;\ i = 1,2,3 \tag{11.14}$$

with $\partial_0 := \partial_t$, i.e.

$$V = (\partial_0 U_1, \partial_0 U_2, \dots, \partial_3 U_3),$$

and the 12×12-matrix $A^r(\nabla U)$, $r = 1, 2, 3$, by

$$A^r(\nabla U) \equiv (A^r_{imjk}(\nabla U))_{imjk}, \quad i, j = 1, 2, 3;\ m, k = 0, 1, 2, 3,$$

where i, m count the rows and j, k count the columns. The element $A^r_{imjk}(\nabla U)$ is given by

$$A^r_{imjk}(\nabla U) := C_{irjk}(\nabla U)\delta_{m0}(1 - \delta_{k0}) + \delta_{k0}\delta_{rm}\delta_{ij}. \tag{11.15}$$

With these notations the differential equation (11.13) resp. (11.8) can be written as the following equation for V:

$$V_t + AV = F(V, \nabla V), \tag{11.16}$$

with initial value

$$V(t = 0) = V^0,$$

where

$$AV := -\sum_{r=1}^{3} A^r(0)\partial_r V,$$

F arises canonically from f, and

$$V^0 := (\partial_r U_i)_{ir}(t = 0) = (U^1, \nabla U^0).$$

In particular we see that V^0 is given in terms of ∇U^0 and U^1.
With (11.16) we have transformed the original equations (11.6) into the general form (11.1). Now we proceed by looking at the general Steps **A**–**E** which will lead to a global existence theorem for small data under certain assumptions on the nonlinearity.

A: Decay for $F \equiv 0$:
If U solves for $t \geq 0$

$$U_{tt} + \mu\nabla \times \nabla \times U - \tau\nabla\nabla' U = 0, \quad U(t = 0) = 0,\ U_t(t = 0) = U^1, \tag{11.17}$$

then we can use explicit representation formulae for the solution U in analogy to the situation in Chapter 2 and one obtains (see F. John [66]):

$$\|(U_t, \nabla U)(t)\|_\infty \leq C(1 + t)^{-1}\|U^1\|_{3,1}, \quad 0 \leq t < \infty. \tag{11.18}$$

Here and in the sequel C denotes a constant that does not depend on t or on the initial data.
The energy is conserved which is expressed by

$$\begin{aligned}\frac{1}{2}\frac{d}{dt}(\|U_t\|_2^2 \;&+\; \mu\|\nabla \times U\|_2^2 + \tau\|\nabla' U\|_2^2)\\ &= \langle U_t, U_{tt} + \mu\nabla \times \nabla \times U - \tau\nabla\nabla' U\rangle\\ &= 0.\end{aligned}$$

We naturally assumed all functions to be real-valued and we dropped the parameter t in $U_t(t,\cdot)$ etc.
Since τ, μ are positive and

$$\begin{aligned}\|\nabla \times U\|_2^2 + \|\nabla' U\|_2^2 &= \langle \nabla \times \nabla \times U - \nabla\nabla' U, U\rangle \\ &= \langle -\Delta U, U\rangle \\ &= \|\nabla U\|_2^2,\end{aligned}$$

where $\Delta U \equiv (\Delta U_1, \Delta U_2, \Delta U_3)$, we obtain

$$\|(U_t, \nabla U)(t)\|_2 \leq C\,\|U^1\|_2. \tag{11.19}$$

Interpolation between (11.18) and (11.19) gives the following estimate for V:

$$\|V(t)\|_q \leq C\,(1+t)^{-(1-\frac{2}{q})}\|V^0\|_{N_p,p}, \quad t \geq 0,\ C = C(q) \tag{11.20}$$

where $2 \leq q \leq \infty$, $1/p + 1/q = 1$, and N_p is not greater than 3.
Remark: In order to describe the decay it would also be possible to make the following ansatz:

$$L^2 = \overline{\nabla W^{1,2}} \oplus \mathcal{D}_0 \quad (\overline{\nabla W^{1,2}} : \text{ closure of } \nabla W^{1,2} \text{ in } L^2) \tag{11.21}$$

is an orthogonal decomposition of $L^2 = (L^2(\mathbb{R}^3))^3$ with

$$\nabla W^{1,2} := \{\nabla\varphi \,|\, \varphi \in W^{1,2}\},$$

and

$$\mathcal{D}_0 := \{W \in L^2 \quad | \quad \forall \varphi \in C_0^\infty : \langle \nabla\varphi, W\rangle = 0\}$$

is the space of vector fields having (weak) divergence zero. (The decomposition (11.21) easily follows from the projection theorem.) The corresponding decomposition of U into

$$U = U^{po} + U^{so}$$

(U^{po}: potential part; U^{so}: solenoidal part) leads to a decomposition of (11.17) into the two systems

$$U_{tt}^{po} - \tau\Delta U^{po} = U_{tt}^{po} - \tau\nabla\nabla' U^{po} = 0,\ U^{po}(t=0) = (U^0)^{po},\ U_t^{po}(t=0) = (U^1)^{po},$$

and

$$U_{tt}^{so} - \mu\Delta U^{so} = U_{tt}^{so} + \mu\nabla \times \nabla \times U^{so} = 0,\ U^{so}(t=0) = (U^0)^{so},\ U_t^{so}(t=0) = (U^1)^{so}.$$

Now one could apply the results from Chapter 2 for the linear wave equation, at least for $2 < q < \infty$. (In order to derive the final decay result (11.20) from the result for U^{so} and U^{po} one has to know that the projections P_{po} and P_{so}

$$P_{po} : L^2 \longrightarrow \overline{\nabla W^{1,2}}, \quad P_{so} : L^2 \longrightarrow \mathcal{D}_0,$$

have the property that

$$\|P_{po}W\|_p \leq C\,\|W\|_p, \quad \|P_{so}W\|_p \leq C\,\|W\|_p, \tag{11.22}$$

which follows from the explicit representation

$$\begin{aligned} P_{po}W &= \mathcal{F}^{-1}(\xi_0\xi_0' \cdot \mathcal{F}W), \\ P_{so}W &= \mathcal{F}^{-1}(-\xi_0 \times \xi_0 \times \mathcal{F}W), \end{aligned}$$

$\xi \in \mathbb{R}^3$ being the variable in Fourier space and $\xi_0 = \frac{\xi}{|\xi|}$. The estimates (11.22) then follow from known theorems on multipliers in Fourier space, see for example Theorem 1.4 in [44].)

B: Local existence and uniqueness:
A change from $C_{imjk}(\nabla U)$ to $\overline{C}_{imjk}(\nabla U)$ with

$$\overline{C}_{imjk}(\nabla U) := C_{imjk}(\nabla U) + c_2^2(\delta_{im}\delta_{jk} - \delta_{ik}\delta_{jm})$$

leaves the differential equation (11.6) invariant but this change will be important for the matrix A^0 (which will be defined below) to be positive definite. We shall write $C_{imjk}(\nabla U)$ again instead of $\overline{C}_{imjk}(\nabla U)$.
Let the matrix A^0 be given by its elements

$$A^0_{imjk}, \quad i,j = 1,2,3;\ m,k = 0,1,2,3,$$

where

$$A^0_{imjk} := (1-\delta_{m0})(1-\delta_{k0})C_{imjk}(\nabla U) + \delta_{ij}\delta_{m0}\delta_{k0},$$

i.e. A^0 essentially depends on V, (formally define $C_{imjk}(\nabla U) := 0$ if $m = 0$ or $k = 0$). Then the differential equations (11.6) (resp. (11.16)) turn into

$$A^0(V)\partial_t V + \sum_{r=1}^{3} A^0(V)A^r(V)\partial_r V = 0,$$

with initial value $V(t=0) = V^0$ and the matrices $A^0(V)$ and $A^0(V)A^r(V)$, $r = 1,2,3$ are symmetric and $A^0(V)$ is positive definite (uniformly with respect to V in each compactum). Hence we can apply Theorem 5.8 and we obtain a unique local solution

$$V \in C^0([0,T], W^{s,2}) \cap C^1([0,T], W^{s-1,2})$$

for some $T > 0$, if $s \geq 3$ and $V^0 \in W^{s,2}$.

C: High energy estimates:
The desired energy estimate of the type (11.4) and the subsequent a priori estimates in **D** and **E**, as well as the final global existence theorem (Theorem 11.1 below) are connected to the behavior of the nonlinearity $F = F(W)$ near $W = 0$. Essentially, F consists of the terms f_i, $i = 1, 2, 3$, where

$$f_i(\nabla U, \nabla^2 U) = \sum_{m,j,k=1}^{3} (C_{imjk}(\nabla U) - C_{imjk}(0))\, \partial_m \partial_k U_j$$

according to (11.8). Hence F vanishes at least of order 2 ($\alpha = 1$) near $W = 0$.
F. John has demonstrated that in the general quadratic case, more precisely, in the so-called "genuinely nonlinear" case, solutions will develop singularities in finite time; see [69] for radially symmetric solutions and also [64] for plane-wave solutions. Recently he investigated in [73] the life span T_∞ of local solutions for the quadratic case and he proved a lower bound for T_∞ in analogy to the situation known for scalar nonlinear wave equations ("almost global existence", cf. Chapter 10 and Theorem 1.2). He used the method of invariant norms adapted to the equations of elasticity, cf. Chapter 9.
In order to obtain a general global existence theorem we therefore assume that, for $i, m, j, k = 1, 2, 3$,

$$|C_{imjk}(\nabla U) - C_{imjk}(0)| = \mathcal{O}(|\nabla U|^2) \quad \text{as } |\nabla U| \longrightarrow 0 \tag{11.23}$$

holds, whence we have

$$F(W) = \mathcal{O}(|W|^3) \quad \text{as } |W| \longrightarrow 0.$$

Cubic nonlinearities turned out to be appropriate for the existence of global solutions to nonlinear wave equations in three space dimensions, see Theorem 1.1. Since the decay behavior of solutions to the linearized equations of elasticity is the same as that of solutions to linear wave equations — compare (11.20) and Theorem 2.3 —, we obtain the corresponding result in this Step **C** and in the following Steps **D, E** in complete analogy to the considerations in the Chapters 6–8. (See also Section 11.3, where the equations of elasticity appear as a special case.)
First we have the following high energy estimate for the local solution:

$$\|V(t)\|_{s,2} \leq C\|V^0\|_{s,2} \exp\{C\int_0^t \|\overline{D}V(r)\|_\infty^2\, dr\}, \quad t \in [0, T],\ C = C(s).$$

D: Weighted a priori estimates:
We have

$$\sup_{0 \leq t \leq T} (1+t)^{2/3} \|V(t, \cdot)\|_{s_1,6} \leq M_0 < \infty,$$

where M_0 is independent of T, s_1 is sufficiently large and

$$\|V^0\|_{s,2} + \|V^0\|_{s,6/5}$$

is sufficiently small ($s > s_1$ being sufficiently large).

E: Final energy estimate:

The following inequality is now easily obtained.

$$\|V(t)\|_{s,2} \leq K\|V^0\|_{s,2}, \quad 0 \leq t \leq T,$$

s being sufficiently large, V^0 being sufficiently small, K being independent of T (and V^0).

Remark: The minimal value of s can be given explicitly as it was done for the wave equation in Chapters 5–8.

Altogether we obtain the following global existence theorem.

Theorem 11.1 *We assume (11.23). Then there exist an integer $s_0 \geq 3$ and a $\delta > 0$ such that the following holds:*
If $V^0 = (U^1, \nabla U^0)$ belongs to $W^{s,2} \cap W^{s,6/5}$ with $s \geq s_0$ and

$$\|V^0\|_{s,2} + \|V^0\|_{s,6/5} < \delta,$$

then there is a unique solution U of the initial value problem to the nonlinear equations of elasticity in the initially isotropic case in $\mathbb{R}^3$ (11.6), (11.7), with

$$(U_t, \nabla U) \in C^0([0,\infty), W^{s,2}) \cap C^1([0,\infty), W^{s-1,2}).$$

Moreover we have

$$\begin{aligned} \|(U_t, \nabla U)(t)\|_\infty + \|(U_t, \nabla U)(t)\|_6 &= \mathcal{O}(t^{-2/3}), \\ \|(U_t, \nabla U)(t)\|_{s,2} &= \mathcal{O}(1) \quad \text{as} \quad t \to \infty. \end{aligned}$$

Remarks: We also have

$$\|(U_t, \nabla U)(t)\|_{s_1,6} = \mathcal{O}(t^{-2/3}) \quad \text{as } t \to \infty,$$

with s_1 given in Step C, but we shall not stress this and the corresponding results in the theorems in the following Sections 11.2 – 11.7.

In Theorem 11.1 the smoothness of the nonlinearity, i.e. C_{imjk} being C^∞, was assumed, the reason being the applicability of Theorem 5.8. But this can be relaxed as we mentioned in Chapter 5 and in Chapter 8 discussing the wave equation.

11.1.2 Initially cubic media in $\mathbb{R}^2$

As a second example from elasticity we study (initially) cubic media, which are the next more complicated ones following the (initially) isotropic media. This will be done for media filling the whole of $\mathbb{R}^2$. Considering two dimensions provides a technical simplification compared to the three-dimensional case, but nevertheless the essential features are shown as well. There will appear greater differences to the isotropic case than might be anticipated. This is of course expressed in the main theorem, Theorem 11.4 below, but is mainly given by the different rates of decay of solutions to the linearized problem. These decay rates will not be obtained using explicit representation formulae in terms of surface or volume integrals (as it was done for isotropic elasticity or for the wave equation). There exist such kinds of representations, see G.F.D. Duff [28], but they are rather complicated and do not seem to be appropriate for calculating decay rates. Instead, we shall apply the Fourier transform with respect to $x \in \mathbb{R}^2$ and the solution will be given as a Fourier integral, essentially an integral over the characteristic manifold (wave cone). The proof of the decay rates follows H. Pecher, see [138], and relies on L^∞–L^∞-estimates of oscillatory integrals over manifolds going back to W. Littman [111]. In case of n space dimensions and at most k vanishing principal curvatures of the characteristic manifold of the differential operator the decay rate $t^{-\frac{n-k}{2}(1-\frac{2}{q})}$, $2 \le q < \infty$, is obtained. This leads to the known decay rates for isotropic elasticity or for the wave equation where $k = 1$ (other examples: $k = 0$ for the Klein–Gordon equation (compare Section 11.5) and for the plate equation (compare Section 11.7)).
In the case of cubic media in $\mathbb{R}^2$ there are flat points on the wave cone, i.e. points where all principal curvatures vanish. Hence the method mentioned above does not directly lead to a decay result since $n = k = 2$. This requires a refined analysis of the method of stationary phase, which has been done by M. Stoth [177] and which will be roughly described below. The set of flat points is a one-dimensional submanifold on the two-dimensional wave cone. The order of vanishing of the principal curvatures at the flat points determines the decay rates which can be obtained by this method. For example we shall get $t^{-\frac{1}{3}(1-\frac{2}{q})}$ for nickel and copper, and $t^{-\frac{1}{2}(1-\frac{2}{q})}$ for aluminium, $2 \le q < \infty$.
Remark: To our knowledge there has not yet been given a *physical* explanation of these kinds of (weaker) decay rates; it still might be hidden in the *mathematical* technique.
A similar phenomenon has been observed by O. Liess in [107, 108] in connection with a system from crystal optics, where singular points and flat points appear ($n = 3$, $k = 2$).
The equations for the displacement vector $U = (U_1, U_2) = U(t, x)$, $t \ge 0$, $x \in \mathbb{R}^2$, are the same as those in (11.6) from Subsection 11.1.1, now with the indices running from 1 to 2, i.e.

$$\partial_t^2 U_i = \sum_{m,j,k=1}^{2} C_{imjk}(\nabla U)\partial_m \partial_k U_j \quad , \quad i = 1, 2, \tag{11.24}$$

with prescribed initial values

$$U(t=0)=U^0, \quad U_t(t=0)=U^1. \tag{11.25}$$

They are written as

$$\begin{aligned} \partial_t^2 U_i - \sum_{m,j,k=1}^{2} C_{imjk}(0)\partial_m\partial_k U_j &= \sum_{m,j,k=1}^{2} (C_{imjk}(\nabla U) - C_{imjk}(0))\partial_m\partial_k U_j \quad (11.26)\\ &=: f_i(\nabla U, \nabla^2 U), \quad i=1,2. \end{aligned}$$

If we assume that the medium is initially cubic, then we have that $(C_{imjk}(0))_{imjk}$ is characterized by three constants λ, μ and τ:

$$(C_{imjk}(0))_{imjk} = \begin{pmatrix} \tau & 0 & 0 & \lambda \\ 0 & \mu & \mu & 0 \\ 0 & \mu & \mu & 0 \\ \lambda & 0 & 0 & \tau \end{pmatrix},$$

(i, m: rows, j, k: columns)
Remark: The *isotropic* case is characterized by

$$\tau = 2\mu + \lambda,$$

(cf. (11.11) – (11.13), and a *weakly coupled* system by

$$\mu = -\lambda$$

(cf. R. Leis [98]). The transformation to a first-order system

$$V_t + AV = F(V, \nabla V), \quad V(t=0) = V^0, \tag{11.27}$$

is analogous to the procedure in the previous subsection (compare (11.14) – (11.16)) with

$$AV = -\sum_{r=1}^{2} A^r(0)\partial_r V,$$

where

$$A^1(\nabla U) \; := \begin{pmatrix} 0 & 0 & (C_{11jk}(\nabla U))_{jk} \\ 0 & 0 & (C_{21jk}(\nabla U))_{jk} \\ 1 & 0 & 0 \; \cdots \; 0 \\ 0 & 1 & 0 \; \cdots \; 0 \\ 0 & 0 & 0 \; \cdots \; 0 \\ 0 & 0 & 0 \; \cdots \; 0 \end{pmatrix},$$

$$A^2(\nabla U) := \begin{pmatrix} 0 & 0 & (C_{21jk}(\nabla U))_{jk} \\ 0 & 0 & (C_{22jk}(\nabla U))_{jk} \\ 0 & 0 & 0 \cdots 0 \\ 0 & 0 & 0 \cdots 0 \\ 1 & 0 & 0 \cdots 0 \\ 0 & 1 & 0 \cdots 0 \end{pmatrix},$$

and F arises from $f = (f_1, f_2)$ canonically.

Now we shall follow the general steps **A**–**E** as before, mainly discussing part **A**, in which the essential differences to the isotropic case appear. (For details not presented here see [177].)

The nonlinearity is at least quadratic. We shall assume

$$|C_{imjk}(\nabla U) - C_{imjk}(0)| = \mathcal{O}(|\nabla U|^\alpha) \quad \text{as } |\nabla U| \to 0, \quad i, m, j, k = 1, 2, \tag{11.28}$$

for some $\alpha \in \mathbb{N}$, hence we have

$$F(W) = \mathcal{O}(|W|^{\alpha+1}) \quad \text{as } |W| \to 0.$$

A: Decay for $F \equiv 0$.

If U solves the linearized equations ((11.26) with $f = 0$), then U satisfies

$$U_{tt} - \mathcal{D}' S \mathcal{D} U = 0, \tag{11.29}$$

$$U(t = 0) = U^0, \quad U_t(t = 0) = U^1,$$

where the matrix S is given by

$$S := \begin{pmatrix} \tau & \lambda & 0 \\ \lambda & \tau & 0 \\ 0 & 0 & \mu \end{pmatrix}.$$

S should be positive definite, i.e.

$$\tau, \mu > 0 \quad , \quad \tau > |\lambda| \tag{11.30}$$

(cf. [98]).

The formal differential symbol $\mathcal{D}$ is given by

$$\mathcal{D} := \begin{pmatrix} \partial_1 & 0 \\ 0 & \partial_2 \\ \partial_2 & \partial_1 \end{pmatrix}.$$

Hence we may rewrite (11.29) as

$$U_{tt} - A_1 U = 0, \tag{11.31}$$

where

$$A_1 \;:\; W^{2,2} \subset L^2 \longrightarrow L^2,$$

$$A_1 U := \begin{pmatrix} \tau\partial_1^2 + \mu\partial_2^2 & (\mu+\lambda)\partial_1\partial_2 \\ (\mu+\lambda)\partial_1\partial_2 & \mu\partial_1^2 + \tau\partial_2^2 \end{pmatrix} U,$$

is a self-adjoint operator.
Applying the Fourier transform, $\hat{U}(y) = (\mathcal{F}U)(y)$, we have

$$\hat{U}_{tt}(y) + \hat{A}_1(y)\hat{U}(y) = 0,$$

$$\hat{U}(t=0) = \hat{U}^0, \quad \hat{U}_t(t=0) = \hat{U}^1,$$

where

$$\hat{A}_1(y) := \begin{pmatrix} \tau y_1^2 + \mu y_2^2 & (\mu+\lambda) y_1 y_2 \\ (\mu+\lambda) y_1 y_2 & \mu y_1^2 + \tau y_2^2 \end{pmatrix}.$$

$\hat{A}_1(y)$ is symmetric and positive definite, so

$$\hat{A}_1(y) = \beta_1^2(y) P_1(y) + \beta_2^2(y) P_2(y),$$

where $\beta_j^2(y)$, $j = 1, 2$, denote the positive eigenvalues of $\hat{A}_1(y)$ and $P_j(y)$, $j = 1, 2$, denote the corresponding projections into the eigenspaces.
$\hat{U} = \hat{U}(t, y)$ is then given by

$$\hat{U}(t,y) = \sum_{j=1}^{2} \left\{ \cos(t\beta_j(y)) P_j(y) \hat{U}^0(y) + \frac{\sin(t\beta_j(y))}{\beta_j(y)} P_j(y) \hat{U}^1(y) \right\} \tag{11.32}$$

and

$$U(t,x) = (\mathcal{F}^{-1}\hat{U}(t,\cdot))(x).$$

In order to be able to describe the asymptotic behavior of $U(t,\cdot)$ as $t \to \infty$ it is necessary to discuss the eigenvalues and the characteristic manifolds.
It holds, for $j = 1, 2$,

$$2\beta_j^2(y) = (y_1^2 + y_2^2)(\mu + \tau) + (-1)^j \left((y_1^2 - y_2^2)^2 (\mu - \tau)^2 + 4 y_1^2 y_2^2 (\lambda + \mu)^2 \right)^{1/2}. \tag{11.33}$$

In particular we have the following cases:

isotropic media ($\tau = 2\mu + \lambda$): $\beta_1^2(y) = \mu |y|^2, \;\; \beta_2^2(y) = \tau |y|^2$,

weakly coupled ($\mu = -\lambda$): $\beta_1^2(y) = \mu y_1^2 + \tau y_2^2, \;\; \beta_2^2(y) = \tau y_1^2 + \mu y_2^2$,

$\tau = \mu$: $\beta_1^2(y) = \tau |y|^2 - (\lambda + \mu) y_1 y_2, \; \beta_2^2(y) = \tau |y|^2 + (\lambda + \mu) y_1 y_2$.

The eigenvalues $\beta_j(\cdot)$, the projections $P_j(\cdot)$, and the corresponding eigenfunctions $\nu_j(\cdot)$, $j = 1, 2$, are functions (of their argument y) in $C^\infty(\mathbb{R}^2 \setminus \{0\})$ and the following homogeneity relations hold:
Let $s > 0$, $y \in \mathbb{R}^2$, $j = 1, 2$. Then

$$\begin{aligned} \hat{A}_1(sy) &= s^2 \hat{A}_1(y), \\ \beta_j(sy) &= s\beta_j(y), \\ P_j(sy) &= P_j(y), \\ (\nabla\beta_j)(sy) &= (\nabla\beta_j)(y). \end{aligned} \tag{11.34}$$

The *characteristic manifold (wave cone)* is given by

$$K := \left\{ (y_1, y_2, \tau) \in \mathbb{R}^3 \mid p(y_1, y_2, \tau) = 0 \right\},$$

where p is the characteristic polynomial of $\hat{A}_1$, given by

$$p(y_1, y_2, \alpha) := \det\left(\hat{A}_1(y) - \alpha^2\, id\right) = \alpha^4 - \alpha^2|y|^2(\mu+\tau) - y_1^2 y_2^2(\lambda^2 + 2\lambda\mu - \tau^2) + (y_1^4 + y_2^4)\tau\mu.$$

The *Fresnel surface* S is the intersection of K with the plane $\{\alpha = 1\}$, given by

$$S := \left\{ (y_1, y_2) \in \mathbb{R}^2 \mid p(y_1, y_2, 1) = 0 \right\}$$

(*Augustin Jean Fresnel*, 1.5.1788 – 14.7.1827). Since it is a curve in two space dimensions, it will also be called the *Fresnel curve.*
By definition, S is also given by

$$S = \{ y \in \mathbb{R}^2 \mid \beta_j(y) = 1,\ j = 1,\ \text{or } j = 2 \}.$$

The wave cone is a two-sheet hypersurface and $\nabla\beta_j(y) \neq 0$ for $y \neq 0$. The two sheets intersect only in the weakly coupled case ($\lambda = -\mu$) or if $\tau = \mu$. The Fresnel curve S is the union of the two curves S_j, $j = 1, 2$, where

$$S_j := \left\{ \frac{y}{\beta_j(y)} \,\middle|\, |y| = 1 \right\},$$

hence a regular parametrization is given by

$$\begin{aligned} d : [0, 2\pi] &\longrightarrow \mathbb{R}^2, \\ \phi &\longmapsto \begin{pmatrix} \frac{\cos\phi}{\beta_j(\cos\phi, \sin\phi)} \\ \frac{\sin\phi}{\beta_j(\cos\phi, \sin\phi)} \end{pmatrix}. \end{aligned} \tag{11.35}$$

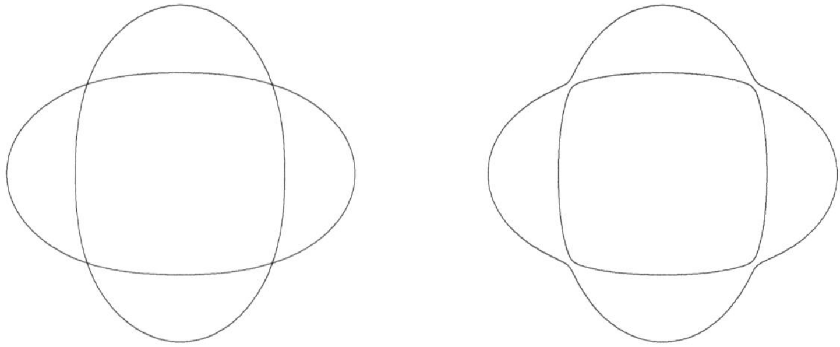

Figure 11.1: $\lambda = -1$ (weakly coupled) $\lambda = -0.8$

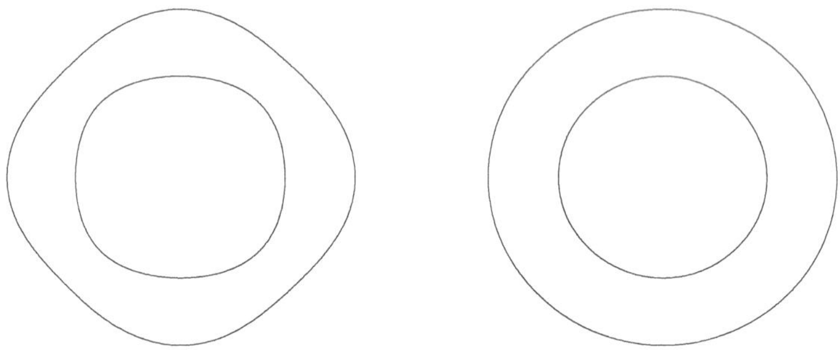

Figure 11.2: $\lambda = \sqrt{12} - 3 = 0.464\ldots$ $\lambda = 1$ (isotropic medium)

Since d is analytic and because d is not a straight line, the curvature may vanish in at most finitely many points; moreover, the curvature vanishes at most of order two (compare the definition of d via a fourth-order polynomial).

The Figures 11.1 – 11.3 (cf. [177]) show typical examples of the Fresnel curves for the cases: $\tau = 3$, $\mu = 1$ and increasing λ.

The curvature of the inner Fresnel curves never vanishes, neither that of the outer Fresnel curve in the isotropic case or in the weakly coupled case.

The cases $\lambda = \sqrt{6} - 1$ and $\lambda = 2$ are shown in Figure 11.3 in enlarged form where those points are easy to find where the curvature vanishes. In general one has that the curvature of the outer Fresnel curve vanishes of second order only if it vanishes for $\phi = 0$ or $\phi = \pi/4$.

We distinguish the following three cases:

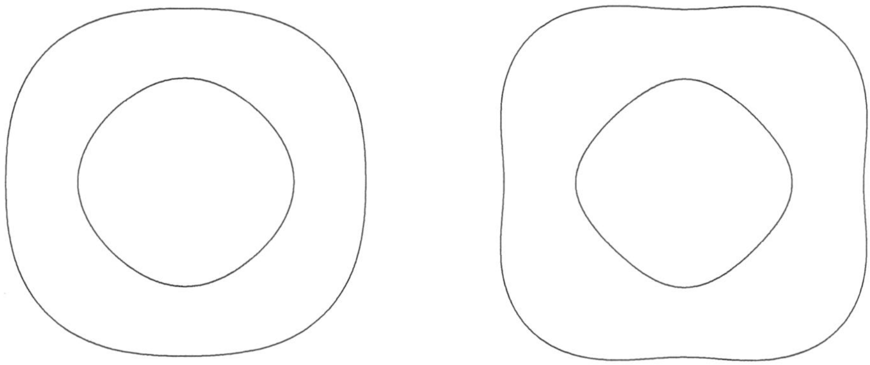

Figure 11.3: $\lambda = \sqrt{6} - 1 = 1.449\ldots$ $\lambda = 2$

#1: The curvature of S never vanishes.

#2: There are points where the curvature of S vanishes and it vanishes of order 1.

#3: There is a point where the curvature vanishes of order 2.

For example in the isotropic or in the weakly coupled case we are in case #1, also for cubic media which are sufficiently close to the isotropic ones, namely for which λ is sufficiently close to $(\tau - 2\mu)$.

Going back to the definition of d in (11.35) and the explicit representation for β_j in (11.33), one can easily obtain characterizations of the cases #1, #2, #3 in terms of the derivatives of d and the coefficients λ, μ, τ, respectively.

As examples we have:

Case #1: isotropic $(\tau = 2\mu + \lambda)$ or weakly coupled $(\lambda = -\mu)$

Case #3: $\tau = 3$, $\mu = 1$, $\lambda = -1 + \sqrt{6} = 1.449\ldots$ (compare Figure 11.3).

For aluminium we have case #1 and for nickel and copper we have case #2, which can be shown with the help of Table 11.1 taken from Miller & Musgrave [126].

	τ	λ	μ
aluminium	9.5	4.9	2.8
nickel	25.5	15.4	12.2
copper	17	12.3	7.55

Table 11.1: Typical elastic moduli

The decay rates for the derivatives of U will follow from the following two Lemmata.

Lemma 11.2 *Let $v \in C_0^\infty(\mathbb{R}^2)$, $\operatorname{supp} v \subset \{x \in \mathbb{R}^2 \mid 1/2 \le |x| \le 2\}$, and $P \in C^\infty$ (neighbourhood of $\operatorname{supp} v$); let $\beta := \beta_j$, $j = 1$ or 2, be one of the eigenvalues of $\hat{A}_1(y)$, $\beta = \beta(y)$. Then we have for all $t > 0$:*

$$\|\mathcal{F}^{-1}\left(\mathrm{e}^{it\beta(\cdot)}v(\cdot)\right)\|_\infty \le ct^{-\varrho/2}\|v\|_{1,\infty},$$

where the constant c may depend on bounds on the derivatives of β on $\operatorname{supp} v$, and ϱ is given by

$$\varrho = \begin{cases} 1 & \text{in case \#1,} \\ 2/3 & \text{in case \#2,} \\ 1/2 & \text{in case \#3.} \end{cases}$$

This Lemma is a modified version of the corresponding Theorem by W. Littman [111]. We sketch the Proof of Lemma 11.2.
In order to estimate

$$\mathcal{F}^{-1}(\mathrm{e}^{it\beta(\cdot)}v(\cdot))(y) = \frac{1}{2\pi}\int_{\mathbb{R}^2} \mathrm{e}^{i(xy+t\beta(x))}v(x)dx$$

we observe that for y, t with

$$\frac{|y|}{t} < \frac{1}{4}\inf_{\{1/2<|x|<2\}}|\nabla\beta(x)| = 2\varepsilon_0 > 0,$$

(implicitly defining ε_0; without loss of generality, $\varepsilon_0 < 1/2$) a partial integration yields

$$|\mathcal{F}^{-1}(\mathrm{e}^{it\beta(\cdot)}v(\cdot))(y)| \le \frac{c}{t}\|v\|_{1,\infty}. \tag{11.36}$$

To deal with the points y, t with

$$\frac{|y|}{t} \ge 2\varepsilon_0,$$

a transformation into distorted polar co-ordinates is used, namely

$$\begin{aligned} T : (0,\infty)\times[0,2\pi) &\longrightarrow \mathbb{R}^2\setminus\{0\}, \\ (r,\phi) &\longmapsto \frac{r}{\beta(x_0)}x_0, \end{aligned}$$

where

$$x_0 := \begin{pmatrix} \cos\phi \\ \sin\phi \end{pmatrix}.$$

Then we get

$$\det\left(\nabla T(r,\phi)\right) = \frac{r}{\beta^2}$$

and the Fourier integral turns into

$$\int_0^\infty \int_0^{2\pi} \exp\left\{ ir \begin{pmatrix} \frac{\cos\phi}{\beta} \\ \frac{\sin\phi}{\beta} \\ 1 \end{pmatrix} \cdot \begin{pmatrix} y_1 \\ y_2 \\ t \end{pmatrix} \right\} v\left(\frac{rx_0}{\beta(x_0)}\right) \frac{r}{\beta^2(x_0)} d\phi dr.$$

We may assume that the Fresnel curve S, which is parametrized by $(\cos\phi, \sin\phi)/\beta$ according to (11.35), is parametrized with the arc length as parameter and we denote the new parametrization by d.
With the notation

$$z := \sqrt{y_1^2 + y_2^2 + t^2}, \quad \xi := \frac{1}{z}\begin{pmatrix} y_1 \\ y_2 \\ t \end{pmatrix}, \quad \varphi_\xi(s) := \begin{pmatrix} d(s) \\ 1 \end{pmatrix} \xi$$

we have to estimate the integral

$$\int_0^\infty r \int_a^b e^{irz\varphi_\xi(s)} w(r,s) ds dr$$

where w stands for all the terms that appear behind the exponential term ($v, 1/\beta^2$, a term from the transformation to arc length).
Since w vanishes outside a fixed interval, say $[a,b] \subset (0,\infty)$, we are now interested in the asymptotic behavior of

$$I = I(z) := \int_a^b e^{iz\varphi_\xi(s)} w(s) ds \quad , \quad w \in C_0^\infty([a,b])$$

as $z \to \infty$, uniformly in ξ. ξ varies in the set

$$\mathcal{M} := \left\{ \xi \in S^2 \mid \xi_3 > 0, \ |(\xi_1, \xi_2)| \geq \varepsilon_0 > 0 \right\}.$$

The behavior of the integral I is determined by the behavior of φ_ξ in its points of stationary phase: $\varphi_\xi'(s) = 0$, and hence it is determined by the behavior of the curvature of d.
The sets

$$F_1 := \{ s \mid d''(s) = 0, d'''(s) \neq 0 \},$$

$$F_2 := \{ s \mid d''(s) = d'''(s) = 0 \}$$

characterize the cases #1, #2, #3, namely

Case #1: $F_1 = F_2 = \emptyset$.

Case #2: $F_1 \neq \emptyset, \ F_2 = \emptyset$.

Case #3: $F_2 \neq \emptyset$.

In case #1 one knows that

$$|I(z)| \leq \frac{1}{z^{1/2}} \left(\frac{2\pi}{\inf |\varphi_\xi''(s)|} \right)^{1/2} \|w\|_{1,\infty}, \tag{11.37}$$

holds, where the infimum is to be taken over the points of stationary phase (see e.g. B.R. Vainberg [185]). (Observe that $|\varphi_\xi''(s)| \geq \varepsilon_0 |d''(s)|$.)
In case #2 one has to consider a neighbourhood V_1 of a point where the curvature vanishes. The integral over $[a,b] \setminus V_1$ leads to a behavior like that in (11.37), the integral over V_1 denoted by I_1 has to be discussed separately. An expansion of $\varphi_\xi(s)$ into powers of s and appropriate partial integrations give the estimate

$$|I_1(z)| \leq \frac{c}{z^{1/3}} \left(1 + \frac{1}{\|\varphi_\xi'''\|_{L^\infty(V_1)}} \right) \|w\|_{1,\infty}, \tag{11.38}$$

where the power $z^{1/3}$ naturally arises by the possible expansion around the point of stationary phase. Similarly, one obtains the estimate

$$|I_2(z)| \leq \frac{c}{z^{1/4}} \left(1 + \frac{1}{\|\varphi_\xi^{iv}\|_{L^\infty(V_2)}} \right) \|w\|_{1,\infty}, \tag{11.39}$$

where I_2 denotes the integral over a neighbourhood V_2 of a point where $\varphi'(s) = \varphi''(s) = \varphi'''(s) = 0$. The estimates (11.36) – (11.39) prove Lemma 11.2.

Q.E.D.

Having proved the L^∞–L^∞-estimate in Lemma 11.2 we may now apply the following Lemma which is a slightly modified version of Theorem 2.2 from H. Pecher [138] (see also Lemma 11.16 in Section 11.7).

Lemma 11.3 *Let $\gamma \geq 0$, $m \in \mathbb{N}$. Let $\beta, Q \in C^\infty(\mathbb{R}^n \setminus \{0\})$ satisfy*

$$\forall s > 0 \quad \forall y \in \mathbb{R}^n : \quad \beta(sy) = s^m \beta(y), \; Q(sy) = Q(y).$$

Assume that for this β an L^∞–L^∞-estimate like in Lemma 11.2 is given with corresponding ϱ. Then there is a constant $c > 0$ such that for all $v \in C_0^\infty(\mathbb{R}^n)$ and all $t \geq 0$ the estimate

$$\left\| \mathcal{F}^{-1} \left(\frac{e^{it\beta(\cdot)}}{|\cdot|^{2m\gamma}} Q(\cdot)(\mathcal{F}v)(\cdot) \right) \right\|_q \leq c(1+t)^{-\frac{n}{m}(\frac{1}{p}-\frac{1}{q})+2\gamma} \|v\|_p$$

holds, provided

$$1 < p \leq 2 \leq q < \infty, \; 1/p + 1/q = 1, \; 1/p - 1/q \geq (2m\gamma)/n, \; (1/p - 1/2)(2n - m\varrho) \leq 2m\gamma.$$

The last two Lemmata easily lead to the following L^p–L^q-estimate (for $1 < p \leq 2 \leq q < \infty$, no estimate for the L^∞-norm).

Theorem 11.4 *Let U be a solution to the linearized equation (11.31) with initial values $U(t=0)=0$, $U_t(t=0)=U^1 \in W^{N,p}$ where*

$$1 < p \leq 2 \leq q < \infty \quad , \quad 1/p + 1/q = 1 \quad \text{and}$$

$$2(1/p - 1/q) \leq N < 2(1/p - 1/q) + 1.$$

Then there is a constant $c > 0$ such that U satisfies

$$\|DU(t)\|_q \leq c(1+t)^{-\frac{\varrho}{2}(\frac{1}{p}-\frac{1}{q})}\|U^1\|_{N,p}$$

where

$$\varrho = \begin{cases} 1 & \text{in case \#1,} \\ 2/3 & \text{in case \#2,} \\ 1/2 & \text{in case \#3.} \end{cases}$$

The constant c depends only on q, τ, μ, and λ.

Proof: By (11.32) we have

$$\begin{aligned} \partial_t \hat{U}(t,y) &= \sum_{j=1}^{2} \cos(t\beta_j(y)) P_j(y) \hat{U}^1(y), \\ \widehat{\partial_k U}(t,y) &= \sum_{j=1}^{2} \frac{iy_k \sin(t\beta_j(y))}{\beta_j(y)} P_j(y) \hat{U}^1(y), \end{aligned}$$

and hence all terms in the components of the vector $DU(t,x)$ are of the form

$$w(t,x) = \mathcal{F}^{-1}\left(\mathrm{e}^{it\beta(\cdot)} Q(\cdot)\hat{g}(\cdot)\right)(x).$$

β equals β_1 or β_2 and is in $C^\infty(\mathbb{R}^2 \setminus \{0\})$ and it is homogeneous of order $m = 1$. The scalar function Q is either a component of P_j or $Q(y) = P_j^{lm}(y) iy_k/\beta_j(y)$ for some $l, m, j, k \in \{1,2\}$, where P_j^{lm} is a component of the matrix P_j. In any case Q is in $C^\infty(\mathbb{R}^2 \setminus \{0\})$ and it is homogeneous of order 0. (For the homogeneity assertion see (11.34).) g essentially equals a component of U^1.
Therefore we may apply Lemma 11.3 in the following way (with $\gamma = s$ to be defined below, $m = 1$, $n = 2$):

$$\begin{aligned} \|w(t,\cdot)\|_q &= \left\|\mathcal{F}^{-1}\left(\mathrm{e}^{it\beta(\cdot)} Q(\cdot)\hat{g}(\cdot)\right)\right\|_q = \left\|\mathcal{F}^{-1}\left(\frac{\mathrm{e}^{it\beta(\cdot)}}{|\cdot|^{2s}} Q(\cdot)|\cdot|^{2s}\hat{g}(\cdot)\right)\right\|_q \qquad (11.40) \\ &= \left\|\mathcal{F}^{-1}\left(\frac{\mathrm{e}^{it\beta(\cdot)}}{|\cdot|^{2s}} Q(\cdot)\mathcal{F}(\Delta^s g)(\cdot)\right)\right\|_q \\ &\leq ct^{-2(1/p-1/q)+2s}\|\Delta^s g\|_p \qquad \text{(by Lemma 11.3)} \\ &\leq ct^{-2(1/p-1/q)+2s}\|g\|_{N,p} \end{aligned}$$

where

$$2s \le N \le 2s + 1.$$

The application of the last Lemma is possible if s satisfies

$$\left(\frac{1}{p} - \frac{1}{2}\right)\left(2 - \frac{\varrho}{2}\right) \le s \le \frac{1}{p} - \frac{1}{q}$$

or equivalently

$$\frac{1}{2}\left(\frac{1}{p} - \frac{1}{q}\right)\left(2 - \frac{\varrho}{2}\right) \le s \le \frac{1}{p} - \frac{1}{q}.$$

For $t \ge 1$ we choose the smallest possible s,

$$s := \frac{1}{2}\left(2 - \frac{\varrho}{2}\right)\theta$$

where

$$\theta := \frac{1}{p} - \frac{1}{q}.$$

This implies for the exponent of t in (11.40):

$$-2\theta + 2s = -\theta\frac{\varrho}{2}$$

hence

$$\|w(t,\cdot)\|_q \le ct^{-\theta\varrho/2}\|g\|_{M,p} \tag{11.41}$$

where

$$\left(2 - \frac{\varrho}{2}\right)\theta \le M < \left(2 - \frac{\varrho}{2}\right)\theta + 1.$$

For $0 \le t \le 1$ we choose the largest possible s, $s := \theta$, and arrive at

$$\|w(t,\cdot)\|_q \le c\|g\|_{N,p} \tag{11.42}$$

with

$$2\theta \le N < 2\theta + 1.$$

The estimates (11.41) and (11.42) prove the theorem.

Q.E.D.

Remark: In (11.40) the Laplace operator Δ in $\mathbb{R}^2$ is used with possibly real powers (*Pierre Simon Laplace*, 28.3.1749 – 5.3.1827). Essentially we only used the property

$$\mathcal{F}(\Delta^s g(\cdot))(y) = |y|^{2s}(\mathcal{F}g)(y)$$

— which can be regarded as a definition for Δ^s — and that $\|\Delta^s g\|_p$ can be estimated by $\|g\|_{N,p}$ if $N \ge 2s$. We refer to the Besov spaces in Appendix A for these questions and we mention that the proof of Lemma 11.3 as given in [138] also relies on the theory of Besov spaces.

We finish Step **A** of the general scheme with the remark that the situation for cubic media in *three* space dimensions will essentially show the same difficulties. To demonstrate this we present a picture of the outer Fresnel surface on page 136 (there are two more hidden inside) for cubic media in $\mathbb{R}^3$ with parameters $\tau = 1$, $\mu = \lambda = 0.7$; see Figure 11.4 (cf. [177]). One may guess where flat points are located.

B: Local existence and uniqueness:
A local existence result is obtained in the same way as that in isotropic elasticity in the previous subsection. Changing $C_{imjk}(\nabla U)$ to $\bar{C}_{imjk}(\nabla U)$ with

$$\bar{C}_{imjk}(\nabla U) := C_{imjk}(\nabla U) + \nu(\delta_{im}\delta_{jk} - \delta_{ik}\delta_{jm}),$$

$$i, m, j, k = 1, 2 \quad , \quad \nu \in \mathbb{R}, \text{ arbitrary, fixed,}$$

leaves the differential equation (11.24) invariant and we write $C_{imjk}(\nabla U)$ again instead of $\bar{C}_{imjk}(\nabla U)$.
We choose ν with

$$0 < \nu < \min(2\mu, \tau - \lambda)$$

which is possible because of the relations (11.30). Then the following matrix $A^0 = A^0(\nabla U) = A^0(V)$,

$$A^0 := \begin{pmatrix} 1 & 0 & 0 & 0 \\ 0 & 1 & 0 & 0 \\ 0 & 0 & & \\ 0 & 0 & \multicolumn{2}{c}{(C_{imjk})_{imjk}} \end{pmatrix},$$

is positive definite and the differential equation (11.24) (resp. (11.27)) turns into

$$A^0(V)\partial_t V + \sum_{r=1}^{2} A^0(V)A^r(V)\partial_r V = 0,$$

with initial value

$$V(t = 0) = V^0.$$

The matrices $A^0(V)$ and $A^0(V)A^r(V)$, $r = 1, 2$, are symmetric and $A^0(V)$ is positive definite (uniformly with respect to V in each compact set). Thus we can apply Theorem 5.8 and we obtain a unique local solution

$$V \in C^0([0, T], W^{s,2}) \cap C^1([0, T], W^{s-1,2})$$

for some $T > 0$, if $s \geq 2$ and $V^0 \in W^{s,2}$.

C: High energy estimates:
Analogously to the result in Chapter 6 we obtain

$$\|V(t)\|_{s,2} \leq C\|V^0\|_{s,2} \exp\left\{C \int_0^t \|\bar{D}V(r)\|_\infty^\alpha dr\right\}, \quad t \in [0, T],\ C = C(s).$$

D: Weighted a priori estimates:
We have

$$\sup_{0\le t\le T}(1+t)^{\varrho/2(1-2/q)}\|V(t)\|_{s_1,q}\le M_0<\infty,$$

where M_0 is independent of T, and ϱ is given in Theorem 11.4, provided

$$q=2\alpha+2,$$

$$\frac{1}{\alpha}\left(1+\frac{1}{\alpha}\right)<\frac{\varrho}{2},$$

s_1 is sufficiently large and

$$\|V^0\|_{s,2}+\|V^0\|_{s,\frac{2\alpha+2}{2\alpha+1}}$$

is sufficiently small ($s>s_1$ being sufficiently large).
This is proved in analogy to the proofs in Chapter 7.

E: Final energy estimate:
As in Chapter 8 we now easily obtain the inequality

$$\|V(t)\|_{s,2}\le K\|V^0\|_{s,2},\quad 0\le t\le T,$$

s being sufficiently large, V^0 being sufficiently small, K being independent of T (and V^0).
Summarizing we obtain the following global existence theorem.

Theorem 11.5 *We assume (11.28) with* $\frac{1}{\alpha}\left(1+\frac{1}{\alpha}\right)<\frac{\varrho}{2}$*, where* ϱ *is given in Theorem 11.4. Then there exist an integer* $s_0\ge 2$ *and a* $\delta>0$ *such that the following holds: If* $V^0=(U^1,\nabla U^0)$ *belongs to* $W^{s,2}\cap W^{s,p}$ *with* $s\ge s_0$ *and* $p=\frac{2\alpha+2}{2\alpha+1}$ *and*

$$\|V^0\|_{s,2}+\|V^0\|_{s,p}<\delta,$$

then there is a unique solution U *of the initial value problem to the nonlinear equations of elasticity in the initially cubic case in* $\mathbb{R}^2$ *(11.24), (11.25), with*

$$(U_t,\nabla U)\in C^0([0,\infty),W^{s,2})\cap C^1([0,\infty),W^{s-1,2}).$$

Moreover, we have

$$\begin{aligned}\|(U_t,\nabla U)(t)\|_\infty+\|(U_t,\nabla U)(t)\|_{2\alpha+2}&=\mathcal{O}\left(t^{-\frac{\varrho}{2}\frac{\alpha}{\alpha+1}}\right),\\ \|(U_t,\nabla U)(t)\|_{s,2}&=\mathcal{O}(1)\quad as\quad t\to\infty.\end{aligned}$$

	#1	#2	#3
ϱ	1	2/3	1/2
α	3	4	5
p	8/7	10/9	12/11
q	8	10	12

Table 11.2: Typical parameter values

The parameter ϱ takes three different values corresponding to the three cases #1, #2, #3 (see Theorem 11.4); then α has a corresponding minimal value determined by the condition

$$\frac{1}{\alpha}\left(1+\frac{1}{\alpha}\right) < \frac{\varrho}{2}.$$

Table 11.2 shows the values of ϱ, the minimal values of α and the corresponding values of p and q.

The discussion of (initially) isotropic and cubic media demonstrates that many interesting problems appear already in the step of the simplest to the next more difficult situation. This underscores the necessity of a lot of further research on these problems despite the general scheme that is available.

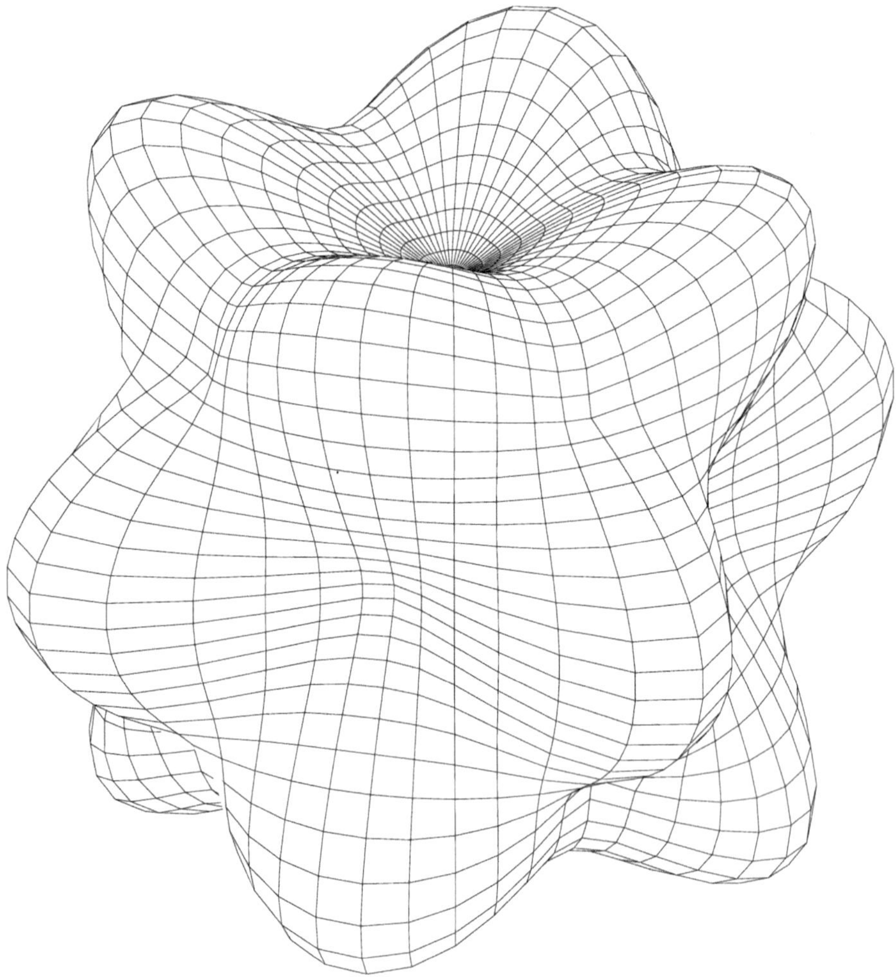

Figure 11.4: A Fresnel surface for a cubic medium in $\mathbb{R}^3$, $\tau = 1$, $\mu = \lambda = 0.7$

11.2 Heat equations

We consider the following type of initial value problems:

$$u_t - \Delta u = F(u, \nabla u, \nabla^2 u), \tag{11.43}$$

$$u(t=0) = u_0, \tag{11.44}$$

for a real-valued function $u = u(t,x)$, $t \geq 0, x \in \mathbb{R}^n$. F is a smooth function satisfying

$$F(w) = \mathcal{O}(|w|^{\alpha+1}) \quad \text{as} \quad |w| \to 0 \quad \text{for some} \quad \alpha \in \mathbb{N}. \tag{11.45}$$

We now cover the steps for this problem.

A: Decay for $F \equiv 0$:
The linear initial value problem

$$u_t - \Delta u = 0, \quad u(t=0) = u_0, \tag{11.46}$$

is solved by

$$\begin{aligned} u(t,x) &:= (4\pi t)^{-n/2} \int_{\mathbb{R}^n} e^{-\frac{|x-y|^2}{4t}} u_0(y)dy \qquad (t>0) \\ &= (4\pi)^{-n/2} \int_{\mathbb{R}^n} e^{-\frac{|z|^2}{4}} u_0(x - \sqrt{t}z)dz \quad (t \geq 0) \end{aligned} \tag{11.47}$$

(for appropriate u_0, e.g. $u_0 \in C_0^\infty$).
We obtain from (11.46)

$$\frac{d}{dt}\|u(t)\|_2^2 + 2\|\nabla u(t)\|_2^2 = 0 \tag{11.48}$$

or

$$\|u(t)\|_2^2 + 2\int_0^t \|\nabla u(r)\|_2^2\, dr = \|u_0\|_2^2, \tag{11.49}$$

which implies

$$\|u(t)\|_2 \leq \|u_0\|_2 \quad \text{for all} \quad t \geq 0. \tag{11.50}$$

For $t > 0$ we obtain from (11.47)

$$|u(t,x)| \leq ct^{-n/2}\|u_0\|_1. \tag{11.51}$$

c will denote various positive constants not depending on t (or u_0).
Let k_t be defined as

$$k_t(z) := (4\pi t)^{-n/2}\, e^{-|z|^2/(4t)}.$$

Then

$$\|k_t\|_1 = (4\pi)^{-n/2} \int_{\mathbb{R}^n} e^{-|y|^2/4}\, dy = 1.$$

This implies, using Sobolev's imbedding theorem,

$$\|u(t)\|_\infty \;\leq\; \|k_t\|_1\|u_0\|_\infty \leq \|u_0\|_\infty \leq c\,\|u_0\|_{n,1}\,. \tag{11.52}$$

From (11.51) and (11.52) we conclude

$$\|u(t)\|_\infty \leq c(1+t)^{-n/2}\|u_0\|_{n,1}, \quad t\geq 0, \tag{11.53}$$

$c=c(n)$ being independent of t.
By interpolation we get from (11.50) and (11.53):

$$\|u(t)\|_q \leq c(1+t)^{-\frac{n}{2}(1-\frac{2}{q})}\|u_0\|_{N_p,p}, \quad t\geq 0, c=c(q,n),$$

where

$$2\leq q\leq\infty, 1/p+1/q=1 \quad \text{and}$$

$$N_p > n(1-2/q) \quad (N_p=n(1-2/q) \text{ if } q\in\{2,\infty\}).$$

Remark: Solutions of the linear heat equation have the property that the L^2-norm decays with a rate too, that there hold L^2–L^∞-decay estimates, and that derivatives decay with a faster rate. This is expressed in the following lemma. The corresponding interpolated versions also hold but they are not recorded here.
Let u_0 belong to C_0^∞ for simplicity.

Lemma 11.6 *Let $m\in\mathbb{N}_0$. There is a constant c only depending on m and n such that the following estimates hold for all $t\geq 0$:*

(i) $\|\nabla^m u(t)\|_2 \;\leq\; c(1+t)^{-(n+m)/4}\|u_0\|_{[\frac{n}{2}]+1+m,1},$

(ii) $\|\nabla^m u(t)\|_\infty \;\leq\; c(1+t)^{-(n+m)/4}\|u_0\|_{[\frac{n}{2}]+1+m,2}.$

(iii) $\|\nabla^m u(t)\|_\infty \;\leq\; c(1+t)^{-(n+m)/2}\|u_0\|_{n+m,1}.$

PROOF: First let $m=0$.
For $t>0$ we have

$$\begin{aligned}\|u(t)\|_2 \;&\leq\; \|k_t\|_2\|u_0\|_1 \qquad \text{(cf. inequality (4.5))}\\ &\leq\; ct^{-n/4}\|u_0\|_1\end{aligned}$$

because

$$\|k_t\|_2^2 = (4\pi t)^{-n}\int\limits_{\mathbb{R}^n} \mathrm{e}^{-\frac{1}{2}|\frac{z}{\sqrt{t}}|^2}dz = ct^{-n/2}.$$

For $t\geq 0$ we get from (11.50) and Sobolev's imbedding theorem

$$\|u(t)\|_2 \leq \|u_0\|_2 \leq c\,\|u_0\|_{[\frac{n}{2}]+1,1}. \tag{11.54}$$

Combining (11.53) and (11.54) we have proved (i) (for $m = 0$).
Let $t > 0$ again. Then using the representation (11.47) we obtain

$$|u(t,x)|^2 \;\leq\; ct^{-n} \int\limits_{\mathbb{R}^n} \mathrm{e}^{-\frac{|x-y|^2}{2t}} dy \, \|u_0\|_2^2 \leq ct^{-n/2} \|u_0\|_2^2. \tag{11.55}$$

For $t \geq 0$ we get

$$\|u(t)\|_\infty \leq c\,\|u_0\|_\infty \leq c\,\|u_0\|_{[\frac{n}{2}]+1,2}. \tag{11.56}$$

Combining (11.55) and (11.56) we have proved (ii) (for $m = 0$).
For $m = 0$ the assertion (iii) is given in (11.53).
The assertions (i), (ii), (iii) now easily follow for $m \geq 1$ observing that each differentiation yields a factor $t^{-1/2}$ which essentially follows from

$$\nabla \mathrm{e}^{-\frac{|x-y|^2}{4t}} = t^{-1/2} \left\{ \tfrac{y-x}{2\sqrt{t}} \mathrm{e}^{-(\frac{|x-y|}{2\sqrt{t}})^2} \right\}.$$

Q.E.D.

B: Local existence and uniqueness:
Theorem 5.8 does not apply here. Parabolic equations like the heat equation have smoothing properties. (The solution u to the linearized equation is C^∞ for $t > 0$ even if $u_0 \in L^1$ only, cf. the representation (11.47).) But there is no finite propagation speed, cf. the remarks at the end of Section 3.1. This is a feature standing in contrast to the situation encountered for hyperbolic systems; see Section 3.1. The results there cannot be used here. For a local existence theorem we refer to Theorem C.4 in Appendix C yielding a solution

$$u \in C^0([0,T], W^{s,2}) \cap C^1([0,T], W^{s-2,2})$$

if $u_0 \in W^{s,2}$ for some $T > 0$, if $s > [n/2] + 3$, and $\|u(t)\|_{2,\infty} < \eta < 1$ holds for all $t \in [0,T]$ if $\|u_0\|_{s,2} < \delta$ is sufficiently small ($\delta = \delta(\eta)$).

C: High energy estimates:
The local solution satisfies

$$\|u(t)\|_{s,2} \leq c\,\|u_0\|_{s,2} \exp\{c \int\limits_0^t \|u(r)\|_{3,\infty}^\alpha dr\}, \quad t \in [0,T], c = c(s). \tag{11.57}$$

We shall give a short proof for the case $\alpha = 1$ because there is an additional consideration necessary compared to those in Chapter 6. Namely, terms of the type

$$c_\varepsilon \int\limits_0^t \|\nabla u(r)\|_{s,2}^2 dr,$$

which arise on the right-hand side in considering the highest derivatives in the nonlinearity, must have sufficiently small factors c_ε in order to be able to be compensated with terms being present on the left-hand side.

Let u be the local solution to

$$u_t - \Delta u = F(u, \nabla u, \nabla^2 u), \tag{11.58}$$

$$u(t=0) = u_0,$$

$$F(0) = 0, \quad \frac{\partial F}{\partial w}(0) = 0,$$

where

$$w := (u, \nabla u, \nabla^2 u).$$

We write F in the form

$$F(w) = \int_0^1 \frac{d}{dr} F(rw)dr = F^0(w)u + \sum_{j=1}^n F_j^1(w)\partial_j u + \sum_{i,j=1}^n F_{ij}^2(w)\partial_i\partial_j u,$$

where

$$F^0(w) := \int_0^1 \frac{\partial F(rw)}{\partial u}\, dr,$$

$$F_j^1(w) := \int_0^1 \frac{\partial F(rw)}{\partial(\partial_j u)}\, dr,$$

$$F_{ij}^2(w) := \int_0^1 \frac{\partial F(rw)}{\partial(\partial_i\partial_j u)}\, dr, \quad i,j = 1,\ldots,n,$$

and

$$F^0(0) = F_j^1(0) = F_{ij}^2(0) = 0, \quad i,j = 1,\ldots,n.$$

Remark: This kind of expansion emphasizes that the choice of the special nonlinearity in the proof of Theorem 1.1, $f = \sum_{i,j=1}^n a_{ij}(Dy)\partial_i\partial_j y$, is no essential restriction.

Differentiation (∇^β) of both sides of (11.58) and taking the inner product in L^2 with $\nabla^\beta u(t)$, we obtain for $0 \le |\beta| \le s$

$$\frac{1}{2}\frac{d}{dt}\|\nabla^\beta u(t)\|_2^2 + \|\nabla^\beta \nabla u(t)\|_2^2 = \langle \nabla^\beta F(w), \nabla^\beta u\rangle(t).$$

This implies

$$\|\nabla^\beta u(t)\|_2^2 + 2\int_0^t \|\nabla^\beta \nabla u(r)\|_2^2\, dr = \|\nabla^\beta u_0\|_2^2 + 2\int_0^t \langle \nabla^\beta F(w), \nabla^\beta u\rangle(r)dr. \tag{11.59}$$

We shall drop the parameter t mostly and we only consider the most difficult terms of the type

$$\langle \nabla^\beta F_{ij}^2(w)\partial_i\partial_j u, \nabla^\beta u\rangle,$$

where $i, j \in \{1, \ldots, n\}$ are arbitrary but fixed in the sequel (the lower-order terms can be handled as in Chapter 6).
First let $|\beta| = 0$. Then

$$\begin{aligned} |\langle F_{ij}^2(w)\partial_i\partial_j u, u\rangle| &\leq \|F_{ij}^2(w)\|_\infty \|\partial_i\partial_j u\|_2 \|u\|_2 \\ &\leq c\,\|w\|_\infty \|u\|_{2,2}\|u\|_2 \\ &\leq c\,\|u\|_{2,\infty}\|u\|_{s,2}^2, \end{aligned} \tag{11.60}$$

(c denotes various constants not depending on t or u).
Now let $k := |\beta| > 0$. Then, writing ∇^k symbolically, we obtain

$$\begin{aligned} \langle \nabla^k(F_{ij}^2(w)\partial_i\partial_j u), \nabla^k u\rangle &= -\langle \nabla^{k-1}(F_{ij}^2(w)\partial_i\partial_j u), \nabla^{k+1}u\rangle \\ &= -\langle F_{ij}^2(w)\nabla^{k-1}\partial_i\partial_j u, \nabla^{k+1}u\rangle + \\ &\quad \langle F_{ij}^2(w)\nabla^{k-1}\partial_i\partial_j u - \nabla^{k-1}(F_{ij}^2(w)\partial_i\partial_j u), \nabla^{k+1}u\rangle \\ &\equiv I + II. \end{aligned} \tag{11.61}$$

The first term I is estimated as follows:

$$\begin{aligned} |I| &\leq c\,\|w\|_\infty \|\nabla u\|_{s,2}^2 \leq c\,\|u\|_{2,\infty}\|\nabla u\|_{s,2}^2 \\ &\leq c\eta\|\nabla u\|_{s,2}^2 \end{aligned} \tag{11.62}$$

where η is small if T resp. $\|u_0\|_{s,2}$ is chosen appropriately small (according to Step **B**).
The second term II is estimated with the help of the Lemmata 4.8, 4.9:

$$\begin{aligned} |II| &\leq c\,(\|\nabla F_{ij}^2(w)\|_\infty \|\nabla^{k-2}\partial_i\partial_j u\|_2 + \|\nabla^{k-1}F_{ij}^2(w)\|_2\|\partial_i\partial_j u\|_\infty)\|\nabla^{k+1}u\|_2 \\ &\leq c\,\|u\|_{3,\infty}\|u\|_{s,2}\|\nabla u\|_{s,2} + c\,\|\nabla^{k-1}w\|_2\|u\|_{2,\infty}\|\nabla u\|_{s,2} \\ &\leq c\|u\|_{3,\infty}\|u\|_{s,2}^2 + c\eta\|\nabla u\|_{s,2}^2. \end{aligned} \tag{11.63}$$

The inequalities (11.60) – (11.63) imply (together with the easier estimates for the lower-order terms which we omit)

$$\begin{aligned} \left| 2\int_0^t \langle \nabla^\beta F(w), \nabla^\beta u\rangle(r)dr \right| &\leq c_1 \int_0^t \|u(r)\|_{3,\infty}\|u(r)\|_{s,2}^2\,dr \\ &\quad + c_2(\eta)\int_0^t \|\nabla u(r)\|_{s,2}^2\,dr, \end{aligned} \tag{11.64}$$

where $c_1, c_2(\eta)$ are positive constants depending at most on η.

Choosing η (resp. T or $\|u_0\|_{s,2}$) — once — sufficiently small we can achieve that

$$c_2(\eta) \leq 1$$

holds.
Therefore we obtain, combining (11.59) and (11.64),

$$\|u(t)\|_{s,2}^2 + \int_0^t \|\nabla u(r)\|_{s,2}^2\, dr \leq \|u_0\|_{s,2}^2 + c\int_0^t \|u(r)\|_{3,\infty}\|u(r)\|_{s,2}^2\, dr.$$

The desired estimate (11.57) now follows immediately using Gronwall's inequality, Lemma 4.1.

D: Weighted a priori estimates:
We have

$$\sup_{0\leq t\leq T} (1+t)^{\frac{n}{2}(1-\frac{2}{q})}\|u(t)\|_{s_1,q} \leq M_0 < \infty,$$

where M_0 is independent of T, provided

$$q = 2\alpha + 2,$$

$$\frac{1}{\alpha}\left(1+\frac{1}{\alpha}\right) < \frac{n}{2},$$

s_1 is sufficiently large and

$$\|u_0\|_{s,2} + \|u_0\|_{s,\frac{2\alpha+2}{2\alpha+1}}$$

is sufficiently small ($s > s_1$ being sufficiently large).
This is proved in analogy to the proof of Theorem 7.1 and uses the representation

$$u(t) = \mathrm{e}^{-At}u_0 + \int_0^t \mathrm{e}^{-A(t-r)}F(w(r))dr.$$

Here A is the Laplace operator realized as a self-adjoint map from $W^{s,2} \subset L^2$ into L^2. $\mathrm{e}^{-tA}u_0$ is the solution to the linear initial value problem (11.46) (given explicitly by the representation (11.47)).
As an easy consequence we then conclude:

E: Final energy estimate:

$$\|u(t)\|_{s,2} \leq K\|u_0\|_{s,2}, \quad 0 \leq t \leq T,$$

s being sufficiently large, u_0 being sufficiently small, K being independent of T (and u_0).
Altogether we obtain the following existence theorem.

Theorem 11.7 *We assume (11.45) with* $\frac{1}{\alpha}(1+\frac{1}{\alpha}) < \frac{n}{2}$. *Then there exist an integer* $s_0 > n/2+3$ *and a* $\delta > 0$ *such that the following holds:*
If u_0 *belongs to* $W^{s,2} \cap W^{s,p}$ *with* $s \geq s_0$ *and* $p = \frac{2\alpha+2}{2\alpha+1}$ *and*

$$\|u_0\|_{s,2} + \|u_0\|_{s,p} < \delta,$$

then there is a unique solution u *of the initial value problem to the nonlinear heat equation (11.43), (11.44) with*

$$u \in C^0([0,\infty), W^{s,2}) \cap C^1([0,\infty), W^{s-2,2}).$$

Moreover, we have

$$\begin{aligned} \|u(t)\|_\infty + \|u(t)\|_{2\alpha+2} &= \mathcal{O}(t^{-\frac{n}{2}\frac{\alpha}{\alpha+1}}), \\ \|u(t)\|_{s,2} &= \mathcal{O}(1) \quad \text{as} \quad t \to \infty. \end{aligned}$$

It was already mentioned in Chapter 1 and in Chapter 8 that the general framework does not lead to optimal results in each case. Here it is possible to use the better decay results expressed in Lemma 11.6 and the dissipation expressed in the energy equality (11.49) (or (11.59)) to improve the result in Theorem 11.7. Actually this will be done in connection with the equations of thermoelasticity in Section 11.3.
We mention that for dissipative systems the technique of A. Matsumura is especially appropriate. This technique consists in considering the differential equation for v^k, where

$$v^k(t,x) := (1+t)^k u(t,x), \quad k \in \mathbb{N},$$

and then deriving the classical energy estimate for v^k, which turns into a weighted a priori estimate for u, see [118] or the paper of S. Zheng [203]. The sharp small data results for solutions of the nonlinear heat equation (11.43) are collated in Table 11.3 (see S. Zheng [203] or G. Ponce [141]).

α \ n	1	2	3	4	5	6	⋯
1	–	–	*	*	+	+	⋯
2	–	+	+	⋯			
3	+	+	⋯				
4	+	⋯					
⋮	⋮	⋱					

Table 11.3: Global existence for nonlinear heat equations

According to Theorem 11.7 global solutions exist for combinations (α, n) where a "+" is written. A "*" indicates that the improved methods lead to global solutions and a "–"

means that one has to expect the development of singularities. In Section 11.3 quadratic nonlinearities in $\mathbb{R}^3$ are discussed. The arguments there justify the cases "∗" in Table 11.3.
In the case that F does not depend on u, i.e. $F = F(\nabla u, \nabla^2 u)$, global small solutions exist for all $\alpha, n \in \mathbb{N}$. For the discussion of blow-up results in the semilinear case $F = F(u)$ we refer the reader to the paper of H.A. Levine [103] and the references therein.

11.3 Equations of thermoelasticity

The equations of thermoelasticity describe the elastic and the thermal behavior of elastic, heat conductive media, in particular the reciprocal actions between elastic stresses and temperature differences. They are a coupling of the equations of elasticity discussed in Section 11.1 and of the heat equation which was discussed in Section 11.2. Hence we have to deal with a hyperbolic-parabolic coupled system for which indeed both hyperbolic and parabolic effects are encountered. We shall consider the initial value problem in $\mathbb{R}^3$ for a homogeneous, initially isotropic medium, but also one-dimensional models are reviewed. The differential equations are equations for the displacement vector $U = U(t,x)$ (compare Section 11.1) and for the temperature difference $\theta = \theta(t,x) := T_a(t,x) - T_0$, where T_a denotes the absolute temperature and T_0 is a fixed reference temperature. The interesting question which arises is whether the behavior will be predominated by the hyperbolic part — mainly the equations of elasticity for U plus coupling terms — or by the parabolic part — mainly a heat equation for θ plus coupling terms. We know from Section 11.1 that in the case of pure elasticity there are global, small solutions if the nonlinearity degenerates up to order two, i.e. if the nonlinearity in the final setting is cubic. Moreover, F. John has shown that in the general "genuinely nonlinear" case a blow-up has to be expected; this was proved for plane waves and for radial solutions, cf. [64, 69]. On the other hand we know from the previous section that quadratic nonlinearities in $\mathbb{R}^3$ still lead to global, small solutions of the heat equation. The question remains whether the dissipative influence through heat conduction is strong enough to prevent solutions from blowing up at least for small data.
The answer to this question will be positive if one excludes purely quadratic nonlinearities in the displacement. This perfectly corresponds to the fact that for these nonlinearities one has to expect a blow-up as was shown in [144]; see below. Thus, we admit all possible cubic nonlinearities (in the final setting) or those quadratic terms which involve θ, which guarantees that a damping effect (dissipation) is present in each equation.
We mention that for *one-dimensional* models global, small solutions always exist. This was shown for the Cauchy problem by Kawashima & Okada in [85, 84]. They proved a global existence theorem for small solutions only using the L^2-energy method. A similar

theorem was obtained by Zheng & Shen in [208] with the method studied here. Again with the L^2-energy method Hrusa & Tarabek proved an existence theorem in [50]. Moreover, important results are found in the initial paper of M. Slemrod [175] for a bounded domain with special mixed boundary conditions, corresponding results by S. Zheng [202], the investigations of S. Jiang [56] for the half-axis, the treatment of the Dirichlet problem for a bounded domain by Shibata & Racke [151] and the improvement of this result by Shibata, Zheng & Racke [152] (*Johann Peter Gustav Lejeune Dirichlet,* 13.2.1805 – 5.5.1859). The Dirichlet problem for the half-axis was discussed by S. Jiang [60] and the Neumann problem for a bounded domain by Y. Shibata [164] and S. Jiang [61], who also discussed the half-line (*Carl Neumann,* 7.5.1832 – 27.3.1925). Periodic solutions are studied in [31] by E. Feireisl and in [152]. Large data lead to the development of singularities which was shown by Dafermos & Hsiao in [24] (for special nonlinearities) and by Hrusa & Messaoudi in [49].

For small initial values the one-dimensional model is predominated by the heat conduction, and there is only one type of elastic waves. In three space dimensions there are two types of elastic waves and the coupling is more complicated. It is interesting to notice that there is a significant difference between the problem treated here and the case of compressible viscous and heat conductive fluids considered by Matsumura & Nishida in [119]; this is the decay of solutions to the linearized system. In [119] the decay of all variables was similar to the pure parabolic linear case. In our situation this is not the case. In fact, the divergence-free part U^{so} of the displacement U in the linearized system behaves asymptotically like solutions of the linear wave equation and does not experience any damping.

We shall now derive the equations of thermoelasticity and then transform to a suitable first-order system. The results presented in this section are taken mainly from our results in [143, 144] and from the joint work with G. Ponce [142].

The equations describing the thermoelastic behavior of a three-dimensional body $\mathcal{B}$ with reference configuration $\mathbb{R}^3$ are those of balance of linear momentum and balance of energy, given by

$$\varrho X_{tt} = \nabla' \tilde{S} + \varrho b, \tag{11.65}$$

$$\varrho \varepsilon_t = \operatorname{tr}\left\{\tilde{S} F_t\right\} + \nabla' q + \varrho r, \tag{11.66}$$

where we use the following notation:

X_i, $i = 1, 2, 3$, are the co-ordinates at time t, $0 \le t < \infty$, of that material point of $\mathcal{B}$ which has co-ordinates x_i, $i = 1, 2, 3$, when $\mathcal{B}$ is in the fixed undistorted reference configuration. The deformation gradient F is given by

$$F^{ij}(t, x) = \frac{\partial}{\partial x_j} X_i(t, x).$$

ϱ is the material density, $\tilde{S}$ is the Piola–Kirchhoff stress tensor, (*Gabrio Piola,* 15.7.1794 – 9.11.1850), b is a specific extrinsic body force, ε is the specific internal energy, $-q$ is the heat flux vector, r is a specific extrinsic heat supply, and tr denotes the trace operator (cf. D.E. Carlson [15] for extended considerations). Furthermore, we denote by η the specific entropy, by T_a the absolute temperature, and by $\psi = \varepsilon - T_a\eta$ the specific Helmholtz free energy (*Hermann Ludwig Ferdinand Helmholtz,* 31.8.1821 – 8.9.1894).

Remark: ψ is also called Helmholtz potential and its existence in relation to the elastic moduli C_{imjk} below is assured (cf. [175]). If one neglected the heat conduction effects this would imply that the medium is hyperelastic; compare section 11.1.

$X, \tilde{S}, \varepsilon, \eta, T_a, q, \psi, b,$ and r are understood to be smooth functions of t and x. For a *homogeneous* medium which we consider we may take $\varrho \equiv 1$ without loss of generality. The constitutive assumption in thermoelasticity is now that $\tilde{S}, q, \psi$ and η are functions of the present values of $F, T_a,$ and ∇T_a. The Clausius–Duhem inequality implies the relations (cf. [15])

$$\psi = \psi(F, T_a), \quad \eta = \eta(F, T_a) = -\frac{\partial \psi(F, T_a)}{\partial T_a},$$

$$\tilde{S} = \tilde{S}(F, T_a) = \frac{\partial \psi(F, T_a)}{\partial F}, \quad q\nabla T_a \geq 0,$$

(*Rudolf Julius Emanuel Clausius,* 2.1.1822 – 24.8.1888; *Pierre Maurice Marie Duhem,* 10.6.1861 – 14.9.1916).

We introduce the displacement vector

$$U := X - x$$

and the temperature difference

$$\theta := T_a - T_0.$$

For simplicity we assume the forces r and b to be zero.

Changing variables from (F, T_a) to $(\nabla U, \theta)$ we obtain from (11.65), (11.66)

$$\partial_t^2 U_i = \sum_{m,j,k=1}^{3} C_{imjk}(\nabla U, \theta)\partial_m \partial_k U_j + \sum_{m=1}^{3} \tilde{C}_{im}(\nabla U, \theta)\partial_m \theta, \quad i = 1, 2, 3, \tag{11.67}$$

$$a(\nabla U, \theta)\partial_t \theta = \frac{1}{f(\theta)}\nabla' q(\nabla U, \theta, \nabla\theta) + \operatorname{tr}\left\{\left(\tilde{C}_{km}(\nabla U, \theta)\right)'_{km} \cdot (\partial_t \partial_s U_r)_{rs}\right\}, \tag{11.68}$$

where

$$C_{imjk} = \frac{\partial \tilde{S}^{im}}{\partial(\partial_k U_j)}, \quad \tilde{C}_{im} = \frac{\partial \tilde{S}^{im}}{\partial \theta}, \tag{11.69}$$

$$\tilde{S}^{im} = \frac{\partial \psi}{\partial(\partial_m U_i)}, \quad a = -\frac{\partial^2 \psi}{\partial \theta^2} \geq a_0 > 0 \tag{11.70}$$

for some positive constant a_0. f is an arbitrary C^∞-function such that $f(\theta) = \theta + T_0$ for $|\theta| \leq T_0/2$ and $0 < f_1 \leq f(\theta) \leq f_2 < \infty$ for $-\infty < \theta < \infty$ with constants f_1, f_2. The equation (11.68) is derived from (11.66) for small values of $|\theta|$, i.e. for $|\theta| \leq T_0/2$ which is *a posteriori* justified by the smallness of the solution which will be obtained later. Additionally, one has prescribed initial conditions

$$U(t=0) = U^0, \quad U_t(t=0) = U^1, \quad \theta(t=0) = \theta^0. \tag{11.71}$$

The medium is assumed to be initially isotropic, that is (compare (11.9) in Section 11.1)

$$C_{imjk}(0,0) = \left(c_1^2 - 2c_2^2\right)\delta_{im}\delta_{jk} + c_2^2\left(\delta_{ij}\delta_{km} + \delta_{jm}\delta_{ik}\right), \quad i,m,j,k = 1,2,3,$$

where the constants c_1, c_2 satisfy

$$c_1 > c_2 > 0.$$

They are related to the Lamé constants λ, μ by

$$c_1^2 = \lambda + 2\mu, \quad c_2^2 = \mu.$$

Moreover, we assume

$$\tilde{C}_{im}(0,0) = -\gamma\delta_{im} \quad \text{with} \quad \gamma \in \mathbb{R}\setminus\{0\},$$

($\gamma = 0$: linearly uncoupled case),

$$\frac{\partial q^i(0,0,0)}{\partial(\partial_j\theta)} = \kappa\delta_{ij}, \quad \kappa > 0 \text{ (heat conduction coefficient)}.$$

$$\frac{\partial q^i(0,0,0)}{\partial(\partial_m U_j)} = \frac{\partial q^i(0,0,0)}{\partial\theta} = 0 \quad , \quad i,j,m = 1,2,3.$$

The equations (11.67), (11.68) are now written as

$$\begin{aligned} \partial_t^2 U_i \quad &- \sum_{m,j,k=1}^{3} C_{imjk}(0,0)\partial_m\partial_k U_j - \sum_{m=1}^{3} \tilde{C}_{im}(0,0)\partial_m\theta \\ &= f_i^1\left(\nabla U, \nabla^2 U, \theta, \nabla\theta\right), \quad i = 1,2,3, \end{aligned} \tag{11.72}$$

$$\begin{aligned} \partial_t\theta \quad &- \kappa\Delta\theta - \operatorname{tr}\left\{(\tilde{C}_{km}(0,0))'_{km} \cdot (\partial_t\partial_s U_r)_{rs}\right\} \\ &= f^2\left(\nabla U, \nabla U_t, \nabla^2 U, \theta, \nabla\theta, \nabla^2\theta\right), \end{aligned} \tag{11.73}$$

where

$$\begin{aligned} f_i^1\left(\nabla U, \nabla^2 U, \theta, \nabla\theta\right) : \quad &= \sum_{m,j,k=1}^{3} (C_{imjk}(\nabla U, \theta) - C_{imjk}(0,0))\partial_m\partial_k U_j \\ &+ \sum_{m=1}^{3} (\tilde{C}_{im}(\nabla U, \theta) - \tilde{C}_{im}(0,0))\partial_m\theta, \quad i = 1,2,3 \end{aligned} \tag{11.74}$$

and

$$f^2\left(\nabla U, \nabla U_t, \nabla^2 U, \theta, \nabla\theta, \nabla^2\theta\right) := \tag{11.75}$$

$$\frac{1}{a(\nabla U,\theta)f(\theta)}\Big[\sum_{i,j,m=1}^{3} \frac{\partial q^i(\nabla U,\theta,\nabla\theta)}{\partial(\partial_m U_j)}\partial_m\partial_i U_j$$

$$+ \sum_{i=1}^{3} \frac{\partial q^i(\nabla U,\theta,\nabla\theta)}{\partial\theta}\partial_i\theta + \sum_{i,j=1}^{3}\left(\frac{\partial q^i(\nabla U,\theta,\nabla\theta)}{\partial(\partial_j\theta)} - \frac{\partial q^i(0,0,0)}{\partial(\partial_j\theta)}\right)\partial_i\partial_j\theta$$

$$+ (1-a(\nabla U,\theta)f(\theta))\kappa\Delta\theta + f(\theta)\,\mathrm{tr}\left\{(\tilde{C}_{km}(\nabla U,\theta) - \tilde{C}_{km}(0,0)\right.$$

$$\left.+ \tilde{C}_{km}(0,0) - a(\nabla U,\theta)\tilde{C}_{km}(0,0))'_{km}\cdot(\partial_t\partial_s U_r)_{rs}\right\}\Big].$$

Introducing $f_i^1 = (f_1^1, f_2^1, f_3^1)$ and the formal differential symbol $\mathcal{D}$ with

$$\mathcal{D} := \begin{pmatrix} \partial_1 & 0 & 0 \\ 0 & \partial_2 & 0 \\ 0 & 0 & \partial_3 \\ 0 & \partial_3 & \partial_2 \\ \partial_3 & 0 & \partial_1 \\ \partial_2 & \partial_1 & 0 \end{pmatrix}$$

as well as the matrix S containing the elastic moduli with

$$S := \begin{pmatrix} 2\mu+\lambda & \lambda & \lambda & 0 & 0 & 0 \\ \lambda & 2\mu+\lambda & \lambda & 0 & 0 & 0 \\ \lambda & \lambda & 2\mu+\lambda & 0 & 0 & 0 \\ 0 & 0 & 0 & \mu & 0 & 0 \\ 0 & 0 & 0 & 0 & \mu & 0 \\ 0 & 0 & 0 & 0 & 0 & \mu \end{pmatrix},$$

assuming S to be positive definite, i.e.

$$\mu > 0, \quad 2\mu + 3\lambda > 0,$$

we arrive at the following simpler representation of the differential equations (11.72), (11.73):

$$U_{tt} - \mathcal{D}'S\mathcal{D}U + \gamma\nabla\theta = f^1\left(\nabla U, \nabla^2 U, \theta, \nabla\theta\right), \tag{11.76}$$

$$\theta_t - \kappa\Delta\theta + \gamma\nabla' U_t = f^2\left(\nabla U, \nabla U_t, \nabla^2 U, \theta, \nabla\theta, \nabla^2\theta\right), \tag{11.77}$$

where we shall assume without loss of generality in the sequel that $T_0 = 1$ and $a(0,0) = 1$. The assumption on the nonlinearity will be

$$\left.\begin{array}{ll} \multicolumn{2}{l}{\text{There are no purely quadratic terms only involving } \nabla U, \nabla U_t, \nabla^2 U} \\ \multicolumn{2}{l}{\text{and additionally one of the following two cases is given:}} \\ \text{Case I:} & \text{Only quadratic terms appear.} \\ \text{Case II:} & \text{Only at least cubic terms appear and one quadratic} \\ & \text{term of the type } \theta\Delta\theta. \end{array}\right\} \tag{11.78}$$

Remark: The specific quadratic nonlinearity of the type $\theta\Delta\theta$ arises from the term $(1 - a(\nabla U, \theta)f(\theta))\kappa\Delta\theta$ in (11.75) and is due to the special function f. This quadratic term cannot be assumed to vanish by any assumption on the general nonlinearities. (See Step **C** below for the typical nonlinearities that may arise; the terms excluded are typically $\nabla U \nabla^2 U$ and $\nabla U \nabla U_t$.) As mentioned above the appearance of purely quadratic terms in ∇U, ∇U_t, $\nabla^2 U$ may lead to the development of singularities.
The transformation to a suitable first-order system is given by

$$V(t) := (S\mathcal{D}U, U_t, \theta)(t) = \left(V^1, V^2, V^3\right)(t),$$

$$V^0 := V(t=0) = \left(S\mathcal{D}U^0, U^1, \theta^0\right).$$

This transformation has turned out to be very useful for the — in general non-homogeneous and anisotropic — linear case, cf. [98].
To recover ∇U from a known function $V^1 = S\mathcal{D}U$, we define the operator $\tilde{B}^j$ by

$$\tilde{B}^j : S\mathcal{D}W^{1,2} \longrightarrow L^2, \quad \tilde{B}^j Z := \partial_j (S\mathcal{D})^{-1} Z, \quad j = 1,2,3,$$

where $S\mathcal{D}W^{1,2} := \{S\mathcal{D}Z | Z \in W^{1,2}\}$.
By Korn's first inequality (cf. [98]) $\tilde{B}^j$ can be continuously extended to a bounded operator

$$B^j : \overline{S\mathcal{D}W^{1,2}} \longrightarrow L^2.$$

(*Arthur Korn*, 20.5.1870 – 22.12.1945)
Let $B^\nabla := (B^1, B^2, B^3)$. Observe that ∇^k, $k \in \mathbb{N}$, commutes with B^∇. The system of equations (11.76), (11.77) now turns into

$$V_t + AV = F\left(V, \nabla V, \nabla^2 V^3, B^\nabla V^1, \nabla B^\nabla V^1\right),$$

$$V(t=0) = V^0,$$

with the nonlinearity

$$F\left(V, \nabla V, \nabla^2 V^3, B^\nabla V^1, \nabla B^\nabla V^1\right) := \begin{pmatrix} 0 \\ f^1\left(B^\nabla V^1, \nabla B^\nabla V^1, V^3, \nabla V^3\right) \\ f^2\left(B^\nabla V^1, \nabla V^2, \nabla B^\nabla V^1, V^3, \nabla V^3, \nabla^2 V^3\right) \end{pmatrix}$$

and A is the differential operator formally given by

$$A_f = \begin{pmatrix} 0 & -S\mathcal{D} & 0 \\ -\mathcal{D}' & 0 & \gamma\nabla \\ 0 & \gamma\nabla' & -\kappa\Delta \end{pmatrix}.$$

$-A$ is the generator of a contraction semigroup in the Hilbert space $\mathcal{H} := \overline{SDW^{1,2}} \times L^2 \times L^2$ $(6+3+1$ components) with domain $D(A) := \{V \in \mathcal{H} \mid A_f V \in L^2\}$, see [98]. The inner product in $\mathcal{H}$ is a weighted L^2-inner product:

$$\langle W, Z\rangle_{\mathcal{H}} := \langle S^{-1}W^1, Z^1\rangle + \langle W^2, Z^2\rangle + \langle W^3, Z^3\rangle.$$

Remark: In the sequel we shall write $V^1, \nabla V^1, \ldots$ instead of $B^\nabla V^1 = \nabla U, \nabla B^\nabla V^1 = \nabla^2 U, \ldots$, i.e. we shall not distinguish between ∇U and $S\mathcal{D}U$; that is, we shall not distinguish between $V(t)$ and $\bar{V}(t) := (B^\nabla V^1(t), V^2(t), V^3(t))$. This is justified since

(i) we have from the representation formula for V,

$$V(t) = \mathrm{e}^{-tA}V^0 + \int_0^t \mathrm{e}^{-(t-s)A}F(\bar{V},\ldots)(s)ds$$

that $\bar{V}$ satisfies

$$\bar{V}(t) = B^\nabla \mathrm{e}^{-tA}V^0 + \int_0^t B^\nabla \mathrm{e}^{-(t-s)}F(\bar{V},\ldots)(s)ds$$

and

(ii) the decay properties in Step **A** below for V,

$$\|V(t)\|_q \le c\cdot(1+t)^{-d}\|V^0\|_{N,p}$$

carry over to $\bar{V}$,

$$\|\bar{V}(t)\|_q \le c\cdot(1+t)^{-d}\|V^0\|_{N,p}.$$

To understand the latter argument one should notice that the operator $\nabla\circ\mathcal{D}^{-1}$ on $\mathcal{D}W^{1,2}$ turns into a bounded multiplication in Fourier space and hence does not change the arguments in Step **A**.
We shall now go through the Steps **A**–**E** as before, of course with necessary modifications due to the hyperbolic-parabolic coupled type of the system. At the very end this example underlines the power of the general method.

A: Decay for $F \equiv 0$:

Let V be a solution of

$$V_t + AV = 0, \quad V(t=0) = V^0,$$

where $V^0 = (SDU^0, U^1, \theta^0)$.
We use a decomposition of U into its curl-free part U^{po} and its divergence-free part U^{so} according to the decomposition (11.21)

$$\begin{aligned} L^2 &= \overline{\nabla W^{1,2}} \oplus \mathcal{D}_0, \\ U &= U^{po} + U^{so}, \end{aligned}$$

which implies a decomposition of V into

$$V = V^{po} + V^{so},$$

where

$$\begin{aligned} V^{po} &= \left(SDU^{po}, U_t^{po}, \theta\right), \\ V^{so} &= \left(SDU^{so}, U_t^{so}, 0\right). \end{aligned}$$

The linear system for (U, θ) (cf. (11.76), (11.77)) is

$$U_{tt} + (\mu\nabla\times\nabla\times -(2\mu+\lambda)\nabla\nabla')U + \gamma\nabla\theta = 0,$$

$$\theta_t - \kappa\Delta\theta + \gamma\nabla' U_t = 0,$$

hence

$$U_{tt}^{po} - (2\mu+\lambda)\Delta U^{po} + \gamma\nabla\theta = U_{tt}^{po} - (2\mu+\lambda)\nabla\nabla' U^{po} + \gamma\nabla\theta = 0,$$

$$U_{tt}^{so} - \mu\Delta U^{so} = U_{tt}^{so} + \mu\nabla\times\nabla\times U^{so} = 0.$$

That is, the linearized system for (U, θ) decouples into a simpler coupled system for (U^{po}, θ) and a wave equation for the components of U^{so}. U^{so} is no longer coupled to θ.
We know the asymptotic behavior for V^{so} from Chapter 2:

$$\|V^{so}(t)\|_q \le c(1+t)^{-(1-2/q)}\|V^{so}(t=0)\|_{N_p,p}, \quad t \ge 0, \quad c = c(q), \tag{11.79}$$

where $2 \le q \le \infty$, $1/p + 1/q = 1$ and N_p is not greater than three.
The asymptotic behavior of V^{po}, which satisfies

$$V^{po} + AV^{po} = 0, \quad V^{po}(t=0) = V^{0,po} = \left(SDU^{0,po}, U^{1,po}, \theta^0\right),$$

will be described with the help of the Fourier transform. It turns out that V^{po} behaves like a solution to the heat equation, i.e. here the damping effect of θ is apparent.
For the discussion of the Fourier representation of V^{po}, the following elementary properties of the Fourier transform will be frequently used ("$\hat{}$" denoting the Fourier transform):

$$\left\|\mathcal{F}^{-1}\left(\hat{g}_1\hat{g}_2\right)\right\|_2 \le \left\|\frac{\hat{g}_1(\cdot)}{(1+|\cdot|)^j}\right\|_\infty \left\|\mathcal{F}^{-1}\left((1+|\cdot|)^j\hat{g}_2(\cdot)\right)\right\|_2, \tag{11.80}$$

$$\left\|\mathcal{F}^{-1}\left(\hat{g}_1\hat{g}_2\right)\right\|_\infty \le \left\|\frac{\hat{g}_1(\cdot)}{(1+|\cdot|)^j}\right\|_1 \left\|\mathcal{F}^{-1}\left((1+|\cdot|)^j\hat{g}_2(\cdot)\right)\right\|_1, \tag{11.81}$$

($j \in \mathbb{N}_0$, g_1, g_2 such that norms appearing are finite).
One major difference to the previously treated equations is that we shall not only make use of the L^p–L^q-estimates for $2 \leq q \leq \infty$ with $1/p + 1/q = 1$. Here we shall also prove L^2–L^∞- and L^1–L^2-estimates for V^{po} which is possible since V^{po} is a solution of a dissipative system. This will finally allow us to treat the different nonlinearities — in particular the mixed quadratic ones — satisfactorily.
For this purpose we shall also make use of the following elementary inequalities (with j, g_1, g_2 as above):

$$\left\|\mathcal{F}^{-1}\left(\hat{g}_1\hat{g}_2\right)\right\|_\infty \leq \left\|\frac{\hat{g}_1(\cdot)}{(1+|\cdot|)^j}\right\|_2 \left\|(1+|\cdot|)^j\hat{g}_2(\cdot)\right\|_2, \tag{11.82}$$

$$\left\|\mathcal{F}^{-1}\left(\hat{g}_1\hat{g}_2\right)\right\|_2 \leq \left\|\frac{\hat{g}_1(\cdot)}{(1+|\cdot|)^j}\right\|_2 \left\|(1+|\cdot|)^j\hat{g}_2(\cdot)\right\|_\infty. \tag{11.83}$$

For simplicity we shall write V instead of V^{po} until we have found the decay estimate for V^{po}. Then V satisfies

$$\hat{V}_t(t,\xi) + \hat{A}(\xi)\hat{V}(t,\xi) = 0, \quad \hat{V}(t=0) = \hat{V}^0, \tag{11.84}$$

where $\hat{A}(\xi)$ is the Fourier symbol of A.
The solution of (11.84) can be written as

$$\hat{V}(t,\xi) = \hat{G}(t,\xi)\hat{V}^0(\xi),$$

where $\hat{G}(t,\xi)$ is described in the following (cf. [98]).
Let

$$\Xi := \begin{pmatrix} \xi_1 & 0 & 0 \\ 0 & \xi_2 & 0 \\ 0 & 0 & \xi_3 \\ 0 & \xi_3 & \xi_2 \\ \xi_3 & 0 & \xi_1 \\ \xi_2 & \xi_1 & 0 \end{pmatrix}.$$

One has

$$\det(\hat{A}(\xi) - \beta) = -\beta^3(\beta^2 + \mu|\xi|^2)^2\Delta_1(\beta,\xi) \tag{11.85}$$

with

$$\Delta_1(\beta,\xi) := -\beta^3 + \beta^2\kappa|\xi|^2 - \beta(2\mu + \lambda + \gamma^2)|\xi|^2 + \kappa(2\mu + \lambda)|\xi|^4.$$

The factor β^3 in (11.85) corresponds to the null space of A considered in $\mathcal{H} = L^2$ instead of $\mathcal{H} = \overline{\mathcal{SDW}^{1,2}} \times L^2 \times L^2$, and the factor $(\beta^2 + \mu|\xi|^2)^2$ corresponds to the U^{so}-component. For $V = V^{po}$ the third factor is of interest.

Let $\beta_1(\xi)$, $\beta_2(\xi)$, $\beta_3(\xi)$ be the zeros of $\Delta_1(\beta,\xi)$ and

$$J(t,\xi) := \sum_{j=1}^{3} \left(\frac{1}{\prod_{k\neq j} (\beta_k(\xi) - \beta_j(\xi))} \right) \mathrm{e}^{-\beta_j(\xi)t},$$

(taking limits for $\beta_k(\xi) = \beta_j(\xi)$, cf. Lemma 11.8)

$$J_k(t,\xi) := \frac{d^k}{dt^k} J(t,\xi), \quad k = 1, 2.$$

Denoting by Id_k, $k = 1, 3, 6$, the identity on $\mathbb{C}^k$, we obtain that $\hat{G}(t,\xi)$ has the following structure (cf. [98], we omit the pair of parameters (t,ξ) for simplicity):

$$\hat{G} = \begin{pmatrix} (J_2 + \kappa|\xi|^2 J_1 + \gamma^2|\xi|^2 J) Id_6 & (J_1 + \kappa|\xi|^2 J) iS\Xi & J\gamma S\Xi\xi \\ (J_1 + \kappa|\xi|^2 J) i\Xi' & (J_2 + \kappa|\xi|^2 J_1) Id_3 & J_1 i\gamma\xi \\ J\gamma\xi'\Xi' & -J_1 i\gamma\xi' & (J_2 + \tau|\xi|^2 J)\, Id_1 \end{pmatrix}.$$

For the discussion of the asymptotic behavior we need the facts below about the eigenvalues $\beta_j(\xi)$, which we take from Zheng & Shen [208] (cf. also R. Leis [98] for similar results).

Remark: It is interesting to notice that the calculations from [208] for the one-dimensional case are of importance here. The reason is that the behavior of the part V^{po} which is determined through a coupling of the curl-free component U^{po} and the temperature difference θ is in principle one-dimensional while the behavior of the part V^{so}, which only consists of derivatives of U^{so}, is really three-dimensional. By "in principle" we mean that the damping effect of the heat conduction part (dissipation) predominates in V^{po} — just as in the one-dimensional case where no term "U^{so}" appears — while there is no damping for U^{so}, a typical elastic behavior. This kind of coupling and splitting will become of interest for blow-up questions; see below.

Let $\tau := 2\mu + \lambda$. From Lemma 2.1 and Lemma 2.2 in [208] we have

Lemma 11.8

(i) *As* $|\xi| \to 0$*:*

$$\begin{aligned} \beta_1(\xi) &= \frac{\kappa\tau}{\tau+\gamma^2}|\xi|^2 + \mathcal{O}(|\xi|^3), \\ \beta_{2,3}(\xi) &= \frac{\kappa\gamma^2}{2(\tau+\gamma^2)}|\xi|^2 \pm i\sqrt{\tau+\gamma^2}|\xi| + \mathcal{O}(|\xi|^3), \end{aligned}$$

as $|\xi| \to \infty$*:*

$$\begin{aligned} \beta_1(\xi) &= \kappa|\xi|^2 - \frac{\gamma^2}{\kappa} - \frac{\alpha_1}{\kappa^3}|\xi|^{-2} + \mathcal{O}(|\xi|^{-3}), \\ \beta_{2,3}(\xi) &= \frac{\gamma^2}{2\kappa} + \frac{\alpha_1}{2\kappa^3}|\xi|^{-2} + \mathcal{O}(|\xi|^{-4}) \pm i\left(\sqrt{\tau}|\xi| + \frac{\alpha_2}{\kappa^2}|\xi|^{-1} + \mathcal{O}(|\xi|^{-3})\right), \end{aligned}$$

where

$$\alpha_1 := \gamma^2(\gamma^2 - \tau), \quad \alpha_2 := \frac{\gamma^2(4\tau - \gamma^2)}{8\sqrt{\tau}}.$$

(ii) *Except for at most two values of* $|\xi| > 0 : \beta_j(\xi) \neq \beta_k(\xi),\ j \neq k.$

(iii) *For any value of* $\xi \neq 0 : \operatorname{Re}\beta_j(\xi) > 0,\ j = 1, 2, 3.$

(iv) *There are positive constants* r_1, r_2 *and* C_j $(j = 1, 2, 3, 4)$ *depending on* r_1, r_2 *such that*

$$\begin{aligned}
|\xi| \leq r_1 \quad &\Rightarrow -C_1|\xi|^2 \leq -\operatorname{Re}\beta_j(\xi) \leq -C_2|\xi|^2,\\
r_1 \leq |\xi| \leq r_2 \quad &\Rightarrow \quad -\operatorname{Re}\beta_j(\xi) \leq -C_3,\\
r_2 \leq |\xi| \quad &\Rightarrow \quad -\operatorname{Re}\beta_j(\xi) \leq -C_4, \quad (j = 1, 2, 3).
\end{aligned}$$

We shall write $\hat{G}(t,\xi) = \left(\hat{G}^{ij}(t,\xi)\right)_{1\leq i,j\leq 3}$ with

$$\hat{G}^{11}(t,\xi) = (J_2(t,\xi) + |\xi|^2 J_1(t,\xi) + \gamma^2|\xi|^2 J(t,\xi))Id_6$$

and so on. We proceed similarly to [208].
Consider

$$\begin{aligned}
\hat{G}^{11}(t,\xi) &= \hat{g}^{11}(t,\xi)\cdot Id_6 :\\
\hat{g}^{11}(t,\xi) &\equiv a_{11}(\xi)\mathrm{e}^{-\beta_1(\xi)t} + b_{11}(\xi)\mathrm{e}^{-\beta_2(\xi)t} + c_{11}(\xi)\mathrm{e}^{-\beta_3(\xi)t}.
\end{aligned}$$

One has from Lemma 11.8 (we shall mostly omit ξ in the coefficients and the eigenvalues)

$$\left.\begin{aligned}
&\text{as } |\xi| \to 0 :\\
&\qquad a_{11} = \tfrac{\gamma^2}{\tau+\gamma^2} + \mathcal{O}(|\xi|^2), \quad b_{11} = \tfrac{\tau}{2(\tau+\gamma^2)} + \mathcal{O}(|\xi|),\\
&\qquad c_{11} = \tfrac{\tau}{2(\tau+\gamma^2)} + \mathcal{O}(|\xi|),\\
&\text{as } |\xi| \to \infty :\\
&\qquad a_{11} = \mathcal{O}(|\xi|^{-4}), \quad b_{11} = \tfrac{1}{2} + \mathcal{O}(|\xi|^{-2}), \quad c_{11} = \tfrac{1}{2} + \mathcal{O}(|\xi|^{-2}).
\end{aligned}\right\} \tag{11.86}$$

Let

$$\begin{aligned}
\hat{I}_{11} &:= \mathrm{e}^{-\frac{\gamma^2}{2\kappa}t}\cos(\sqrt{\tau}|\xi|t),\\
\hat{Z}_{11} &:= \frac{\gamma^2}{\tau+\gamma^2}\mathrm{e}^{-\frac{\kappa\tau}{\tau+\gamma^2}|\xi|^2 t} + \frac{\tau}{\tau+\gamma^2}\mathrm{e}^{\frac{-\kappa\gamma^2}{2(\tau+\gamma^2)}|\xi|^2 t}\cos(\sqrt{\tau+\gamma^2}|\xi|t),\\
\hat{h}_{11} &:= \hat{I}_{11} + \hat{Z}_{11}.
\end{aligned}$$

To prove the L^2–L^2-estimate we notice that for

$$W_{11}(t) := \mathcal{F}^{-1}\left(\hat{G}^{11}(t,\cdot)\hat{V}^{0,1}(\cdot)\right)$$

one has

$$\begin{aligned} W_{11}(t) &= \mathcal{F}^{-1}\left((\hat{G}^{11}(t,\cdot)-\hat{h}_{11}(t,\cdot))\hat{V}^{0,1}\right)+\mathcal{F}^{-1}(\hat{h}_{11}(t,\cdot)\hat{V}^{0,1}) \qquad (11.87)\\ &\equiv T_1(t)+T_2(t), \end{aligned}$$

and

$$\begin{aligned} \|T_1(t)\|_2 &\le \|\left(\hat{G}^{11}(t,\cdot)-\hat{h}_{11}(t,\cdot)\right)/(1+|\cdot|)^k\|_\infty \|(1+|\cdot|)^k\hat{V}^{0,1}(\cdot)\|_2\\ &\le C_k\|\left(\hat{G}^{11}(t,\cdot)-\hat{h}_{11}(t,\cdot)\right)/(1+|\cdot|)^k\|_\infty\|V^{0,1}\|_{k,2}, \quad k\in\mathbb{N}_0. \end{aligned}$$

(C_k being a constant depending only on k.)
Let $t \ge 1$. Without loss of generality we assume $r_1 < 1 < r_2$.

a) $|\xi| \le r_1$:

$$\begin{aligned} &\left|\frac{\hat{G}^{11}(t,\xi)-\hat{h}_{11}(t,\xi)}{(1+|\xi|)^k}\right| \le c|\hat{G}^{11}(t,\xi)-\hat{h}_{11}(t,\xi)| \qquad (11.88)\\ &\le c\left\{|\hat{G}^{11}(t,\xi)-\hat{Z}_{11}(t,\xi)|+|\hat{I}_{11}(t,\xi)|\right\}\\ &\le c\left\{\left|\left(a_{11}-\frac{\gamma^2}{\tau+\gamma^2}\right)e^{-\beta_1 t}\right|+\left|\left(b_{11}-\frac{\tau}{2(\tau+\gamma^2)}\right)e^{-\beta_2 t}\right|\right.\\ &\quad+\left|\left(c_{11}-\frac{\tau}{2(\tau+\gamma^2)}\right)e^{-\beta_3 t}\right|+\frac{\gamma^2}{\tau+\gamma^2}\left|e^{-\beta_1 t}-e^{-\frac{\kappa\tau}{\tau+\gamma^2}|\xi|^2 t}\right|\\ &\quad+\frac{\tau}{2(\tau+\gamma^2)}\left|e^{-\beta_2 t}-e^{-\left(\frac{\kappa\gamma^2}{2(\tau+\gamma^2)}|\xi|^2+i\sqrt{\tau+\gamma^2}|\xi|\right)t}\right|\\ &\quad\left.+\frac{\tau}{2(\tau+\gamma^2)}\left|e^{-\beta_3 t}-e^{-\left(\frac{\kappa\gamma^2}{2(\tau+\gamma^2)}|\xi|^2-i\sqrt{\tau+\gamma^2}|\xi|\right)t}\right|+|\hat{I}_{11}(t,\xi)|\right\}. \end{aligned}$$

the first term in (11.88) we get by Lemma 11.8 and (11.86)

$$\begin{aligned} \left|\left(a_{11}-\frac{\gamma^2}{\tau+\gamma^2}\right)e^{-\beta_1 t}\right| &\le c(1+t)^{-1}t|\xi|^2 e^{-c|\xi|^2 t}\\ &\le c(1+t)^{-1}. \end{aligned}$$

(c will denote various constants in the sequel which do not depend on t or V^0, but possibly on k.) For the second term we obtain

$$\left|\left(b_{11} - \frac{\tau}{2(\tau+\gamma^2)}\right) e^{-\beta_2 t}\right| \le c(1+t)^{-1/2} \left(\sqrt{t}|\xi| e^{-c(\sqrt{t}|\xi|)^2}\right) \le c(1+t)^{-1/2}$$

and analogously the third one is treated.

The fourth term is estimated by Lemma 11.8 in the following way:

$$\begin{aligned}\left|e^{-\beta_1 t} - e^{-\frac{\kappa\tau}{\tau+\gamma^2}|\xi|^2 t}\right| &\le ce^{-c|\xi|^2 t}|\xi|^3 t \\ &\le c(1+t)^{-1/2};\end{aligned}$$

similarly for the fifth and sixth term.

The last term is estimated by

$$|\hat{I}_{11}(t,\xi)| \le ce^{-ct}.$$

Thus we obtain for $|\xi| \le r_1$

$$\left|\frac{\hat{G}^{11}(t,\xi) - \hat{h}_{11}(t,\xi)}{(1+|\xi|)^k}\right| \le c(1+t)^{-1/2}, \quad k \in \mathbb{N}_0.$$

b) $r_1 \le |\xi| \le r_2$: By Lemma 11.8 we obtain

$$\left|\frac{\hat{G}^{11}(t,\xi) - \hat{h}_{11}(t,\xi)}{(1+|\xi|)^k}\right| \le ce^{-ct}, \quad k \in \mathbb{N}_0.$$

c) $|\xi| \ge r_2$:

$$\begin{aligned}\hat{G}^{11} - \hat{h}_{11} &= \hat{G}^{11} - \hat{I}_{11} - \hat{Z}_{11} \\ &= a_{11}e^{-\beta_1 t} + \left(b_{11} - \frac{1}{2}\right) e^{-\beta_2 t} + \left(c_{11} - \frac{1}{2}\right) e^{-\beta_1 t} \\ &\quad + \frac{1}{2}\left(e^{-\beta_2 t} - e^{\left(-\gamma^2/2 + i\sqrt{\tau}|\xi|\right)t}\right) \\ &\quad + \frac{1}{2}\left(e^{-\beta_3 t} - e^{-\left(\gamma^2/(2\kappa) - i\sqrt{\tau}|\xi|\right)t}\right) - \hat{Z}_{11},\end{aligned} \tag{11.89}$$

$$\left|a_{11}e^{-\beta_1 t}\right| \le c|\xi|^{-4} e^{-c_4 t} \le ce^{-ct}$$

by (11.86), analogously for the second and the third term in (11.89).

$$\left|e^{-\beta_{2,3} t} - e^{-\left(\frac{\gamma^2}{2\kappa} \pm i\sqrt{\tau}|\xi|\right)t}\right| \le ce^{-ct},$$

$$|\hat{Z}_{11}(t,\xi)| \le ce^{-ct}.$$

That is:

$$\left|\frac{\hat{G}^{11}(t,\xi) - \hat{h}_{11}(t,\xi)}{(1+|\xi|)^k}\right| \le c \cdot e^{-ct}, \quad k \in \mathbb{N}_0,$$

and we obtain

$$\forall t \ge 1 \quad \forall k \in \mathbb{N}_0 : \quad \left\|\frac{\hat{G}^{11}(t,\xi) - \hat{h}_{11}(t,\cdot)}{(1+|\cdot|)^k}\right\|_\infty \le c \cdot (1+t)^{-1/2}. \tag{11.90}$$

For $0 \le t \le 1$ we get

$$\left|\frac{\hat{G}^{11}(t,\xi) - \hat{h}_{11}(t,\xi)}{(1+|\xi|)^k}\right| \le c.$$

Thus we have estimated the term $T_1(t)$ from (11.87) by

$$\|T_1(t)\|_2 \le c(1+t)^{-1/2}\|V^{0,1}\|_{k,2}, \quad t \ge 0, \quad k \in \mathbb{N}_0 \text{ arbitrary}. \tag{11.91}$$

For all $t \ge 0$ one has

$$\begin{aligned}\|T_2(t)\|_2 &\le \|\hat{I}_{11}(t,\cdot)\hat{V}^{0,1}(\cdot)\|_2 + \|\hat{Z}_{11}(t,\cdot)\hat{V}^{0,1}(\cdot)\|_2 \\ &\le ce^{-ct}\|V^{0,1}\|_2 + c \cdot \|e^{c\Delta t}V^{0,1}\|_2.\end{aligned}$$

$e^{c\Delta t}V^{0,1}$ is the solution of the heat equation

$$W_t - c\Delta W = 0, \quad W(t=0) = V^{0,1}.$$

Thus by the known results for the heat equation (cf. Section 11.2) we conclude

$$\|T_2(t)\|_2 \le c\|V^{0,1}\|_2. \tag{11.92}$$

From (11.91) and (11.92) we conclude the L^2–L^2-behavior for the first component:

$$\|W_{11}(t)\|_2 \le c\|V^{0,1}\|_2, \quad t \ge 0. \tag{11.93}$$

Remark: The factor $(1+|\xi|)^k$ — which could have been omitted up to now — will become important later on for similar calculations for the remaining estimates. Since we do not carry out all of the details we already wanted to point out how weighting factors are incorporated.
We now turn to the study of the L^1–L^∞-estimate. Notice that by (11.87) we get

$$\|T_1(t)\|_\infty \le c\left\|\frac{\hat{G}^{11}(t,\cdot) - \hat{h}_{11}(t,\cdot)}{(1+|\cdot|)^k}\right\|_1 \|V^{0,1}\|_{k,1}, \quad k \in \mathbb{N}_0.$$

With similar calculations as in (11.88) – (11.90) we get for $t \ge 1$ and $k \ge 3$

$$\left\|\frac{\hat{G}^{11}(t,\cdot) - \hat{h}_{11}(t,\cdot)}{(1+|\cdot|)^k}\right\|_1 \le c_k(1+t)^{-2},$$

e.g.

$$
\begin{aligned}
\int_{|\xi|\le r_1} \left| \left(b_{11}(\xi) - \frac{\tau}{2(\tau+\gamma^2)} \right) \mathrm{e}^{-\beta_2(\xi)t} \right| d\xi &\le c \int_{|\xi|\le r_1} |\xi| \mathrm{e}^{-c|\xi|^2 t} d\xi \\
&= c \int_{|\eta|\le r_1 t^{1/2}} t^{-1/2} |\eta| \mathrm{e}^{-c|\eta|^2} t^{-3/2} d\eta \\
&\le c(1+t)^{-2},
\end{aligned}
$$

$$
\begin{aligned}
\int_{|\xi|\ge r_2} \frac{\left| \mathrm{e}^{-\beta_2(\xi)t} - \mathrm{e}^{\left(-\gamma^2/(2\kappa) + i\sqrt{\tau}|\xi|\right)t} \right|}{(1+|\xi|)^k} d\xi &\le c\mathrm{e}^{-ct} \int_{|\xi|\ge r_2} \frac{|\xi|^{-1}}{(1+|\xi|)^k} d\xi \\
&\le c\mathrm{e}^{-ct} \quad \text{if } k \ge 3.
\end{aligned}
$$

For $0 \le t \le 1$ we obtain

$$
\left\| \frac{\hat{G}^{11}(t,\cdot) - \hat{h}_{11}(t,\cdot)}{(1+|\cdot|)^k} \right\|_1 \le c.
$$

Thus we have estimated $T_1(t)$ by

$$
\|T_1(t)\|_\infty \le c(1+t)^{-2} \|V^{0,1}\|_{k,1}, \quad k \ge 3, \quad t \ge 0.
$$

For $T_2(t)$ we have (cf. the estimates for the heat equation in Section 11.2 or [206])

$$
\|T_2(t)\|_\infty \le c_k \left\| \frac{\hat{h}_{11}(t,\cdot)}{(1+|\cdot|)^k} \right\|_1 \|V^{0,1}\|_{k,1}, \quad k \in \mathbb{N}_0
$$

and with the previous arguments

$$
\left\| \frac{\hat{h}_{11}(t,\cdot)}{(1+|\cdot|)^k} \right\|_1 \le c(1+t)^{-3/2} \text{ if } k \ge 4.
$$

Thus we have obtained the L^1–L^∞-behavior of W_{11}:

$$
\|W_{11}(t)\|_\infty \le c(1+t)^{-3/2} \|V^{0,1}\|_{4,1}, \quad t \ge 0. \tag{11.94}
$$

For the contributions of $\hat{G}^{12}, \ldots, \hat{G}^{33}$ similar lengthy but straightforward calculations are omitted, and we just state the following results:

For $\hat{G}^{12}$ (analogously for $\hat{G}^{21}$):
$\hat{G}^{12}(t,\xi) \equiv \tilde{g}^{12}(t,\xi) iS\Xi$. For the estimates we consider a typical term

$$
\begin{aligned}
\hat{g}^{12}(t,\xi) &:= \eta \tilde{g}^{12}(t,\xi) \\
&\equiv a_{12}(\xi)\mathrm{e}^{-\beta_1(\xi)t} + b_{12}(\xi)\mathrm{e}^{-\beta_2(\xi)t} + c_{12}(\xi)\mathrm{e}^{-\beta_3(\xi)t}, \quad \eta \in \{\xi_1, \xi_2, \xi_3\}.
\end{aligned}
$$

As $|\xi| \to 0$:

$$a_{12} = \mathcal{O}(|\xi|), \quad b_{12} = \mathcal{O}(1), \quad c_{12} = \mathcal{O}(1),$$

as $|\xi| \to \infty$:

$$a_{12} = \mathcal{O}(|\xi|^{-3}), \quad b_{12} = \mathcal{O}(1), \quad c_{12} = \mathcal{O}(1).$$

L^2–L^2-estimate, L^1–L^∞-estimate:

$$\left.\begin{aligned} \|W_{12}(t)\|_2 &\leq c\|V^{0,2}\|_2, \quad t \geq 0, \\ \|W_{12}(t)\|_\infty &\leq c(1+t)^{-3/2}\|V^{0,2}\|_{4,1}, \quad t \geq 0. \end{aligned}\right\} \tag{11.95}$$

(Analogously for $W_{21}(t)$).

$\underline{\hat{G}^{13}}$ (analogously for $\hat{G}^{31}$):

Consider

$$\hat{G}^{13}(t,\xi) \equiv \tilde{g}^{13}(t,\xi)\gamma S \Xi \xi/|\xi|^2$$

and the typical term

$$\begin{aligned} \hat{g}^{13}(t,\xi) &:= \eta_1\eta_2\tilde{g}^{13}(t,\xi) \\ &\equiv a_{13}(t,\xi)\mathrm{e}^{-\beta_1(\xi)t} + b_{13}(\xi)\mathrm{e}^{-\beta_2(\xi)t} + c_{13}(\xi)\mathrm{e}^{-\beta_3(\xi)t}, \\ &\qquad \eta_1, \eta_2 \in \left\{\frac{\xi_1}{|\xi|}, \frac{\xi_2}{|\xi|}, \frac{\xi_3}{|\xi|}\right\}. \end{aligned}$$

As $|\xi| \to 0$:

$$a_{13} = \mathcal{O}(1), \quad b_{13} = \mathcal{O}(1), \quad c_{13} = \mathcal{O}(1),$$

as $|\xi| \to \infty$:

$$a_{13}(\xi) = \mathcal{O}(|\xi|^{-2}), \quad b_{13} = \mathcal{O}(|\xi|^{-1}), \quad c_{13} = \mathcal{O}(|\xi|^{-1}).$$

L^2–L^2-estimate, L^1–L^∞-estimate:

$$\left.\begin{aligned} \|W_{13}(t)\|_2 &\leq c\|V^{0,3}\|_2, \quad t \geq 0, \\ \|W_{13}(t)\|_\infty &\leq c(1+t)^{-3/2}\|V^{0,3}\|_{4,1}, \quad t \geq 0. \end{aligned}\right\} \tag{11.96}$$

(Analogously for $W_{31}(t)$).

$\underline{\hat{G}^{22}}$:

$$\begin{aligned} \hat{G}^{22}(t,\xi) &\equiv \hat{g}^{22}(t,\xi) Id_3, \\ \hat{g}^{22}(t,\xi) &\equiv a_{22}(\xi)\mathrm{e}^{-\beta_1(\xi)t} + b_{22}(\xi)\mathrm{e}^{-\beta_2(\xi)t} + c_{22}(\xi)\mathrm{e}^{-\beta_3(\xi)t}. \end{aligned}$$

As $|\xi| \to 0$:

$$a_{22} = \mathcal{O}(|\xi|^2), \quad b_{22} = \frac{1}{2} + \mathcal{O}(|\xi|), \quad c_{22} = \frac{1}{2} + \mathcal{O}(|\xi|),$$

as $|\xi| \to \infty$:

$$a_{22}(\xi) = \mathcal{O}(|\xi|^{-2}), \quad b_{22} = \frac{1}{2} + \mathcal{O}(|\xi|^{-1}), \quad c_{22} = \frac{1}{2} + \mathcal{O}(|\xi|^{-1}).$$

L^2–L^2-estimate, L^1–L^∞-estimate:

$$\left. \begin{aligned} \|W_{22}(t)\|_2 &\leq c\|V^{0,2}\|_2, \quad t \geq 0, \\ \|W_{22}(t)\|_\infty &\leq c(1+t)^{-3/2}\|V^{0,2}\|_{4,1}, \quad t \geq 0. \end{aligned} \right\} \tag{11.97}$$

$\underline{\hat{G}^{23}}$ (analogously for $\hat{G}^{32}$):

$$\hat{G}^{23}(t,\xi) \equiv \tilde{g}^{23}(t,\xi) i\gamma\xi.$$

Consider the typical term

$$\begin{aligned} \hat{G}^{23}(t,\xi) &:= \eta\tilde{g}^{23}(t,\xi) \\ &\equiv a_{23}(\xi)\mathrm{e}^{-\beta_1(\xi)t} + b_{23}(\xi)\mathrm{e}^{-\beta_2(\xi)t} + c_{23}(\xi)\mathrm{e}^{-\beta_3(\xi)t}, \quad \eta \in \{\xi_1, \xi_2, \xi_3\}. \end{aligned}$$

As $|\xi| \to 0$:

$$a_{23} = \mathcal{O}(|\xi|), \quad b_{23} = \mathcal{O}(1), \quad c_{23} = \mathcal{O}(1),$$

as $|\xi| \to \infty$:

$$a_{23} = \mathcal{O}(|\xi|^{-1}), \quad b_{23} = \mathcal{O}(|\xi|^{-1}), \quad c_{23} = \mathcal{O}(|\xi|^{-1}).$$

L^2–L^2-estimate, L^1–L^∞-estimate:

$$\left. \begin{aligned} \|W_{23}(t)\|_2 &\leq c\|V^{0,3}\|_2, \quad t \geq 0, \\ \|W_{23}(t)\|_\infty &\leq c(1+t)^{-3/2}\|V^{0,3}\|_{4,1}, \quad t \geq 0. \end{aligned} \right\} \tag{11.98}$$

(Analogously for $W_{32}(t)$.)

$\underline{\hat{G}^{33}}$:

$$\begin{aligned} \hat{G}^{33}(t,\xi) &\equiv \hat{g}^{33}(t,\xi) Id_1. \\ \hat{g}^{33}(t,\xi) &\equiv a_{33}(\xi)\mathrm{e}^{-\beta_1(\xi)t} + b_{33}(\xi)\mathrm{e}^{-\beta_2(\xi)t} + c_{33}(\xi)\mathrm{e}^{-\beta_3(\xi)t}. \end{aligned}$$

As $|\xi| \to 0$:

$$a_{33} = \frac{\kappa\tau}{\tau+\gamma^2} + \mathcal{O}(|\xi|), \quad b_{33} = \frac{\gamma^2}{2(\tau+\gamma^2)} + \mathcal{O}(|\xi|), \quad c_{33} = \frac{\gamma^2}{2(\tau+\gamma^2)} + \mathcal{O}(|\xi|),$$

as $|\xi| \to \infty$:

$$a_{33} = 1 + \mathcal{O}(|\xi|^{-1}), \quad b_{33} = \mathcal{O}(|\xi|^{-2}), \quad c_{33} = \mathcal{O}(|\xi|^{-2}).$$

L^2–L^2-estimate, L^1–L^∞-estimate:

$$\left. \begin{aligned} \|W_{33}(t)\|_2 &\leq c\|V^{0,3}\|_2, \quad t \geq 0, \\ \|W_{33}(t)\|_\infty &\leq c(1+t)^{-3/2}\|V^{0,3}\|_{4,1}, \quad t \geq 0. \end{aligned} \right\} \tag{11.99}$$

Summarizing (11.93) – (11.99) we obtain

$$\begin{aligned} \|V^{po}(t)\|_2 &\le c\|V^{0,po}\|_2 \\ \|V^{po}(t)\|_\infty &\le c(1+t)^{-3/2}\|V^{0,po}\|_{4,1}, \end{aligned}$$

and by interpolation the corresponding L^p–L^q-estimate. Since a differentiation of order $|\alpha|$ yields a factor of order $|\xi|^{|\alpha|}$ in Fourier space we can prove in the same way for $\alpha \in \mathbb{N}_0^n$:

$$\|\nabla^\alpha V^{po}(t)\|_q \le c(1+t)^{-(3/2-3/q+|\alpha|/2)}\|V^{0,po}\|_{N^\alpha_{p,p}} \tag{11.100}$$

where $t \ge 0$, $c = c(q,\alpha)$, $2 \le q \le \infty$, $1/p + 1/q = 1$ and N_p^α is not greater than $4 + |\alpha|$. With the help of the differential equation one also obtains a decay rate for $\|\nabla^\alpha \theta_t(t)\|_q$ of order $(2 - 3/q + |\alpha|/2)$. Taking the results for V^{po} in (11.100) and for V^{so} in (11.79) together we have proved

Lemma 11.9 *Let $1/p+1/q = 1$, $2 \le q \le \infty$, $\alpha \in \mathbb{N}_0^3$. Then there exist $N_p \in \mathbb{N}$, $N_p \le 4$, and $c = c(q,\alpha)$ such that for all $t \ge 0$ and all (U^0, U^1, θ^0) with $V^0 = (\nabla U^0, U^1, \theta^0) \in W^{N_p+|\alpha|,p}$ the following estimates hold:*

(i) $\|\nabla^\alpha DU(t)\|_q \le c(1+t)^{-(1-2/q)}\|V^0\|_{N_p+|\alpha|,p}$,

(ii) $\|\nabla^\alpha \theta(t)\|_q \le c(1+t)^{-(3/2-3/q+|\alpha|/2)}\|V^0\|_{N_p+|\alpha|,p}$,

(iii) $\|\nabla^\alpha \theta_t(t)\|_q \le c(1+t)^{-(2-3/q+|\alpha|/2)}\|V^0\|_{N_p+|\alpha|+2,p}$.

(Observe that we do not distinguish between $\mathcal{SD}U^0$ and ∇U^0.)
We have used the elementary properties of the Fourier transform given in (11.80), (11.81) to prove Lemma 11.9. If we use the inequalities given in (11.82), (11.83) we immediately obtain the following L^2–L^∞- resp. L^1–L^2- decay estimates.

Lemma 11.10 *Let $\alpha \in \mathbb{N}_0^3$. There exist a constant $c = c(\alpha)$ and integers $N_1^*, N_2^* \le 4$ such that*

(i) $\forall t \ge 0 \quad \forall V^0 \in W^{N_2^*+|\alpha|,2}$:

$$\|\nabla^\alpha \theta(t)\|_\infty \le c(1+t)^{-(3/4+|\alpha|/2)}\|V^0\|_{N_2^*+|\alpha|,2},$$

(ii) $\forall t \ge 0 \quad \forall V^0 \in W^{N_1^*+|\alpha|,1}$:

$$\|\nabla^\alpha \theta(t)\|_2 \le c(1+t)^{-(3/4+|\alpha|/2)}\|V^0\|_{N_1^*+|\alpha|,1},$$

(iii) $\forall t \ge 0 \quad \forall V^0 \in W^{N_2^*+|\alpha|+2,2}$:

$$\|\nabla^\alpha \theta_t(t)\|_\infty \le c(1+t)^{-(3/4+(|\alpha|+1)/2)}\|V^0\|_{N_2^*+|\alpha|+2,2},$$

(iv) $\forall t \geq 0 \quad \forall V^0 \in W^{N_1^*+|\alpha|+2,1}$:

$$\|\nabla^\alpha \theta_t(t)\|_2 \leq c(1+t)^{-(3/4+(|\alpha|+1)/2)} \|V^0\|_{N_1^*+|\alpha|+2,1}.$$

B: Local existence and uniqueness:

Applying to the hyperbolic part of the differential equations the transformation which we used in Section 11.1.1 to obtain a symmetric hyperbolic system, we end up here with a symmetric hyperbolic-parabolic coupled system for which a local existence theorem is given in Appendix C. We obtain a local solution $V = (\nabla U, U_t, \theta)$ in some time interval $[0,T]$ with $0 < T \leq 1$ provided $V^0 \in W^{s,2}$ with $s \geq 4$ and $\|V^0\|_{s,2} < \delta$ is small enough yielding

$$\|DU(t)\|_\infty + \|\theta(t)\|_{2,\infty} < \eta < 1 \text{ for all } t \in [0,T]$$

($\delta = \delta(\eta)$, η arbitrary).
V satisfies

$$\begin{aligned} DU &\in C^0([0,T], W^{s,2}) \cap C^1([0,T], W^{s-1,2}), \\ \theta &\in C^0([0,T], W^{s,2}) \cap C^1([0,T], W^{s-2,2}). \end{aligned}$$

Remark: T. Mukoyama directly investigated the second-order system for (U,θ) and obtained a similar result, see [134].

C: High energy estimates:

We shall prove the following estimate for the local solution provided η is small enough.

$$\|(DU,\theta)(t)\|_{s,2} \leq C \left\|(\nabla U^0, U^1, \theta^0)\right\|_{s,2} \cdot \tag{11.101}$$

$$\cdot \exp\left\{ C \int_0^t \left(\|DU\|_{1,\infty}^2 + \|\theta\|_{2,\infty} + \|\theta_t\|_\infty \right)(r) dr \right\},$$

$$t \in [0,T], \quad C = C(s).$$

The crucial point is to obtain the quadratic term $\|DU\|_{1,\infty}^2$ in the exponent, a linear term $\|DU\|_{1,\infty}$ would not be sufficient because of the weak decay rates for DU. The energy estimates in the previous sections in their simple form lead to a linear term $\|DU\|_{1,\infty}$ since there are quadratic nonlinearities involving DU. Here the elementary estimate

$$abc \leq \frac{\varepsilon}{2} a^2 + \frac{1}{2\varepsilon} b^2 c^2$$

for positive real numbers a, b, c, ε will produce the quadratic term $\|DU\|_{1,\infty}^2$. One may think for example of a being a θ-term, b being the DU-term and c being a DU- or θ-term.

Then the term εa^2 can be incorporated into the left-hand side of the inequality to which abc denotes the right-hand side. In the previous section abc was just estimated by

$$abc \leq \frac{b}{2}(a^2 + c^2).$$

Due to our assumption (11.78) the following typical nonlinearities appear (up to unessential constants), cf. (11.74), (11.75):

$$f^1 = \text{ quadratic terms } f^{1,q} + \text{ cubic terms } f^{1,c} + \text{ higher-order terms}$$

where

$$f^{1,q} = \theta\nabla^2 U + \nabla\theta\nabla U + \theta\nabla\theta,$$

($\theta\nabla^2 U$ is to be read symbolically for a typical term $\theta\partial_i\partial_m U_k$, and so on),

$$f^{1,c} = (\nabla U)^2\nabla^2 U + \nabla U\theta\nabla^2 U + \theta^2\nabla^2 U + (\nabla U)^2\nabla\theta + \nabla U\theta\nabla\theta + \theta^2\nabla\theta. \tag{11.102}$$

Analogously

$$f^2 = f^{2,q} + f^{2,c} + \text{ higher-order terms},$$

where

$$\begin{aligned} f^{2,q} &= \theta\nabla^2 U + \nabla\theta\nabla^2 U + \nabla U\nabla\theta + \theta\nabla\theta \\ &\quad +\nabla\theta\nabla\theta + \nabla^2\theta\nabla U + \theta\nabla^2\theta + \nabla\theta\nabla^2\theta + \theta\nabla U_t, \end{aligned}$$

$$\begin{aligned} f^{2,c} &= (\nabla U)^2\nabla^2 U + \nabla U\theta\nabla^2 U + \theta^2\nabla^2 U + \nabla U\nabla\theta\nabla^2 U \\ &\quad +(\nabla\theta)^2\nabla^2 U + \theta\nabla\theta\nabla^2 U + (\nabla U)^2\theta + \nabla U\theta\nabla\theta \\ &\quad +\theta^2\nabla\theta + \nabla U\nabla\theta\nabla\theta + (\nabla\theta)^2\nabla\theta + (\nabla U)^2\nabla^2\theta \\ &\quad +(\nabla U)\theta\nabla^2\theta + \theta^2\nabla^2\theta + \nabla U\nabla\theta\nabla^2\theta + (\nabla\theta)^2\nabla^2\theta \\ &\quad +\theta\nabla\theta\nabla^2\theta + (\nabla U)^2\nabla U_t + \nabla U\theta\nabla U_t + \theta^2\nabla U_t. \end{aligned}$$

The terms of higher order will be neglected in the sequel since they are always easier to deal with than the quadratic and cubic terms. We take ∇^α on both sides of (11.76), (11.77) and multiply (11.76) by $\nabla^\alpha U_t$ and (11.77) by $\nabla^\alpha\theta$ in L^2, $|\alpha| \leq s$. Now we have to discuss $3 + 6 + 9 + 20$ terms numbered from ($\underline{1}$) to ($\underline{38}$). Since a lot of calculations recur we shall pick out the characteristic ones.
We get from (11.76) (dropping the argument t)

$$\begin{aligned} &\frac{1}{2}\frac{d}{dt}\left(\|\nabla^\alpha U_t\|_2^2 + \langle\nabla^\alpha S\mathcal{D}U, \nabla^\alpha\mathcal{D}U\rangle\right) + \gamma\langle\nabla^\alpha\nabla\theta, \nabla^\alpha U_t\rangle \\ &\quad = \sum_{i=1}^{9}\langle\nabla^\alpha(\underline{i}), \nabla^\alpha U_t\rangle, \end{aligned} \tag{11.103}$$

$$\begin{aligned}\langle\nabla^\alpha(\underline{1}),\nabla^\alpha U_t\rangle &= \langle\nabla^\alpha(\theta\nabla^2 U),\nabla^\alpha U_t\rangle = \langle\theta\nabla^2\nabla^\alpha U,\nabla^\alpha U_t\rangle\\ &\quad+\langle\nabla^\alpha(\theta\nabla^2 U)-\theta\nabla^\alpha\nabla^2 U,\nabla^\alpha U_t\rangle \equiv R_1+R_2,\end{aligned}$$

$$\begin{aligned}R_1 &= -\langle\nabla\theta\nabla^\alpha\nabla U,\nabla^\alpha U_t\rangle - \langle\theta\nabla^\alpha\nabla U,\nabla^\alpha\nabla U_t\rangle\\ &= -\langle\nabla\theta\nabla^\alpha\nabla U,\nabla^\alpha U_t\rangle - \frac{1}{2}\frac{d}{dt}\langle\theta\nabla^\alpha\nabla U,\nabla^\alpha\nabla U\rangle\\ &\quad+\frac{1}{2}\langle\theta_t\nabla^\alpha\nabla U,\nabla^\alpha\nabla U\rangle.\end{aligned}$$

The second term in the last right-hand side will be incorporated into the left-hand side of (11.103) after integration with respect to t using $\sup\limits_{0\le t\le T}\|\theta(t)\|_\infty<\eta$ and Korn's inequality for $\langle\nabla^\alpha S\mathcal{D}U,\nabla^\alpha\mathcal{D}U\rangle$. Using the inequalities on composite functions from Chapter 4 we obtain

$$\begin{aligned}|R_2| &\le c\left(\|\nabla\theta\|_\infty\|\nabla^\alpha\nabla U\|_2+\|\nabla^2 U\|_\infty\|\nabla^\alpha\theta\|_2\right)\|\nabla^\alpha U_t\|_2\\ &\le c\left(\|\nabla\theta\|_\infty\left(\|\nabla^\alpha\nabla U\|_2^2+\|\nabla^\alpha U_t\|_2^2\right)+\varepsilon_1\|\nabla^\alpha\theta\|_2^2+\frac{1}{\varepsilon_1}\|\nabla^2 U\|_\infty^2\|\nabla^\alpha U_t\|_2^2\right),\end{aligned}$$

where $\varepsilon_1>0$ is arbitrary.
Without loss of generality we assume $|\alpha|>0$ because $R_2=0$ for $\alpha=0$. The term $\varepsilon_1\|\nabla^\alpha\theta\|_2^2$ can be incorporated into the left-hand side choosing $\varepsilon_1>0$ small enough after adding the left-hand side arising from (11.77), where the term $\kappa\|\nabla^\alpha\nabla\theta\|_2^2$ will appear; see the following formula.
We obtain from (11.77)

$$\frac{1}{2}\frac{d}{dt}\|\nabla^\alpha\theta\|_2^2+\kappa\|\nabla^\alpha\nabla\theta\|_2^2+\gamma\langle\nabla^\alpha\nabla' U_t,\nabla^\alpha\theta\rangle=\sum_{i=10}^{38}\langle\nabla^\alpha(\underline{i}),\nabla^\alpha\theta\rangle. \tag{11.104}$$

$$\begin{aligned}\langle\nabla^\alpha(\underline{10}),\nabla^\alpha\theta\rangle &= \langle\nabla^\alpha(\theta\nabla^2 U),\nabla^\alpha\theta\rangle=\langle\theta\nabla^\alpha\nabla^2 U,\nabla^\alpha\theta\rangle\\ &\quad+\langle\nabla^\alpha(\theta\nabla^2 U)-\theta\nabla^2\nabla^\alpha U,\nabla^\alpha\theta\rangle\equiv R_3+R_4,\end{aligned}$$

$$\begin{aligned}|R_3| &= |\langle\nabla\theta\nabla^\alpha\nabla U,\nabla^\alpha\theta\rangle+\langle\theta\nabla^\alpha\nabla U,\nabla^\alpha\nabla\theta\rangle|\\ &\le \|\nabla\theta\|_\infty\left(\|\nabla^\alpha\nabla U\|_2^2+\|\nabla^\alpha\theta\|_2^2\right)+\varepsilon_1\|\nabla^\alpha\nabla\theta\|_2^2+\frac{1}{\varepsilon_1}\|\theta\|_\infty^2\|\nabla^\alpha\nabla U\|_2^2.\end{aligned}$$

$$\begin{aligned}|R_4| &\le c\left(\|\nabla\theta\|_\infty\|\nabla^\alpha\nabla U\|_2+\|\nabla^2 U\|_\infty\|\nabla^\alpha\theta\|_2\right)\|\nabla^\alpha\theta\|_2\\ &\le c\left(\|\nabla\theta\|_\infty\left(\|\nabla^2\nabla U\|_2^2+\|\nabla^\alpha\theta\|_2^2\right)+\varepsilon_1\|\nabla^\alpha\theta\|_2^2+\frac{1}{\varepsilon_1}\|\nabla^2 U\|_\infty^2\|\nabla^\alpha\theta\|_2^2\right).\end{aligned}$$

$$\begin{aligned}
\langle \nabla^\alpha(\underline{18}), \nabla^\alpha \theta\rangle &= \langle \nabla^\alpha(\theta \nabla U_t), \nabla^\alpha \theta\rangle \\
&= \langle \theta \nabla \nabla^\alpha U_t, \nabla^\alpha \theta\rangle + \langle \nabla^\alpha(\theta\nabla U_t) - \theta \nabla\nabla^\alpha U_t, \nabla^\alpha\theta\rangle \\
&\equiv R_5 + R_6,
\end{aligned}$$

$$\begin{aligned}
|R_5| &= |\langle \nabla\theta\nabla^\alpha U_t, \nabla^\alpha\theta\rangle + \langle \theta\nabla^\alpha U_t, \nabla\nabla^\alpha\theta\rangle| \\
&\leq \|\nabla\theta\|_\infty \left(\|\nabla^\alpha U_t\|_2^2 + \|\nabla^\alpha\theta\|_2^2\right) + \varepsilon_1\|\nabla^\alpha\nabla\theta\|_2^2 + \frac{1}{\varepsilon_1}\|\theta\|_\infty^2\|\nabla^\alpha U_t\|_2^2.
\end{aligned}$$

$$\begin{aligned}
|R_6| &\leq c\left(\|\nabla\theta\|_\infty\|\nabla^\alpha U_t\|_2 + \|\nabla U_t\|_\infty\|\nabla^\alpha\theta\|_2\right)\|\nabla^\alpha\theta\|_2 \\
&\leq c\left(\|\nabla\theta\|_\infty\left(\|\nabla^\alpha U_t\|_2^2 + \|\nabla^\alpha\theta\|_2^2\right) + \varepsilon_1\|\nabla^\alpha\theta\|_2^2 + \frac{1}{\varepsilon_1}\|\nabla U_t\|_\infty^2\|\nabla^\alpha\theta\|_2^2\right).
\end{aligned}$$

Up to now we have considered typical quadratic terms. The remaining quadratic terms can be handled in a similar fashion. We shall now deal with two typical cubic terms.

$$\begin{aligned}
\langle \nabla^\alpha(\underline{4}), \nabla^\alpha U_t\rangle &= \langle \nabla^\alpha((\nabla U)^2\nabla^2 U), \nabla^\alpha U_t\rangle \\
&= \langle (\nabla U)^2\nabla^\alpha\nabla^2 U, \nabla^\alpha U_t\rangle + \langle \nabla^\alpha((\nabla U)^2\nabla^2 U) - (\nabla U)^2\nabla^\alpha\nabla^2 U, \nabla^\alpha U_t\rangle \\
&\equiv R_7 + R_8.
\end{aligned}$$

$$\begin{aligned}
R_7 &= -\langle \nabla(\nabla U)^2\nabla^\alpha\nabla U, \nabla^\alpha U_t\rangle - \langle (\nabla U)^2\nabla^\alpha\nabla U, \nabla^\alpha\nabla U_t\rangle \\
&= -\langle \nabla(\nabla U)^2\nabla^\alpha\nabla U, \nabla^\alpha U_t\rangle - \frac{1}{2}\frac{d}{dt}\langle (\nabla U)^2\nabla^\alpha\nabla U, \nabla^\alpha\nabla U\rangle \\
&\quad + \langle \partial_t(\nabla U)^2\nabla^\alpha\nabla U, \nabla^\alpha\nabla U\rangle.
\end{aligned}$$

The second term in the last right-hand side is again incorporated into the left-hand side (compare R_1 above) and the inequality

$$\|\nabla(\nabla U)^2\|_\infty + \|\partial_t(\nabla U)^2\|_\infty \leq c\|DU\|_{1,\infty}^2$$

leads to the desired quadratic term in the exponent.

$$\begin{aligned}
|R_8| &\leq c\left(\|\nabla(\nabla U)^2\|_\infty\|\nabla^\alpha\nabla U\|_2 + \|\nabla^2 U\|_\infty\|\nabla^\alpha(\nabla U)^2\|_2\right)\|\nabla^\alpha U_t\|_2 \\
&\leq c\left(\|DU\|_{1,\infty}^2\|\nabla^\alpha DU\|_2^2 + \|\nabla^2 U\|_\infty\|\nabla U\|_\infty\|\nabla^\alpha\nabla U\|_2\|\nabla^\alpha U_t\|_2\right) \\
&\leq c\|DU\|_{1,\infty}^2\|\nabla^\alpha DU\|_2^2.
\end{aligned}$$

$$\begin{aligned}\langle \nabla^\alpha (\underline{22}), \nabla^\alpha \theta\rangle &= \langle \nabla^\alpha (\nabla U \nabla\theta \nabla^2 U), \nabla^\alpha \theta\rangle \\ &= \langle \nabla U \nabla\theta \nabla^\alpha \nabla^2 U, \nabla^\alpha\theta\rangle + \langle \nabla^\alpha(\nabla U\nabla\theta\nabla^2 U) - \nabla U\nabla\theta\nabla^\alpha\nabla^2 U, \nabla^\alpha\theta\rangle \\ &\equiv R_9 + R_{10},\end{aligned}$$

$$\begin{aligned}|R_9| &= |\langle \nabla(\nabla U\nabla\theta)\nabla^\alpha\nabla U, \nabla^\alpha\theta\rangle + \langle \nabla U\nabla\theta\nabla^\alpha\nabla U, \nabla^\alpha\nabla\theta\rangle| \\ &\leq \|\nabla\theta\|_{1,\infty}\left(\|\nabla^\alpha\nabla U\|_2^2 + \|\nabla^\alpha\theta\|_2^2\right) + \varepsilon_1\|\nabla^\alpha\nabla\theta\|_2^2 + \frac{1}{\varepsilon_1}\|\nabla\theta\|_\infty\|\nabla^\alpha\nabla U\|_2^2.\end{aligned}$$

$$\begin{aligned}|R_{10}| &\leq c\left(\|\nabla\theta\|_{1,\infty}\|\nabla^\alpha\nabla U\|_2 + \|\nabla^2 U\|_\infty\|\nabla^\alpha(\nabla U\nabla\theta)\|_2\right)\|\nabla^\alpha\theta\|_2 \\ &\leq c\left(\|\nabla\theta\|_{1,\infty}\left(\|\nabla^\alpha\nabla U\|_2^2 + \|\nabla^\alpha\theta\|_2^2\right) + \varepsilon_1\|\nabla^\alpha\theta\|_2^2 + \frac{1}{\varepsilon_1}\|\nabla U\|_{1,\infty}^2\|\nabla^\alpha\theta\|_2^2\right).\end{aligned}$$

The remaining cubic terms are handled in a similar fashion. Adding (11.103) and (11.104), summing over all $|\alpha| \leq s$, and integrating with respect to t, we obtain the desired special energy estimate (11.101) using Gronwall's inequality, Lemma 4.1, and choosing η (from Step **B**) and ε_1 sufficiently small. (Observe that the third terms on the left-hand side of (11.103) and (11.104), respectively, cancel.)

D: Weighted a priori estimates:

According to the assumption (11.78) on the nonlinearity we first discuss case I (no cubic terms).

Let s_0, k, k', l be arbitrary integers satisfying $l \geq 6$, $k' \geq l+1$, $k \geq k'+7$, $s_0 \geq k+7$ and let $0 < \varepsilon < 1/8$. We define for the local solution:

$$\begin{aligned}M_\varepsilon(T) := \sup_{0\leq t\leq T}\Big\{&(1+t)^{3/4-\varepsilon}\|\nabla^3\theta(t)\|_{k,2};\quad (1+t)^{3/2-2\varepsilon}\|\nabla^3\theta(t)\|_{k',2}; \\ &(1+t)^{5/4}\|\nabla\theta(t)\|_{1,2};\quad (1+t)^{3/4}\|\theta(t)\|_{2,2};\quad (1+t)^{3/2-\varepsilon}\|\theta(t)\|_{2,\infty}; \\ &(1+t)^{5/4}\|\theta_t(t)\|_\infty;\quad (1+t)^{3/4-\varepsilon}\|DU(t)\|_{l,\infty}\Big\}.\end{aligned}$$

We shall use the integral equation for V:

$$V(t) = \mathrm{e}^{-tA}V^0 + \int_0^t \mathrm{e}^{-(t-r)A}F(\ldots)(r)dr, \tag{11.105}$$

$(F = (0, f^1, f^2))$. We first estimate terms involving θ, then the term in M_ε involving U. These estimates use two basic technical ingredients.

First of all we shall of course use the decay estimates obtained in Lemma 11.9 and Lemma 11.10, in particular we shall use the L^2–L^∞- resp. L^1–L^2-estimates in Lemma 11.10 for

θ and if necessary we shall split integrals for an L^∞-estimate, $\{L^2$-estimate$\}$ $\|\int_0^t \ldots\|_{\infty\{2\}}$ into $\|\int_0^{t/2} \ldots + \int_{t/2}^t \ldots\|_{\infty\{2\}}$ where the L^1–L^∞-estimate $\{L^1$–L^2-estimate$\}$ is used in the first integral and the L^2–L^∞-estimate $\{L^2$–L^2-estimate$\}$ is used in the second integral. The second idea is to rewrite the nonlinear term F in the following form. The quadratic terms in which $\nabla\theta$ or $\nabla^2\theta$ appear remain unchanged, and the terms where only θ appears, i.e. $\theta\nabla^2 U$ and $\theta\nabla U_t$, will be written as $\nabla(\theta\nabla U)-\nabla\theta\nabla U$ and $\nabla(\theta U_t)-\nabla\theta U_t$, respectively. This has the advantage that either the better decay rate of $\nabla\theta$ is available or a ∇ can be put in front of the semigroup also leading to better decay (for θ-terms). Thus the quadratic terms F^q in F can be expressed as $F^q = F_1^q + F_2^q$ where $F_1^q = \nabla(\theta\ldots) \equiv \nabla\tilde{F}_1^q$ and $F_2^q = \nabla\theta(\ldots) + \nabla^2\theta(\ldots)$. The integral equation (11.105) then turns into

$$V(t) = \mathrm{e}^{-tA}V^0 + \int_0^t \mathrm{e}^{-(t-r)A}\left(\nabla\tilde{F}_1^q + F_2^q\right)(r)dr.$$

The initial data V^0 will be assumed to satisfy

$$\|V^0\|_{s_0,2} + \|V^0\|_{s_0,1} < \delta,$$

where δ will be chosen sufficiently small leading to the desired uniform bound for $M_\varepsilon(T)$. According to the Lemmata 11.9 and 11.10, and, with the technique outlined above, we obtain the following sequence of estimates. (C denotes various constants not depending on T or V^0.)

$$\begin{aligned}\|\nabla^3\theta(t)\|_{k,2} \;\le\;& C(1+t)^{-3/2}\delta + C\int_0^t (1+t-r)^{-3/2}\,\cdot\\ &\cdot\left[\|DU(r)\|_{s_0,2}\|\theta(r)\|_{2,\infty} + \|DU(r)\|_{1,\infty}\|\theta(r)\|_{s_0,2}\right]dr.\end{aligned}$$

Analogously,

$$\begin{aligned}\|\nabla^3\theta(t)\|_{k',2} \;\le\;& C(1+t)^{-3/2}\delta + C\int_0^t (1+t-r)^{-3/2}\,\cdot\\ &\cdot\left[\|DU(r)\|_{s_0,2}\|\theta(r)\|_{2,\infty} + \|DU(r)\|_{1,\infty}\left(\|\theta(r)\|_{2,2} + \|\nabla^3\theta(r)\|_{k,2}\right)\right]dr.\end{aligned}$$

In the following estimates the technique described above becomes relevant for the first time.

$$\begin{aligned}\|\nabla\theta(t)\|_{1,2} \;\le\;& C(1+t)^{-(3/4+1/2)}\delta + C\int_0^{t/2} (1+t-r)^{-(3/4+1)}\,\cdot\\ &\cdot\left[\|DU(r)\|_{s_0,2}\left(\|\theta(r)\|_{2,2} + \|\nabla^3\theta(r)\|_{k',2}\right)\right]dr\end{aligned}$$

$$
\begin{aligned}
&+C\int_{t/2}^{t}(1+t-r)^{-1}\left[\|DU(r)\|_{s_0,2}\|\theta(r)\|_{2,\infty}\right.\\
&\left.+\|DU(r)\|_{\infty}\left(\|\theta(r)\|_{2,2}+\|\nabla^3\theta(r)\|_{k',2}\right)\right]dr\\
&+C\int_{0}^{t}(1+t-r)^{-(3/4+1/2)}\left[\|DU(r)\|_{s_0,2}\left(\|\nabla\theta(r)\|_{1,2}+\|\nabla^3\theta(r)\|_{k',2}\right)\right]dr.
\end{aligned}
$$

$$
\begin{aligned}
\|\theta(t)\|_{2,2} \;\le\;& C(1+t)^{-3/4}\delta+C\int_0^t(1+t-r)^{-(3/4+1/2)}\cdot\\
&\cdot\left[\|DU(r)\|_{s_0,2}\left(\|\theta(r)\|_{2,2}+\|\nabla^3\theta(r)\|_{k',2}\right)\right]dr\\
&+C\int_0^t(1+t-r)^{-3/4}\left[\|DU(r)\|_{s_0,2}\left(\|\nabla\theta(r)\|_{1,2}+\|\nabla^3\theta(r)\|_{k',2}\right)\right]dr.
\end{aligned}
$$

$$
\begin{aligned}
\|\theta(t)\|_{2,\infty} \;\le\;& C(1+t)^{-3/2}\delta+C\int_0^{t/2}(1+t-r)^{-(3/2+1/2)}\cdot\\
&\cdot\left[\|DU(r)\|_{s_0,2}\left(\|\theta(r)\|_{2,2}+\|\nabla^3\theta(r)\|_{k',2}\right)\right]dr\\
&+C\int_{t/2}^{t}(1+t-r)^{-(3/4+1/2)}\left[\|DU(r)\|_{s_0,2}\|\theta(r)\|_{2,\infty}\right.\\
&\left.+\|DU(r)\|_{1,\infty}\left(\|\theta(r)\|_{2,2}+\|\nabla^3\theta(r)\|_{k',2}\right)\right]dr\\
&+C\int_0^{t/2}(1+t-r)^{-3/2}\left[\|DU(r)\|_{s_0,2}\left(\|\nabla\theta(r)\|_{1,2}+\|\nabla^3\theta(r)\|_{k',2}\right)\right]dr\\
&+C\int_{t/2}^{t}(1+t-r)^{-3/2}\left[\|DU(r)\|_{l,\infty}\left(\|\nabla\theta(r)\|_{1,2}+\|\nabla^3\theta(r)\|_{k',2}\right)\right]dr.
\end{aligned}
$$

Since

$$
\theta_t(t)=\frac{\partial}{\partial t}e^{-tA}V^0\Big|_{3^{rd}comp.}+\int_0^t\frac{\partial}{\partial t}e^{(t-r)A}F(\ldots)(r)\Big|_{3^{rd}comp.}dr
$$

we obtain

$$
\begin{aligned}
\|\theta_t(t)\|_{\infty} \;\le\;& C(1+t)^{-(3/2+1/2)}\delta+C\|\theta(t)\|_{2,\infty}\|DU(t)\|_{1,\infty}\\
&+C\int_0^t(1+t-r)^{-(3/4+1/2)}\left[\|DU(r)\|_{s_0,2}\|\theta(r)\|_{2,\infty}\right.\\
&\left.+\|DU(r)\|_{1,\infty}\left(\|\theta(r)\|_{2,2}+\|\nabla^3\theta(r)\|_{k',2}\right)\right]dr.
\end{aligned}
$$

Summarizing the estimates above, using the definition of $M_\varepsilon(T)$ and the high energy estimate from Step **C**, inequality (11.101), we obtain

$$M_\varepsilon(T) \le C\delta + C\delta((M_\varepsilon(T))^2 + M_\varepsilon(T)) \exp\left\{C((M_\varepsilon(T))^2 + M_\varepsilon(T))\right\}. \qquad (11.106)$$

The integrals of the type $\sup\limits_{0\le t\le T} \int\limits_0^t (1+t)^\gamma (1+t-r)^{-\alpha}(1+r)^{-\beta} dr$, which naturally appear in proving (11.106) are uniformly bounded according to Lemma 7.4. The decay exponents in the definition of $M_\varepsilon(T)$ and ε have been suitably chosen such that the assumptions of Lemma 7.4 are satisfied (which is shown by a lengthy but easy calculation).
As in Chapter 7 we are now able to conclude that $M_\varepsilon(T)$ is uniformly bounded, i.e.

$$M_\varepsilon(T) \le M_0 < \infty, \qquad (11.107)$$

where $M_0 = M_0(\varepsilon)$ is independent of T, provided δ is sufficiently small.

Now we deal with case II in assumption (11.78) (no quadratic terms besides $\theta\Delta\theta$).
Let

$$\begin{aligned} M(T) \;&:=\; \sup_{0\le t\le T} \Big\{ (1+t)^{5/9} \|DU(t)\|_{s_1,9/2}; \\ &\quad (1+t)^{27/26}\Big(\|\theta(t)\|_{s_3,13/2} + \|\nabla^2\theta(t)\|_{s_5,26/11} + \|\nabla^2\theta(t)\|_{s_5,18/7}\Big)\Big\}, \end{aligned}$$

where s_1, s_3, s_5 are sufficiently large such that

$$\begin{aligned} \left[\frac{s_1+N_{9/7}}{2}\right] + 4 \;&\le\; \min\{s_1, s_3\}, \quad s_1 + N_{9/7} \le s_5, \\ \left[\frac{s_3+N_{13/11}}{2}\right] + 4 \;&\le\; \min\{s_1, s_3\}, \quad s_3 + N_{13/11} \le s_5, \\ \left[\frac{s_5+N_{18/11}}{2}\right] + 4 \;&\le\; \min\{s_1, s_3\}, \\ \left[\frac{s_5+N_{26/15}}{2}\right] + 4 \;&\le\; \min\{s_1, s_3\}, \end{aligned}$$

($N_{\dots}$ from Lemma 11.66, $N_{\dots} \le 4$).
The initial data V^0 will be assumed to satisfy again

$$\|V^0\|_{s_0,2} + \|V^0\|_{s_0,1} < \delta,$$

where δ will be chosen sufficiently small such that $M(t)$ will be uniformly bounded; s_0 has to be sufficiently large, at least $s_0 \ge \max\{s_1 + N_{9/7} + 3, s_3 + N_{13/11} + 3, s_5 + 4\}$.
Using Lemma 11.9 we conclude

$$\begin{aligned} \|DU(t)\|_{s_1,9/2} \;&\le\; C(1+t)^{-5/9}\delta + C\int_0^t (1+t-r)^{-5/9} \|F(V,\dots)(r)\|_{s_1+N_{9/7},9/7}\, dr, \\ \|\theta(t)\|_{s_3,13/2} \;&\le\; C(1+t)^{-27/26}\delta + C\int_0^t (1+t-r)^{-27/26} \|F(V,\dots)(r)\|_{s_3+N_{13/11},13/11}\, dr. \end{aligned}$$

Let

$$p_1 := 9/7, \quad q_1 := 9/2, \quad \tilde{q}_1 := 18,$$
$$p_2 := 13/11, \quad q_2 := 13/2, \quad \tilde{q}_2 := 26/5.$$

Then

$$1/p_j = 1/2 + 1/q_j + 1/\tilde{q}_j, \quad j = 1, 2,$$
$$18 = \tilde{q}_1 > \tilde{q}_2 > q_2 > q_1 = 9/2.$$

This implies the validity of the following inequality ($N \in \mathbb{N}$):

$$\begin{aligned} \|f_1 f_2 f_3\|_{N,p_j} \le\ & C\Big\{\|f_1\|_{N,2}\|f_2\|_{[N/2]+1,\tilde{q}_j}\|f_3\|_{[N/2]+1,q_j} \\ & +\|f_1\|_{[N/2]+1,q_j}\|f_2\|_{N,2}\|f_3\|_{[N/2]+1,\tilde{q}_j} \\ & +\|f_1\|_{[N/2]+1,q_j}\|f_2\|_{[N/2]+1,\tilde{q}_j}\|f_3\|_{N,2}\Big\}. \end{aligned} \tag{11.108}$$

Remember that F has the form $F = (0, f^1, f^2)$ (cf. (11.76)). Typical terms in $f^1 = f^{1,c}$ have the form given in (11.102).
Using (11.108) we get

$$\|f^1(V,\ldots)\|_{s_1+N_{9/7},9/7} \le C\|V\|_{s_0,2}\left(\|\theta\|_{s_3,13/2} + \|\theta\|^2_{s_3,13/2} + \|\nabla U\|^2_{s_1,9/2}\right).$$

The typical cubic terms in f^2 are dealt with similarly. For the quadratic term $\theta\Delta\theta$ we obtain

$$\|\theta\Delta\theta\|_{s_1+N_{9/7},9/7} \le C\|\theta\|_{s_1+N_{9/7}+2,9/7}\|\nabla^2\theta\|_{s_1+N_{9/7},18/7}.$$

Thus we conclude

$$\begin{aligned} &\|F(V,\ldots)\|_{s_1+N_{9/7},9/7} \\ &\le C\|V\|_{s_0,2}\left(\|\theta\|_{s_3,13/2} + \|\theta\|^2_{s_3,13/2} + \|\nabla U\|^2_{s_1,9/2} + \|\nabla^2\theta\|_{s_5,18/7}\right). \end{aligned} \tag{11.109}$$

Analogously

$$\begin{aligned} &\|F(V,\ldots)\|_{s_3+N_{13/11},13/11} \\ &\le C\|V\|_{s_0,2}\left(\|\theta\|_{s_3,13/2} + \|\theta\|^2_{s_3,13/2} + \|\nabla U\|^2_{s_1,9/2} + \|\nabla^2\theta\|_{s_5,26/11}\right). \end{aligned} \tag{11.110}$$

Finally we notice that the integrands $\|F(V,\ldots)\|_{\ldots}$ appearing in the estimates for $\|\nabla^2\theta(t)\|_{s_5,18/7}$ and $\|\nabla^2\theta(t)\|_{s_5,26/11}$ can be estimated by the same right-hand sides in (11.109) and (11.110), respectively. (Observe for example that the following inequality holds:

$$\begin{aligned} \|\theta\Delta\theta\|_{s_5+N_{18/11},18/11} &\le C\|\theta\|_{s_5+N_{18/11}+2,2}\|\theta\|_{\left[\frac{s_5+N_{18/11}}{2}\right]+3,13/2} \\ &\le C\|\theta\|_{s_0,2}\|\theta\|_{s_3,13/2}.) \end{aligned}$$

Summarizing these estimates we get

$$M(T) \leq C\delta + C\delta((M(T))^2 + M(T)) \exp\left\{C((M(T))^2 + M(T))\right\}.$$

This is the same inequality as that obtained for $M_\varepsilon(T)$ in (11.106). Analogously, $M(T)$ is bounded by M_0 if δ is sufficiently small, M_0 being independent of T:

$$M(T) \leq M_0 < \infty. \tag{11.111}$$

(11.107) and (11.111) are the desired weighted a priori estimates and now lead as usual to the desired energy estimate.

E: Final energy estimate:

$$\|V(t)\|_{s,2} \leq K\|V^0\|_{s,2}, \quad 0 \leq t \leq T,$$

$s \geq s_0$ (s_0 from Step **D**), δ small enough as given in Step **D**, K being independent of T. Altogether we have proved the following global existence theorem.

Theorem 11.11 *Let the nonlinearity satisfy (11.78). Then there exist an integer s_0 and a $\delta > 0$ such that the following holds:*
If $(\nabla U^0, U^1, \theta^0) \in W^{s,2} \cap W^{s,1}$ with $s \geq s_0$ and

$$\|(\nabla U^0, U^1, \theta^0)\|_{s,2} + \|(\nabla U^0, U^1, \theta^0)\|_{s,1} < \delta,$$

then there is a unique solution (U, θ) of the initial value problem to the nonlinear equations of thermoelasticity (11.67), (11.68), (11.71) in $\mathbb{R}^3$ with

$$(\nabla U, U_t) \in C^0([0,\infty), W^{s,2}) \cap C^1([0,\infty), W^{s-1,2}),$$

$$\theta \in C^0([0,\infty), W^{s,2}) \cap C^1([0,\infty), W^{s-2,2}).$$

Moreover, the asymptotic behavior can be described as follows.
Case I (no cubic terms): There exist integers $l < k' < k \leq s$ such that for $\varepsilon < 1/8$ we have

$$\begin{aligned}
\|(\nabla U, U_t, \theta)(t)\|_{s,2} &= \mathcal{O}(1), \\
\|\nabla^3\theta(t)\|_{k,2} + \|(\nabla U, U_t)(t)\|_{l,\infty} &= \mathcal{O}(t^{-3/4+\varepsilon}), \\
\|\nabla^3\theta(t)\|_{k',2} &= \mathcal{O}(t^{-3/2+2\varepsilon}), \\
\|\nabla\theta(t)\|_{1,2} + \|\theta_t(t)\|_\infty &= \mathcal{O}(t^{-5/4}), \\
\|\theta(t)\|_{2,2} &= \mathcal{O}(t^{-3/4}), \\
\|\theta(t)\|_{2,\infty} &= \mathcal{O}(t^{-3/2+\varepsilon}) \quad \text{as} \quad t \to \infty.
\end{aligned}$$

<u>Case II</u> *(no quadratic terms but $\theta\Delta\theta$): There exist integers s_1, s_3, $s_5 \leq s$ with*

$$\begin{aligned}\|(\nabla U, U_t, \theta)(t)\|_{s,2} &= \mathcal{O}(1),\\ \|(\nabla U, U_t)(t)\|_{s_1,9/2} &= \mathcal{O}(t^{-5/9}),\\ \|\theta(t)\|_{s_3,13/2} + \|\nabla^2\theta(t)\|_{s_5,26/11} + \|\nabla^2\theta(t)\|_{s_5,18/7} &= \mathcal{O}(t^{-27/26}) \quad \text{as} \quad t \to \infty.\end{aligned}$$

We remark that beyond the previous theorem there are only local existence theorems for initial-*boundary* value problems in three space dimensions. These cover also non-homogeneous, anisotropic media and both bounded and unbounded domains, see A. Chrzęszczyk [20] and Jiang & Racke [62].

This section will be concluded by an example that shows that one has to expect the development of singularities in finite time if there are purely quadratic nonlinear terms only involving derivatives of the displacement, not involving the temperature, as

$$\nabla U \nabla^2 U, \; \nabla U \nabla U_t.$$

The idea of proving a blow-up is that the hyperbolic part predominates and that solutions to the nonlinear equations of elasticity in general develop singularities in three space dimensions if there are quadratic nonlinearities, see Section 11.1.

The ansatz that we make comes from the observation that solutions (U, θ) to the linearized equations can be decoupled into $(U^{so}, 0) + (U^{po}, \theta)$ where the divergence-free U^{so} is no longer coupled to θ and solves a linear wave equation, see Step **A** above. That is, for $U^{0,so} \neq 0$ or $U^{1,so} \neq 0$ there is always a non-trivial purely hyperbolic part in the (linear) equations which does not involve θ and hence does not experience any damping.

Remark: $U^{so} \equiv 0$ in one space dimension, i.e. the ansatz below cannot work there. Indeed, we know that in one dimension global solutions always exist if the data are sufficiently small without further restrictions on the nonlinearity; see the discussion of one-dimensional models at the beginning of this section.

In the (three-dimensional) nonlinear case it will now be the aim to obtain a decomposition of U for a *plane wave* U into some divergence-free part U^σ and a curl-free part U^π, compatible with the special nonlinearity that we shall choose, which has to be determined such that U^σ satisfies

$$\partial_t^2 U_i^\sigma - \sum_{m,j,k=1}^{3} C_{imjk}(\nabla U^\sigma, 0)\partial_m \partial_k U_j^\sigma = 0, \quad i = 1, 2, 3,$$

$$U^\sigma(t=0) = U^{0,\sigma}, \quad U_t^\sigma(t=0) = U^{1,\sigma}.$$

Actually, U^σ will be a function of x_1 only (plane wave) and U^σ will satisfy

$$\partial_t^2 U_i^\sigma - \sum_{j=1}^{3} B_{ij}(\partial_1 U^\sigma)\partial_1^2 U_j^\sigma, \quad i = 1, 2, 3$$

for some B_{ij}, and this will be a genuinely nonlinear strictly hyperbolic system. Then the general results from T.-P. Liu [112] $\{$or F. John [64]$\}$ will imply that there are nonlinearities which on one hand satisfy at least the basic (physical) properties — as they are given for instance in the global existence theorem above (of course besides the degeneracy requirement (11.78)). On the other hand, the nonlinearities are such that for compactly supported (in x_1), non-vanishing smooth data $(U^{0,\sigma}, U^{1,\sigma})$, which are sufficiently small, a plane-wave solution cannot be of class C^2 $\{$resp. $C^3\}$ for all positive t.

The components (U_2, U_3) will develop the singularities. The coefficients in the differential equations (11.67), (11.68) for (U, θ) are — besides the heat flux vector q — given by a free Helmholtz energy (Helmholtz potential) $\psi = \psi(\nabla U, \theta)$, see (11.69), (11.70) where in particular we should have

$$a \geq a_0 > 0 \quad \text{for some constant } a_0, \tag{11.112}$$

$$(\tilde{C}_{im})_{im} \neq 0 \quad (\text{“really coupled”}). \tag{11.113}$$

We consider plane waves

$$U(t,x) = \tilde{U}(t, \tau \cdot x), \quad \tau \in \mathbb{R}^3 \quad \text{fixed.}$$

We also assume $\theta(t,x) = \tilde{\theta}(t, \tau \cdot x)$ and we write U and θ again instead of $\tilde{U}$ and $\tilde{\theta}$ respectively. For simplicity we choose

$$\tau := (1,0,0)',$$

that is U and θ become functions of x_1 (and t) only. We may decompose U into

$$U = (0, U_2, U_3)' + (U_1, 0, 0)' \equiv U^\sigma + U^\pi,$$

where U^σ, U^π satisfy

$$\nabla' U^\sigma = 0, \quad \nabla \times U^\pi = 0.$$

In this sense it is again a decomposition into divergence-free, respectively, curl-free parts, but this decomposition is no longer in L^2.

Let P^σ denote the corresponding projection

$$U \mapsto P^\sigma U := U^\sigma.$$

As indicated above the aim is to obtain the relation

$$\begin{aligned} 0 &= \partial_t^2 U^\sigma - P^\sigma \left\{ \sum_{m,j,k=1}^{3} C_{imjk}(\nabla U, \theta) \partial_m \partial_k U_j + \sum_{m=1}^{3} \tilde{C}_{im}(\nabla U, \theta) \partial_m \theta \right\}_{i=1,2,3} \qquad (11.114) \\ &= \partial_t^2 U^\sigma - \left\{ \sum_{m,j,k=1}^{3} C_{imjk}(\nabla U^\sigma, 0) \partial_m \partial_k U_j^\sigma \right\}_{i=1,2,3}, \end{aligned}$$

where the coefficient $C_{imjk}(\nabla U, 0)$ does not depend on θ. For this purpose we require at first

$$P^\sigma \left\{ \sum_{m=1}^{3} \tilde{C}_{im}(\nabla U, \theta)\partial_m \theta \right\}_{i=1,2,3} = 0. \tag{11.115}$$

In order to have a nonlinear dependence of C_{imjk} in θ we additionally require

$$\frac{\partial C_{imjk}}{\partial \theta} \neq 0 \text{ at least for one quadrupel } (i, m, j, k). \tag{11.116}$$

To obtain (11.114) we now have to postulate

$$P^\sigma \left\{ \sum_{m,j,k=1}^{3} C_{imjk}(\nabla U, \theta)\partial_m \partial_k U_j \right\}_{i=1,2,3} = \left\{ \sum_{m,j,k=1}^{3} C_{imjk}(\nabla U^\sigma, 0)\partial_m \partial_k U_j^\sigma \right\}_{i=1,2,3}. \tag{11.117}$$

Since we are considering plane waves the final system for U^σ should be

$$\partial_t^2 U_i^\sigma - \sum_{j=1}^{3} B_{ij}(\partial_1 U^\sigma)\partial_1^2 U_j^\sigma = 0, \quad i = 1, 2, 3, \tag{11.118}$$

$$U^\sigma(t=0) = U^{0,\sigma}, \quad U_t^\sigma(t=0) = U^{1,\sigma},$$

where

$$B_{ij}(\partial_1 U^\sigma) = C_{i1j1}(\nabla U^\sigma, 0), \quad i, j = 1, 2, 3. \tag{11.119}$$

Let

$$V(\alpha) := \psi\left(\begin{pmatrix} \alpha_1 & 0 & 0 \\ \alpha_2 & 0 & 0 \\ \alpha_3 & 0 & 0 \end{pmatrix}, 0 \right), \quad \alpha \in \mathbb{R}^3, \tag{11.120}$$

and

$$V_{ij}(\alpha) := B_{ij}\left(\begin{pmatrix} \alpha_1 & 0 & 0 \\ \alpha_2 & 0 & 0 \\ \alpha_3 & 0 & 0 \end{pmatrix} \right). \tag{11.121}$$

Then the system (11.118) is *strictly hyperbolic* if

$$\text{the matrix } (V_{ij}(\alpha))_{ij} \text{ has only positive distinct eigenvalues} \tag{11.122}$$

and it is *genuinely nonlinear* if

$$\left.\frac{d^3(V(s\beta))}{ds^3}\right|_{s=0} \neq 0 \text{ for any right eigenvector } \beta \text{ of } (V_{ij}(\alpha))_{ij}. \tag{11.123}$$

(For the notion of "strictly hyperbolic" and "genuinely nonlinear" see [64].) Then the following theorem will follow from the general result in [112, p. 107] $\big\{$resp. [64, p. 387]$\big\}$

if there is a nonlinearity ψ satisfying all the conditions gathered in the following formula (11.124):

$$(11.112), (11.113), (11.115), (11.116), (11.117), (11.122), (11.123) \tag{11.124}$$

(at least in a neighbourhood of zero).

Theorem 11.12 *There exist nonlinearities satisfying (11.124) such that for compactly supported non-vanishing smooth data $(U^{0,\sigma}, U^{1,\sigma})$ which are sufficiently small, i.e.*

$$\sup_{x_1 \in \mathbf{R}} \left| \partial_1 \left(\partial_1 U_2^0, \partial_1 U_3^0, U_2^1, U_3^1 \right)(x_1) \right|$$

$$\left\{ resp. \quad \sup_{x_1 \in \mathbf{R}} \left| \partial_1^2 \left(\partial_1 U_2^0, \partial_1 U_3^0, U_2^1, U_3^1 \right)(x_1) \right| \right\}$$

is sufficiently small, a plane-wave solution of the nonlinear equations of thermoelasticity (11.67), (11.68), (11.71) in $\mathbb{R}^3$ cannot be of class C^2 $\{$resp. $C^3\}$ for all positive t.

PROOF: We choose the following function ψ:

$$\begin{aligned} \psi(\nabla U, \theta) \;:=\; & 3(\partial_1 U_1)^2 + (\partial_1 U_2)^2 + (\partial_1 U_3)^2 + (\partial_1 U_2)(\partial_1 U_3) \\ & + a_{111}(\partial_1 U_1)^3 + a_{222}(\partial_1 U_2)^3 + a_{333}(\partial_1 U_3)^3 \\ & + a_{223}(\partial_1 U_2)^2(\partial_1 U_3) + a_{233}(\partial_1 U_2)(\partial_1 U_3)^2 \\ & + (\partial_1 U_1)^2\theta + \gamma\theta \sum_{j=1}^{3} (\partial_j U_j) - \theta^2 \end{aligned}$$

with coefficients satisfying

$$a_{111}, a_{222}, a_{333}, a_{223}, a_{233}, \gamma \in \mathbb{R} \setminus \{0\}, \tag{11.125}$$

$$a_{222} + a_{333} + a_{223} + a_{233} \neq 0, \tag{11.126}$$

$$a_{222} - a_{333} - a_{223} + a_{233} \neq 0. \tag{11.127}$$

Now it has become a more or less simple algebraic task to check whether this ψ satisfies the conditions (11.124).

By the definition of C_{imjk}, and $\tilde{C}_{im}$ the relations required in (11.112), (11.113), (11.115), (11.116) are immediately clear. It remains to show (11.117), (11.122) and (11.123).

We have

$$C_{imjk}(\nabla U, \theta) = 0 \quad \text{for} \quad (m, k) \neq (1, 1).$$

The relation (11.117) is then equivalent to

$$0 = \sum_{j=1}^{3} C_{11j1}(\nabla U^\sigma, 0)\partial_1^2 U_j^\sigma \tag{11.128}$$

and

$$\sum_{j=1}^{3} C_{i1j1}(\nabla U,\theta)\partial_1^2 U_j = \sum_{j=1}^{3} C_{i1j1}(\nabla U^\sigma,0)\partial_1^2 U_j^\sigma, \quad i=2,3. \tag{11.129}$$

The relation (11.128) holds because

$$U_1^\sigma = 0 \quad \text{and} \quad C_{1121} = C_{1131} = 0.$$

We have

$$(C_{i1j1}(\nabla U,\theta))_{ij} = \tag{11.130}$$

$$\begin{pmatrix} 6+6a_{111}(\partial_1 U_1)+2\theta & 0 & 0 \\ 0 & 2+6a_{222}(\partial_1 U_2)+2a_{223}(\partial_1 U_3) & 1+2a_{223}(\partial_1 U_2)+2a_{233}(\partial_1 U_3) \\ 0 & 1+2a_{223}(\partial_1 U_2)+2a_{233}(\partial_1 U_3) & 2+6a_{333}(\partial_1 U_3)+2a_{233}(\partial_1 U_2) \end{pmatrix}$$

and

$$C_{i1j1}(\nabla U,\theta) = C_{i1j1}(\nabla U,0) \quad \text{for} \quad (i,j)\neq(1,1).$$

Let $i\in\{2,3\}$. Then we obtain from (11.130) and the definition of U^σ:

$$\begin{aligned} \sum_{j=1}^{3} C_{i1j1}(\nabla U,\theta)\partial_1^2 U_j &= \sum_{j=2}^{3} C_{i1j1}(\nabla U,\theta)\partial_1^2 U_j \\ = \sum_{j=2}^{3} C_{i1j1}(\nabla U,0)\partial_1^2 U_j &= \sum_{j=2}^{3} C_{i1j1}(\nabla U^\sigma,0)\partial_1^2 U_j \\ = \sum_{j=2}^{3} C_{i1j1}(\nabla U^\sigma,0)\partial_1^2 U_j^\sigma &= \sum_{j=1}^{3} C_{i1j1}(\nabla U^\sigma,0)\partial_1^2 U_j^\sigma \end{aligned}$$

which yields (11.129) and thus (11.117).
By (11.119) – (11.121) we obtain

$$(V_{ij}(\alpha))_{ij} = \begin{pmatrix} 6+6a_{111}\alpha_1 & 0 & 0 \\ 0 & 2+6a_{222}\alpha_2+2a_{223}\alpha_3 & 1+2a_{223}\alpha_2+2a_{233}\alpha_3 \\ 0 & 1+2a_{223}\alpha_2+2a_{233}\alpha_3 & 2+6a_{333}\alpha_3+2a_{233}\alpha_2 \end{pmatrix}$$

and it is clear that $(V_{ij}(0))_{ij}$ has the three positive distinct eigenvalues

$$\lambda_1 = 6, \quad \lambda_2 = 3, \quad \lambda_3 = 1.$$

Hence (11.122) is satisfied in a neighbourhood of zero. Moreover,

$$\begin{aligned} \left.\frac{d^3 V(s\beta)}{ds^3}\right|_{s=0} &= 6a_{111}\beta_1^3 + 6a_{222}\beta_2^3 + 6a_{333}\beta_3^3 + 6a_{223}\beta_2^2\beta_3 + 6a_{233}\beta_2\beta_3^2 \\ &=: H(\beta), \end{aligned}$$

where $\beta = \beta(\alpha)$ is a right eigenvector of $(V_{ij}(\alpha))_{ij}$. For $\alpha = 0$ we have the three right eigenvectors

$$\beta^1 = (1,0,0)', \quad \beta^2 = (0,1,1)', \quad \beta^3 = (0,1,-1)'$$

and we obtain from (11.125)

$$H(\beta^1) = 6a_{111} \neq 0, \tag{11.131}$$

and from (11.126), (11.127)

$$H(\beta^2) = 6(a_{222} + a_{333} + a_{223} + a_{233}) \neq 0, \tag{11.132}$$

and

$$H(\beta^3) = 6(a_{222} - a_{333} - a_{223} + a_{233}) \neq 0. \tag{11.133}$$

A combination of the relations (11.131) – (11.133) yields (11.123) in a neighbourhood of zero which completes the proof of Theorem 11.12.

Q.E.D.

We remark that with the results in this section the nonlinear equations of thermoelasticity were *basically* understood, but there are many interesting remaining problems such as problems with anisotropic media (cf. Section 11.1) or boundary value problems in three space dimensions or large data problems. (This remark applies mutatis mutandis to the other systems in Chapter 11, too.)

11.4 Schrödinger equations

(*Erwin Schrödinger,* 12.8.1887 – 4.1.1961)
The following type of differential equation

$$u_t - i\Delta u = F(u, \nabla u), \tag{11.134}$$

with initial value

$$u(t=0) = u_0, \tag{11.135}$$

for a complex-valued function $u = u(t,x), t \geq 0, x \in \mathbb{R}^n, i = \sqrt{-1}$, will be studied. (In quantum mechanics u describes the state of a particle in a quantum mechanical system.) The smooth function F is assumed to satisfy

$$F(w) = \mathcal{O}(|w|^{\alpha+1}) \quad \text{as } |w| \to 0 \quad \text{for some } \alpha \in \mathbb{N}, \tag{11.136}$$

and

$$\frac{\partial F(w)}{\partial(\partial_j u)} \quad \text{is real,} \quad j = 1,...,n, \; w \in \mathbb{R}^{n+1}. \tag{11.137}$$

It is apparent that F does not depend on the second derivatives of u. One reason for this is that there seems to be no local existence theorem covering general second-order nonlinearities, see Step **B** below and Section 11.7, where a similar situation is given for nonlinear plate equations. In connection with this one should notice that if u is a solution of the linear Schrödinger equation, $u = u_1 + iu_2$, with real-valued functions u_1, u_2, then u_1 and u_2 are solutions of the linear plate equation

$$\partial_t^2 u_j + \Delta^2 u_j = 0, \quad j = 1, 2,$$

which easily follows from

$$\partial_t u_1 = -\Delta u_2 \quad \text{and} \quad \partial_t u_2 = \Delta u_1.$$

On the other hand, if the real-valued function v solves

$$v_{tt} + \Delta^2 v = 0$$

then u defined by

$$u := v_t + i\Delta v$$

solves the linear Schrödinger equation,

$$u_t - i\Delta u = 0,$$

a fact which will be used in Section 11.7. This elementary but important relation is expected to have implications for the relation between the corresponding nonlinear systems too.

Now we start following the general Steps **A**–**E**.

A: Decay for $F \equiv 0$:
The linear equation

$$u_t - i\Delta u = 0 \tag{11.138}$$

can formally be obtained from the linear heat equation (11.46) by the transformation $t \to it$ and indeed, the solution formula corresponding to that for the heat equation, formula (11.47), now becomes

$$u(t,x) := (4\pi it)^{-n/2} \int_{\mathbf{R}^n} e^{\frac{i|x-y|^2}{4t}} u_0(y)dy, \tag{11.139}$$

which solves (11.138) for a suitable class of initial data u_0. (The class of admissible data u_0 such that (11.139) defines a classical solution is smaller than the corresponding one for the heat equation due to the regularizing properties of the heat kernel which decays exponentially.) As in Section 11.2 we immediately obtain from (11.139)

$$\|u(t)\|_\infty \leq c(1+t)^{-n/2} \|u_0\|_{n_1,1}, \quad n_1 > n, \tag{11.140}$$

$t \geq 0, c > 0$ being independent of t.
Moreover we get from (11.138)

$$\mathrm{Re}\left[\int_{\mathbb{R}^n} u_t\overline{u} + i\int_{\mathbb{R}^n} \nabla u \nabla \overline{u}\right] = 0$$

($\overline{z}$ denoting the complex conjugate of $z \in \mathbb{C}$),
which implies

$$\frac{d}{dt}\|u(t)\|_2^2 = 0 \tag{11.141}$$

or

$$\|u(t)\|_2 = \|u_0\|_2 \quad \text{for all} \quad t \geq 0. \tag{11.142}$$

By interpolation we obtain from (11.140), (11.142):

$$\|u(t)\|_q \leq c(1+t)^{-\frac{n}{2}(1-\frac{2}{q})}\|u_0\|_{N_p,p}, \tag{11.143}$$

$t \geq 0, c = c(q,n)$, where $2 \leq q \leq \infty, 1/p + 1/q = 1$ and

$$N_p > n(1-2/q) \quad (N_p = n(1-2/q) \quad \text{if } q \in \{2,\infty\})\,.$$

In contrast to the situation in Section 11.2 where we discussed the (linear) heat equation, derivatives of solutions to the linear Schrödinger equation do not decay with a larger decay rate as do solutions to the linear heat equation. Also the L^2-norm $\|u(t)\|_2$ of u here does not decay at all. In addition, one should observe the difference between the energy equation (11.142) and the corresponding energy equation (11.49). The dissipative integral term in (11.49) is missing in the non-dissipative but conservative Schrödinger energy equation (11.142).

B: Local existence and uniqueness:
The nonlinear system (11.134), (11.135) is neither covered by the local existence theorem for symmetric hyperbolic systems, Theorem 5.8, nor by that on symmetric hyperbolic-parabolic coupled systems in Appendix C (although a similar proof to that for Theorem C.4 used in Appendix C gives a unique local solution to the nonlinear Schrödinger system).
A local solution

$$u \in C^0([0,T], W^{s,2}) \cap C^1([0,T], W^{s-2,2})$$

for some $T > 0$ is provided by T. Kato in [83, pp. 70,71] if $u_0 \in W^{s,2}$. If s is odd, s has to be at least $2[\frac{n+2}{4}]+3$; if s is even, s has to be at least $2[\frac{n}{4}]+4$. These restrictions arise in [83] in studying *boundary* value problems. In [83] special second-order nonlinearities are also dealt with. For recent developments see the references in Chapter 13.2.

C: High energy estimates:

$$\|u(t)\|_{s,2} \leq c\|u_0\|_{s,2} \exp\{c\int_0^t \|u(r)\|_{2,\infty}^\alpha dr\}, \quad t \in [0,T],\ c = c(s).$$

We shall give a proof for the case $\alpha = 1$ in order to point out where the second assumption on the nonlinearity F, (11.137), plays a role.

F is written as

$$F(u, \nabla u) = F^0(w)u + \sum_{j=1}^n F_j^1(w)\partial_j u, \ \ w := (u, \nabla u),$$

where

$$\begin{aligned} F^0(w) &:= \int_0^1 \frac{\partial F(rw)}{\partial u}\, dr, \\ F_j^1(w) &:= \int_0^1 \frac{\partial F(rw)}{\partial(\partial_j u)}\, dr, \quad j = 1, \ldots, n. \end{aligned}$$

The first assumption on F, (11.136), implies

$$F^0(0) = F_j^1(0) = 0, \quad j = 1, ..., n.$$

Let $\beta \in \mathbb{N}_0^n, 0 \leq |\beta| \leq s$. We have for the local solution the identity

$$\frac{1}{2}\frac{d}{dt}\|\nabla^\beta u(t)\|_2^2 = \mathrm{Re}\,\langle \nabla^\beta F(w), \nabla^\beta u\rangle(t)$$

which follows directly from the differential equation. Dropping the parameter t, we shall investigate the most difficult terms

$$\mathrm{Re}\,\langle \nabla^\beta (F_j^1(w)\partial_j u), \nabla^\beta u\rangle,$$

$j \in \{1, \ldots, n\}$ arbitrary but fixed in the sequel.

1. $\underline{|\beta| = 0}$:

$$\mathrm{Re}\,\langle F_j^1(w)\partial_j u, u\rangle = -\,\mathrm{Re}\,\langle (\partial_j F_j^1(w))u, u\rangle - \mathrm{Re}\,\langle F_j^1(w)u, \partial_j u\rangle.$$

The assumption (11.137) says that $F_j^1(w)$ is real which implies

$$\mathrm{Re}\,\langle F_j^1(w)\partial_j u, u\rangle = \mathrm{Re}\,\langle F_j^1(w)u, \partial_j u\rangle$$

and hence

$$\begin{aligned} \mathrm{Re}\,\langle F_j^1(w)\partial_j u, u\rangle &= -\frac{1}{2}\,\mathrm{Re}\,\langle (\partial_j F_j^1(w))u, u\rangle \\ &\leq c\,\|w\|_{1,\infty}\|u\|_2^2 \\ &\leq c\,\|u\|_{2,\infty}\|u\|_{s,2}^2. \end{aligned}$$

2. $\underline{0 < |\beta| \leq s-1}$:

$$\mathrm{Re}\,\langle \nabla^\beta (F_j^1(w)\partial_j u),\, \nabla^\beta u\rangle \;=\; \mathrm{Re}\,\langle F_j^1(w)\nabla^\beta \partial_j u,\, \nabla^\beta u\rangle$$

$$+\,\mathrm{Re}\,\langle \nabla^\beta (F_j^1(w)\partial_j u) \;-\; F_j^1(w)\nabla^\beta \partial_j u,\, \nabla^\beta u)\rangle \equiv R_1 + R_2.$$

In analogy to the case 1 ($|\beta| = 0$) we obtain

$$|R_1| \leq c\|u\|_{2,\infty}\|u\|^2_{|\beta|,2} \leq c\|u\|_{2,\infty}\|u\|^2_{s,2}.$$

The second term R_2 is estimated with the help of the commutator estimates in Lemma 4.9:

$$\begin{aligned}
|R_2| &\leq c(\|\nabla F_j^1(w)\|_\infty \|\nabla^{|\beta|-1}\partial_j u\|_2 + \|\nabla^{|\beta|}F_j^1(w)\|_2\|\partial_j u\|_\infty)\|\nabla^{|\beta|}u\|_2 \\
&\leq c\|u\|_{2,\infty}\|u\|^2_{|\beta|,2} + c\|(u,\nabla u)\|_{|\beta|,2}\|u\|_{1,\infty}\|u\|_{|\beta|,2} \\
&\leq c\|u\|_{2,\infty}\|u\|^2_{|\beta|,2} + c\|u\|_{2,\infty}\|u\|_{|\beta|+1,2}\|u\|_{|\beta|,2} \\
&\leq c\|u\|_{2,\infty}\|u\|^2_{s,2}.
\end{aligned}$$

3. $\underline{|\beta| = s}$:

The critical term in case 2 was $\|u\|_{|\beta|+1,2}$ which stemmed from $\mathrm{Re}\,\langle \nabla^s F(u,\nabla u), \nabla^s u\rangle$. According to the cases 1, 2 only the term with the highest derivatives is critical. This term has the form

$$\mathrm{Re}\,\langle a(u,\nabla u)\nabla^s\nabla u, \nabla^s u\rangle$$

with

$$a(u,\nabla u) \in \mathbb{R}$$

because of the assumption (11.137). Thus we obtain (as in case 1 where $s = 0$) that

$$\mathrm{Re}\,\langle a(u,\nabla u)\nabla^s\nabla u, \nabla^s u\rangle \leq c\|u\|_{2,\infty}\|u\|^2_{s,2}.$$

This proves the desired high energy estimate.

D: Weighted a priori estimates:

The following estimate is proved in the standard manner.

$$\sup_{0\leq t\leq T}(1+t)^{n/2(1-2/q)}\|u(t)\|_{s_1,q} \leq M_0 < \infty,$$

where M_0 is independent of T, provided

$$q = 2\alpha + 2,$$

$$\frac{1}{\alpha}(1+\frac{1}{\alpha}) < \frac{n}{2},$$

s_1 is sufficiently large and

$$\|u_0\|_{s,2} + \|u_0\|_{s,\frac{2\alpha+2}{2\alpha+1}}$$

is sufficiently small ($s > s_1$ being sufficiently large).

E: Final energy estimate:

$$\|u(t)\|_{s,2} \le K\|u_0\|_{s,2}, \quad 0 \le t \le T,$$

s being sufficiently large, u_0 being sufficiently small, K being independent of T (and u_0).

Summarizing, we obtain the following global existence theorem.

Theorem 11.13 *We assume (11.136) with $\frac{1}{\alpha}(1+\frac{1}{\alpha}) < \frac{n}{2}$ and (11.137). Then there exist an integer s_0 and a $\delta > 0$ such that the following holds:*
If u_0 belongs to $W^{s,2} \cap W^{s,p}$ with $s \ge s_0$ and $p = \frac{2\alpha+2}{2\alpha+1}$ and

$$\|u_0\|_{s,2} + \|u_0\|_{s,p} < \delta,$$

then there is a unique solution u of the initial value problem to the nonlinear Schrödinger equation (11.134), (11.135) with

$$u \in C^0([0,\infty), W^{s,2}) \cap C^1([0,\infty), W^{s-2,2}).$$

Moreover, we have

$$\|u(t)\|_\infty + \|u(t)\|_{2\alpha+2} = \mathcal{O}(t^{-\frac{n}{2}\frac{\alpha}{\alpha+1}}),$$

$$\|u(t)\|_{s,2} = \mathcal{O}(1) \quad as \quad t \to \infty.$$

Remark: The semilinear case where $F = F(u)$ does not depend on derivatives of u, has found a wide interest and there are already many more results; see for example the papers of Ginibre & Velo [38], Y. Tsutsumi [182] or the book of W. Strauss [179] and the references there.
In analogy to the wave equation (cf. Chapter 4 and the proof of Theorem 1.2 in Chapter 8) invariance properties of the Schrödinger equation have been investigated and optimal L^∞-decay rates have been obtained, see P. Constantin [22].

11.5 Klein–Gordon equations

(*Oskar Benjamin Klein*, 15.9.1894 – 5.2.1977; *Walter Gordon*, 3.8.1893 – 24.12.1939)

The equations of the type

$$u_{tt} - \Delta u + m\,u = f(u, Du, D\nabla u), \tag{11.144}$$

with initial values

$$u(t=0) = u_0, \quad u_t(t=0) = u_1, \tag{11.145}$$

for a real function $u = u(t,x),\ t \geq 0, x \in \mathbb{R}^n$,
where

$$m > 0 \quad \text{is a constant} \quad (\text{"mass"}),$$

and where f is smooth and satisfies

$$f(W) = \mathcal{O}(|\,W|^{\alpha+1}) \quad \text{as} \quad |W| \to 0, \qquad \text{for some} \quad \alpha \in \mathbb{N}, \tag{11.146}$$

are covered to a large extent by the considerations already made for the nonlinear wave equations — by which we denoted the corresponding equations with $m = 0$ — in the Chapters 5–8. Here we find better decay rates and the L^2-norm of $u(t)$ can be estimated easily (cf. the remarks in Chapter 8 for the case $f = f(y, \ldots)$ there), which leads to a global existence result with weaker assumptions on the relation between α and n.

Remarks: The nonlinear Klein–Gordon equation (11.144) is a relativistically invariant (in contrast to the Schrödinger equation (11.134)) equation describing the wave function of a particle with spin zero. Also, the only difference with the discussion on the nonlinear wave equation will be the derivation of better L^p–L^q-decay estimates.

The canonical transformation to a first-order system in t is given by

$$V := (u_t, \nabla u, \sqrt{m}\,u), \quad V^0 := (u_1, \nabla u_0, \sqrt{m}\,u_0).$$

Then V satisfies

$$V_t + AV = F(V), \quad V(t=0) = V^0,$$

where A is given elementarily and

$$F(V) = (f(u, Du, D\nabla u), 0, 0)'.$$

A: Decay for $F \equiv 0$:

Let u solve in $[0,\infty) \times \mathbb{R}^n$

$$u_{tt} - \Delta u + m\,u = 0, \quad u(t=0) = u_0,\, u_t(t=0) = u_1, \tag{11.147}$$

$$u_0, u_1 \in C_0^\infty(\mathbb{R}^n).$$

Then v defined by

$$v(t, x_1, \ldots, x_{n+1}) := e^{-i\sqrt{m}x_{n+1}} u(t, x_1, \ldots, x_n) \tag{11.148}$$

solves

$$\begin{aligned} v_{tt} - \Delta v &= 0, & (11.149) \\ v(0, x_1, \ldots, x_{n+1}) &= e^{-i\sqrt{m}x_{n+1}} u_0(x_1, \ldots, x_n), & (11.150) \\ v_t(0, x_1, \ldots, x_{n+1}) &= e^{-i\sqrt{m}x_{n+1}} u_1(x_1, \ldots, x_n) & (11.151) \end{aligned}$$

in $[0, \infty) \times \mathbb{R}^{n+1}$. Therefore, we shall first study decay rates for solutions to the linear wave equation and then we have to check how the factor $e^{-i\sqrt{m}x_{n+1}}$ affects the calculations.

The following brief estimates for $t \geq 1$ are taken from W. von Wahl's paper [187] and lead to another proof of Theorem 2.3 (which we proved in Chapter 2 even more elementarily for $n = 3$).

The two cases of odd respectively even space dimensions have to be discussed separately.

Case 1: n odd, $n \geq 3$. (For $n = 1$ cf. Chapter 2.)
Let $\varphi \in C^{\frac{n+1}{2}}(\mathbb{R}^n)$, $\psi \in C^{\frac{n-1}{2}}(\mathbb{R}^n)$ be compactly supported and let v solve

$$v_{tt} - \Delta v = 0 \quad \text{in} \quad [0, \infty) \times \mathbb{R}^n, \tag{11.152}$$

$$v(t = 0) = \varphi, \quad v_t(t = 0) = \psi. \tag{11.153}$$

Then v is given by the classical formula

$$\begin{aligned} v(t, x) &= \sum_{j=0}^{(n-3)/2} (j + 1) a_j t^j \left(\frac{\partial^j}{\partial t^j} Q_1 \right)(t, x) & (11.154) \\ &\quad + t \sum_{j=0}^{(n-3)/2} a_j t^j \left(\left(\frac{\partial^{j+1}}{\partial t^{j+1}} Q_1 \right)(t, x) + \left(\frac{\partial^j}{\partial t^j} Q_2 \right)(t, x) \right), \end{aligned}$$

where the coefficients $a_j, j = 0, \ldots, (n - 3)/2$, are constants depending only on n, and Q_1, Q_2 are given by

$$\begin{aligned} Q_1(t, x) &:= \frac{1}{\omega_n} \int_{S^{n-1}} \varphi(x + t\xi) d\xi, & (11.155) \\ Q_2(t, x) &:= \frac{1}{\omega_n} \int_{S^{n-1}} \psi(x + t\xi) d\xi, & (11.156) \end{aligned}$$

where ω_n denotes the surface measure of the unit sphere S^{n-1} in $\mathbb{R}^n$, (see [187] resp. Courant & Hilbert [23, pp. 681–691]).

Example: For $n = 3$ and $\varphi = 0$ one has $a_0 = 1$ and we obtain

$$v(t, x) = \frac{t}{4\pi} \int_{S^2} \psi(x + t\xi) d\xi.$$

This is Kirchhoff's formula which we used in Chapter 2.

Using the elementary formula

$$\int_{S^{n-1}} \xi_j f(t\xi)d\xi = t^{-(n-1)} \int_{K(0,t)} (\partial_j f)(x)dx$$

for $f \in C^1(K(0,t)), t > 0,\ K(0,t) = \{x \in \mathbb{R}^n \mid |x| \leq t\}$, we obtain for $j \geq 1$:

$$\begin{aligned}
\frac{\partial^j}{\partial t^j} Q_1(t,x) &= \frac{1}{\omega_n} \sum_{|\alpha|=j} \int_{S^{n-1}} (\nabla^\alpha \varphi)(x+t\xi)\xi^\alpha d\xi \\
&\leq \frac{1}{\omega_n} \sum_{|\alpha|=j} \sum_{|\beta|=1,\alpha\geq\beta} t^{-|\alpha-\beta|} \left| \int_{S^{n-1}} (\nabla^\alpha \varphi)(x+t\xi)(t\xi)^{\alpha-\beta}\xi^\beta d\xi \right| \\
&= \frac{1}{\omega_n} \sum_{|\alpha|=j,|\beta|=1,\alpha\geq\beta} t^{-|\alpha-\beta|-(n-1)} \left| \int_{K(0,t)} (\nabla^\alpha \nabla^\beta \varphi)(x+y)y^{\alpha-\beta}dy \right| \\
&\leq \frac{1}{\omega_n} \sum_{|\alpha|=j+1,|\gamma|=|\alpha|-2} t^{-(n-1)}t^{-|\gamma|} \left| \int_{K(0,t)} (\nabla^\alpha \varphi)(x+y)y^\gamma dy \right|,
\end{aligned}$$

analogously for Q_2.

Thus we get from (11.154)

$$\begin{aligned}
|v(t,x)| \leq\ & c\Big\{|Q_1(t,x)| \\
&+ \sum_{j=1}^{(n-3)/2} t^{j-(n-1)} \sum_{|\alpha|=j+1,|\gamma|=|\alpha|-2} t^{-|\gamma|}\{| \int_{K(0,t)} (\nabla^\alpha \varphi)(x+y)y^\gamma dy| \\
&+ t| \int_{K(0,t)} (\nabla^\alpha \psi)(x+y)y^\gamma dy|\} + t|Q_2(t,x)| \\
&+ \sum_{j=1}^{(n-3)/2} t^{j+1-(n-1)} \sum_{|\alpha|=j+2,|\gamma|=|\alpha|-2} t^{-|\gamma|} | \int_{K(0,t)} (\nabla^\alpha \varphi)(x+y)y^\gamma dy|\Big\},
\end{aligned} \tag{11.157}$$

where c denotes a constant which at most depends on n (the symbol c will also be used in the sequel).

Moreover we have

$$\begin{aligned}
|Q_1(t,x)| &\leq \frac{1}{t\omega_n} |\sum_{j=1}^n \int_{S^{n-1}} \varphi(x+t\xi)t\xi_j\xi_j d\xi| \\
&\leq \frac{c}{\omega_n} t^{-n} \sum_{0\leq|\alpha|\leq 1,\, 0\leq|\gamma|\leq 1} | \int_{K(0,t)} (\nabla^\alpha \varphi)(x+y)y^\gamma dy|,
\end{aligned} \tag{11.158}$$

analogously for Q_2.
In order to estimate the right-hand side of (11.157) we consider the typical term

$$t^{-|\gamma|} \int\limits_{K(0,t)} (\nabla^\alpha \varphi)(x+y) y^\gamma dy.$$

Let $\lambda, \mu \geq 0$ be such that

$$\mu < 1, \quad \lambda + \mu \leq 1.$$

Since

$$(1-\mu) + (1-\lambda) = 1 + (1-\lambda-\mu)$$

we may apply the convolution inequality (4.5) to get

$$\begin{aligned} & \|t^{-|\gamma|} \int\limits_{K(0,t)} (\nabla^\alpha \varphi)(\cdot + y) y^\gamma dy\|_{L^{1/(1-\lambda-\mu)}(\mathbf{R}^n)} \qquad (11.159) \\ & \leq \|\nabla^\alpha \varphi\|_{L^{1/(1-\mu)}(\mathbf{R}^n)} t^{-|\gamma|} \| |\cdot^\gamma| \|_{L^{1/(1-\lambda)} K(0,t)} \\ & \leq c\, t^{n(1-\lambda)} \|\nabla^\alpha \varphi\|_{L^{1/(1-\mu)}(\mathbf{R}^n)}. \end{aligned}$$

The inequalities (11.157) – (11.159) imply for $v(t) = v(t,\cdot)$ and $t \geq 1$:

$$\begin{aligned} \|v(t)\|_{L^{1/(1-\lambda-\mu)}(\mathbf{R}^n)} \leq\ & c\, t^{n(1-\lambda)} t^{-\frac{n-1}{2}} \{ t^{-\frac{n-1}{2}} \|\varphi\|_{L^{1/(1-\mu)}(\mathbf{R}^n)} \\ & + \sum_{1 \leq |\alpha| \leq (n-3)/2+2} \|\nabla^\alpha \varphi\|_{L^{1/(1-\mu)}(\mathbf{R}^n)} + t^{-\frac{n-3}{2}} \|\psi\|_{L^{1/(1-\mu)}(\mathbf{R}^n)} \\ & + \sum_{1 \leq |\alpha| \leq (n-3)/2+1} \|\nabla^\alpha \psi\|_{L^{1/(1-\mu)}(\mathbf{R}^n)} \}. \end{aligned}$$

In particular we have for $\mu = 0, \lambda = 1$:

$$\|v(t)\|_\infty \leq c\, t^{-\frac{n-1}{2}} \{ \|\varphi\|_{\frac{n+1}{2},1} + \|\psi\|_{\frac{n-1}{2},1} \}, \qquad (11.160)$$

which is the desired L^1–L^∞-decay estimate for $t \geq 1$ and for n odd.

Case 2: n even.
In this case the solution v of (11.152), (11.153), is explicitly given by

$$\begin{aligned} v(t,x) =\ & \sum_{j=0}^{(n-2)/2} (j+1) b_j t^j \left(\frac{\partial^j}{\partial t^j} G_1 \right)(t,x) \qquad (11.161) \\ & + t \sum_{j=0}^{(n-2)/2} b_j t^j \left(\left(\frac{\partial^{j+1}}{\partial t^{j+1}} G_1 \right)(t,x) + \left(\frac{\partial^j}{\partial t^j} G_2 \right)(t,x) \right), \end{aligned}$$

where the b_j, $j = 0, \ldots, (n-2)/2$, are constants depending only on n, and G_1, G_2 are given by

$$G_1(t,x) = \frac{1}{\omega_{n+1}} \int\limits_{S^n} \varphi(x_1 + t\xi_1, \ldots, x_n + t\xi_n) d\xi, \tag{11.162}$$

$$G_2(t,x) = \frac{1}{\omega_{n+1}} \int\limits_{S^n} \psi(x_1 + t\xi_1, \ldots, x_n + t\xi_n) d\xi. \tag{11.163}$$

φ, ψ are here assumed to satisfy $\varphi \in C^{\frac{n+2}{2}}(\mathbb{R}^n)$, $\psi \in C^{\frac{n}{2}}(\mathbb{R}^n)$, both having compact support.

Using the representations

$$G_1(t,x) = \frac{2\Gamma(\frac{n+1}{2})}{\sqrt{\pi}\Gamma(\frac{n}{2})t^{n-1}} \int\limits_0^t \frac{r^{n-1}}{\omega_n (t^2 - r^2)^{1/2}} \int\limits_{S^{n-1}} \varphi(x + r\xi) d\xi \, dr, \tag{11.164}$$

$$G_2(t,x) = \frac{2\Gamma(\frac{n+1}{2})}{\sqrt{\pi}\Gamma(\frac{n}{2})t^{n-1}} \int\limits_0^t \frac{r^{n-1}}{\omega_n (t^2 - r^2)^{1/2}} \int\limits_{S^{n-1}} \psi(x + r\xi) d\xi \, dr, \tag{11.165}$$

and dividing the integral $\int\limits_0^t \ldots$ into $\int\limits_0^{t-\varepsilon} \ldots + \int\limits_{t-\varepsilon}^t$ for $0 < \varepsilon < t$, one obtains after a lengthy but straightforward calculation similar to that in case 1 above for $t \geq 1$:

$$\|v(t)\|_\infty \leq c t^{-\frac{n-1}{2}} \{\|\varphi\|_{\frac{n+2}{2},1} + \|\psi\|_{\frac{n}{2},1}\}. \tag{11.166}$$

This is the desired L^1–L^∞-decay estimate for $t \geq 1$ and n even.

It remains to get estimates for $\|v(t)\|_\infty$ if $0 \leq t \leq 1$. Let $0 \leq t \leq 1$.

<u>Case 1</u>: n odd, $n \geq 3$.

Looking at the representations (11.154), (11.155), (11.156), we consider the typical term

$$t^k \int\limits_{S^{n-1}} (\nabla^\alpha h)(x + t\xi)\xi^\alpha \, d\xi,$$

where $|\alpha| = k - 1$, $0 \leq k \leq (n-3)/2$ for $h = \psi$ and $|\alpha| = k$, $0 \leq k \leq (n-3)/2 + 1$ for $h = \varphi$.

We have for $m \in \mathbb{N}_0$

$$|t^k \int\limits_{S^{n-1}} (\nabla^\alpha h)(x + t\xi)\xi^\alpha \, d\xi| = t^k \left| \int\limits_{S^{n-1}} \int\limits_t^\infty \frac{(s-t)^m}{m!} \frac{d^{m+1}}{ds^{m+1}} \nabla^\alpha h(x + s\xi)\xi^\alpha ds \, d\xi \right|$$

$$= \frac{t^k}{m!} \left| \int\limits_{S^{n-1}} \int\limits_t^\infty \frac{s^{n-1}}{s^{n-1}} \frac{(s-t)^m}{s^{|\alpha|+m+1}} (s\xi)^\alpha \underbrace{(s\xi) \cdot \ldots \cdot (s\xi)}_{(m+1) \text{ times}} (\underbrace{\nabla \cdot \ldots \cdot \nabla}_{m+1 \text{ times}} \nabla^\alpha h)(x + s\xi) ds \, d\xi \right|$$

$$\leq \frac{1}{m!} \int\limits_{|z|>t} \frac{t^k(|z|-t)^m |z|^{|\alpha|+m+1}}{|z|^{|\alpha|+m+n}} |(\underbrace{\nabla \cdot \ldots \cdot \nabla}_{m+1+|\alpha| \text{ times}} h)(x+z)| dz$$

$$\leq c \int\limits_{|z|>t} \frac{|z|^{m+k}}{|z|^{n-1}} |(\underbrace{\nabla \cdot \ldots \cdot \nabla}_{m+1+|\alpha| \text{ times}} h)(x+z)| dz.$$

Choosing $m := n-1-k$ we obtain

$$|t^k \int\limits_{S^{n-1}} (\nabla^\alpha h)(x+t\xi)\xi^\alpha \, d\xi| \leq c \begin{cases} \|h\|_{n-1,1}, & \text{if} \quad h=\psi, \\ \|h\|_{n,1}, & \text{if} \quad h=\varphi. \end{cases}$$

(Observe that $m \geq 0$ since $k \leq (n-3)/2+1$.)
Thus we have proved

$$\|v(t)\|_\infty \leq c\{\|\varphi\|_{n,1} + \|\psi\|_{n-1,1}\} \tag{11.167}$$

for $0 \leq t \leq 1$ and $\varphi \in C^n(\mathbb{R}^n), \psi \in C^{n-1}(\mathbb{R}^n)$ having compact support.

Case 2: n even.
Looking at the representations (11.161), (11.164), (11.165) we obtain in the same way as in case 1, using

$$\int\limits_0^t \left(\frac{r}{t}\right)^{n-1} \frac{dr}{\sqrt{t^2-r^2}} = \int\limits_0^1 \frac{s^{n-1}}{\sqrt{1-s^2}} ds < \infty,$$

the estimate

$$\|v(t)\|_\infty \leq c\,\{\|\varphi\|_{n,1} + \|\psi\|_{n-1,1}\} \tag{11.168}$$

for $0 \leq t \leq 1$ and $\varphi \in C^n(\mathbb{R}^n), \psi \in C^{n-1}(\mathbb{R}^n)$ having compact support.
Summarizing the estimates (11.160), (11.166), (11.167), (11.168) we have obtained the following L^1–L^∞-estimates for solutions of the linear wave equation

$$\|v(t)\|_\infty \leq c\,(1+t)^{-\frac{n-1}{2}} \{\|\varphi\|_{n,1} + \|\psi\|_{n-1,1}\}, \quad t \geq 0.$$

Now we consider a solution u of the linear Klein–Gordon equation in $[0,\infty) \times \mathbb{R}^n$ with initial values $u(t=0) = u_0$, $u_t(t=0) = u_1$, i.e. u is a solution to (11.147). v defined by (11.148) solves the linear wave equation (11.149) in $[0,\infty) \times \mathbb{R}^{n+1}$ with initial values (11.150), (11.151).
Since the factor $e^{-i\sqrt{m}x_{n+1}}$ disturbs the integrability properties of the initial data one has to show directly that this factor does not affect the estimates above. This is carried out in detail in [187] for $t \geq 1$ and goes along the lines of the considerations above. (11.160), (11.166) then lead to the following estimate for $t \geq 1$:

$$\|u(t)\|_\infty \leq c\,t^{-n/2} \left\{\|u_0\|_{\frac{N_1+1}{2},1} + \|u_1\|_{\frac{N_1-1}{2},1}\right\}, \tag{11.169}$$

where

$$N_1 := \begin{cases} n+2, & \text{if } n \text{ is odd}, \\ n+1, & \text{if } n \text{ is even}. \end{cases}$$

For $0 \le t \le 1$ we use the representation for v in (11.154) resp. (11.161) and the reduction argument for reducing an integral over S^{k+1}, $k \in \mathbb{N}$, to an integral over S^k, described in the formulae (11.162), (11.163), (11.164), (11.165), and we obtain

$$\|u(t)\|_\infty \le c \left\{\|u_0\|_{n,1} + \|u_1\|_{n-1,1}\right\}. \tag{11.170}$$

Combining the formulae (11.169), (11.170) we have found the following L^1–L^∞-estimate for solutions to the linear Klein–Gordon equation:

$$\|u(t)\|_\infty \le c\,(1+t)^{-n/2} \left\{\|u_0\|_{N_2,1} + \|u_1\|_{N_2-1,1}\right\} \tag{11.171}$$

where

$$N_2 := \begin{cases} n & \text{if } n \ge 2, \\ 2 & \text{if } n = 1. \end{cases} \tag{11.172}$$

Remarks: Further more detailed estimates may be found in the papers by S. Klainerman [89, 92], J. Shatah [158] and T.C. Sideris [173].
u_t is also a solution of the linear Klein–Gordon equation with initial values

$$u_t(t=0) = u_1, \quad u_{tt}(t=0) = \Delta u_0 - m u_0\,;$$

hence we obtain

$$\|u_t(t)\|_\infty \le c\,(1+t)^{-n/2} \left\{\|u_1\|_{N_2,1} + \|u_0\|_{N_2+1,1}\right\}. \tag{11.173}$$

Analogously, for ∇u with initial values

$$\nabla u(t=0) = \nabla u_0,\ \nabla u_t(t=0) = \nabla u_1,$$

we have

$$\|\nabla u(t)\|_\infty \le c(1+t)^{-n/2} \left\{\|u_0\|_{N_2+1,1} + \|u_1\|_{N_2,1}\right\}. \tag{11.174}$$

Summarizing (11.171), (11.173), (11.174) we obtain the following L^1–L^∞-decay estimate for $V = (u_t, \nabla u, u)$ with $V(t=0) = V^0 = (u_1, \nabla u_0, u_0)$:

$$\|V(t)\|_\infty \le c(1+t)^{-n/2} \|V^0\|_{N,1}, \quad t \ge 0, \tag{11.175}$$

where

$$N := \begin{cases} n & \text{if } n \ge 2, \\ 2 & \text{if } n = 1. \end{cases} \tag{11.176}$$

The corresponding L^2–L^2-estimate for V easily follows by multiplication of the differential equation for u with u_t in $L^2(\mathbb{R}^n)$ leading to

$$\frac{d}{dt}\left\{\|u_t(t)\|_2^2 + \|\nabla u(t)\|_2^2 + m\|u(t)\|_2^2\right\} = 0,$$

whence we conclude

$$\|V(t)\|_2 = \|V^0\|_2, \quad t \geq 0. \tag{11.177}$$

The estimates (11.175) and (11.177) give us the L^p–L^q-decay estimates by interpolation:

$$\|V(t)\|_q \leq c(1+t)^{-\frac{n}{2}(1-\frac{2}{q})}\|V^0\|_{N_p,p}, \quad t \geq 0,\, c = c(q,n),$$

where $2 \leq q \leq \infty$, $1/p + 1/q = 1$ and $N_p > N(1-2/q)$, N being defined in (11.176) ($N_p = N(1-2/q)$ if $q \in \{2,\infty\}$).

B: Local existence and uniqueness:

Transforming the nonlinear Klein–Gordon equation (11.144) in the standard way to a first-order symmetric hyperbolic system (cf. Chapters 3, 5) for V, we obtain a unique local solution by Theorem 5.8,

$$V \in C^0([0,T], W^{s,2}) \cap C^1([0,T], W^{s-1,2})$$

for some $T > 0$, if $s > n/2 + 1$ and $V^0 \in W^{s,2}$.
The following Steps **C**–**E** are analogous to those for the wave equation. We obtain

C: High energy estimates:

$$\|V(t)\|_{s,2} \leq C\|V^0\|_{s,2} \exp\left\{C\int_0^t \|\overline{D}V(r)\|_\infty^\alpha\, dr\right\}, \quad t \in [0,T],\, C = C(s).$$

D: Weighted a priori estimates:

$$\sup_{0\leq t\leq T}(1+t)^{\frac{n}{2}(1-\frac{2}{q})}\|V(t,\cdot)\|_{s_1,q} \leq M_0 < \infty,$$

where M_0 is independent of T, provided

$$q = 2\alpha + 2,$$

$$\frac{1}{\alpha}\left(1 + \frac{1}{\alpha}\right) < \frac{n}{2},$$

s_1 is sufficiently large and

$$\|V^0\|_{s,2} + \|V^0\|_{s,\frac{2\alpha+2}{2\alpha+1}}$$

is sufficiently small ($s > s_1$ being sufficiently large).

E: Final energy estimate:

$$\|V(t,\cdot)\|_{s,2} \leq K\|V^0\|_{s,2}, \quad 0 \leq t \leq T,$$

s being sufficiently large, V^0 being sufficiently small, K being independent of T (and V^0).
Altogether we obtain the following global existence theorem.

Theorem 11.14 *We assume (11.146) with* $\frac{1}{\alpha}(1+\frac{1}{\alpha}) < \frac{n}{2}$. *Then there exist an integer* $s_0 > \frac{n}{2}+1$ *and a* $\delta > 0$ *such that the following holds:*
If $V^0 = (u_1, \nabla u_0, \sqrt{m}\, u_0)$ *belongs to* $W^{s,2} \cap W^{s,p}$ *with* $s \geq s_0$ *and* $p = \frac{2\alpha+2}{2\alpha+1}$ *and*

$$\|V^0\|_{s,2} + \|V^0\|_{s,p} < \delta,$$

then there is a unique solution u of the initial value problem to the nonlinear Klein–Gordon equation (11.144), (11.145) with

$$(u_t, \nabla u, u) \in C^0([0,\infty), W^{s,2}) \cap C^1([0,\infty), W^{s-1,2}).$$

Moreover, we have

$$\begin{aligned} \|(u_t, \nabla u, u)(t)\|_\infty + \|(u_t, \nabla u, u)(t)\|_{2\alpha+2} &= \mathcal{O}(t^{-\frac{n}{2}\frac{\alpha}{\alpha+1}}), \\ \|(u_t, \nabla u, u)(t)\|_{s,2} &= \mathcal{O}(1) \quad as \quad t \to \infty. \end{aligned}$$

In analogy to the situation for the wave equation ("$m = 0$") one can use the special invariance properties of $\partial_t^2 - \Delta$ to improve the foregoing result, namely it is possible to obtain global solutions for the cases $\alpha = 1$ and $n = 3$ or $n = 4$ (while Theorem 11.14 would require $n > 4$); see S. Klainerman [89]. This may also be obtained with different methods, see the papers by J. Shatah [159] and T.C. Sideris [173]. (For the former see also the remarks in Chapter 9.)

11.6 Maxwell equations

(James Clerk Maxwell, 13.6.1831 – 5.11.1879)
The Maxwell equations in $\mathbb{R}^3$ are given by

$$\partial_t D - \nabla \times H = 0, \tag{11.178}$$

$$\partial_t B + \nabla \times E = 0. \tag{11.179}$$

Additionally one has the initial conditions

$$D(t=0) = D^0, \quad B(t=0) = B^0, \tag{11.180}$$

and the restriction

$$\nabla' D = 0, \quad \nabla' B = 0. \tag{11.181}$$

The following notation is used:

D: dielectric displacement, $D = (D_1, D_2, D_3) = D(t,x) \in \mathbb{R}^3, t \geq 0, x \in \mathbb{R}^3$,

H: magnetic field, $H = (H_1, H_2, H_3) = H(t,x)$,

B: magnetic induction, $B = (B_1, B_2, B_3) = B(t,x)$,

E: electric field, $E = (E_1, E_2, E_3) = E(t,x)$.

Remark: D ist not used for $(\partial_t, \nabla)'$ in this section.
Electric and magnetic currents are assumed to be zero. Then (11.178) – (11.181) describe electro-magnetic waves without damping, i.e. the electric conductivity is assumed to be zero.
The unknown fields are

$$D = \varepsilon(E) \quad \text{and} \quad B = \mu(H)$$

(respectively $E = \varepsilon^{-1}(D)$ and $H = \mu^{-1}(B)$), where we assume that

$$\varepsilon, \mu : \quad \mathbb{R}^3 \longrightarrow \mathbb{R}^3$$

are smooth bijections and the derivatives $\partial\varepsilon/\partial E$, $\partial\mu/\partial H$ are uniformly positive definite with respect to their arguments in each compact set.
The main assumptions on the nonlinearities are

$$\varepsilon(E) = \varepsilon_0 E + \mathcal{O}(|E|^3) \quad \text{as} \quad |E| \to 0 \tag{11.182}$$

and

$$\mu(H) = \mu_0 H + \mathcal{O}(|H|^3) \quad \text{as} \quad |H| \to 0, \tag{11.183}$$

where ε_0, μ_0 are positive constants (ε_0: dielectric constant; μ_0: permeability).
The inverse relation

$$\varepsilon^{-1}(D) = \varepsilon_0^{-1} D + \mathcal{O}(|D|^3) \quad \text{as} \quad |D| \to 0 \tag{11.184}$$

and

$$\mu^{-1}(B) = \mu_0^{-1} B + \mathcal{O}(|B|^3) \quad \text{as} \quad |B| \to 0 \tag{11.185}$$

follows immediately. Thus we may write (11.178), (11.179) in the following form:

$$D_t - \mu_0^{-1} \nabla \times B = \nabla \times F_1(B),$$

$$B_t + \varepsilon_0^{-1} \nabla \times D = \nabla \times F_2(D),$$

where F_1, F_2 are smooth vector-valued functions. The relations (11.184), (11.185) imply

$$F_j(W) = \mathcal{O}(|W|^3) \quad \text{as} \quad |W| \to 0, \quad j = 1, 2.$$

Without loss of generality we assume

$$\varepsilon_0 = \mu_0 = 1.$$

A: Decay for $F_1, F_2 = 0$:

From the linearized initial value problem

$$D_t - \nabla \times B = 0, \tag{11.186}$$

$$B_t + \nabla \times D = 0, \tag{11.187}$$

$$D(t=0) = D^0,\ B(t=0) = B^0,$$

$$\nabla' D = 0,\ \nabla' B = 0, \tag{11.188}$$

we obtain by differentiation

$$\begin{aligned} D_{tt} + \nabla \times \nabla \times D &= 0, \\ B_{tt} + \nabla \times \nabla \times B &= 0. \end{aligned}$$

Using the formula

$$\Delta = \nabla\nabla' - \nabla \times \nabla \times$$

and (11.188) we obtain the equations

$$\begin{aligned} D_{tt} - \Delta D &= 0, \\ B_{tt} - \Delta B &= 0. \end{aligned}$$

Therefore we can apply the same technique as in Chapter 2 (cf. also Section 11.4) leading to

$$\|(D,B)(t)\|_\infty \le c(1+t)^{-1}\|(D^0,B^0)\|_{3,1}, \quad t \ge 0, \tag{11.189}$$

where c is independent of t.
In contrast to Section 2 it is now easy to get an L^2–L^2-estimate for D and B. Multiplying both sides of (11.186) with D in L^2 and both sides of (11.187) with B in L^2 we end up with

$$\begin{aligned} \frac{1}{2}\frac{d}{dt}\|D(t)\|_2^2 &- \langle \nabla \times B, D\rangle = 0, \\ \frac{1}{2}\frac{d}{dt}\|B(t)\|_2^2 &+ \langle \nabla \times D, B\rangle = 0. \end{aligned}$$

Adding the last two equations we get

$$\|(D,B)(t)\|_2 = \|(D^0,B^0)\|_2, \quad t \ge 0. \tag{11.190}$$

By interpolation we obtain from (11.189), (11.190)

$$\|(D,B)(t)\|_q \le c(1+t)^{-(1-2/q)}\|(D^0,B^0)\|_{N_p,p}, \tag{11.191}$$

$$t \ge 0,\ c = c(q), \quad \text{where} \quad 2 \le q \le \infty,\ 1/p + 1/q = 1 \quad \text{and}$$

$$N_p > 3(1-2/q) \quad (N_p = 3(1-2/q) \quad \text{if} \quad q \in \{2,\infty\}).$$

Further estimates similar to those obtained for solutions of the wave equation using invariance properties (cf. Chapter 8) are proved by D. Christodoulou and S. Klainerman in [19].

B: Local existence and uniqueness:
We want to apply Theorem 5.8. For this purpose we write the equations (11.178), (11.179) in the form

$$\frac{\partial \varepsilon(E)}{\partial E}\partial_t E - \nabla \times H = 0, \tag{11.192}$$

$$\frac{\partial \mu(H)}{\partial H}\partial_t H + \nabla \times E = 0. \tag{11.193}$$

Introducing

$$U := (E, H),$$

$$A^0(U) := \begin{pmatrix} \partial\varepsilon/\partial E & 0 \\ 0 & \partial\mu/\partial H \end{pmatrix},$$

$$A^1 := \begin{pmatrix} 0 & 0 & 0 & 0 & 0 & 0 \\ 0 & 0 & 0 & 0 & 0 & 1 \\ 0 & 0 & 0 & 0 & -1 & 0 \\ 0 & 0 & 0 & 0 & 0 & 0 \\ 0 & 0 & -1 & 0 & 0 & 0 \\ 0 & 1 & 0 & 0 & 0 & 0 \end{pmatrix},$$

$$A^2 := \begin{pmatrix} 0 & 0 & 0 & 0 & 0 & -1 \\ 0 & 0 & 0 & 0 & 0 & 0 \\ 0 & 0 & 0 & 1 & 0 & 0 \\ 0 & 0 & 1 & 0 & 0 & 0 \\ 0 & 0 & 0 & 0 & 0 & 0 \\ -1 & 0 & 0 & 0 & 0 & 0 \end{pmatrix},$$

and

$$A^3 := \begin{pmatrix} 0 & 0 & 0 & 0 & 1 & 0 \\ 0 & 0 & 0 & -1 & 0 & 0 \\ 0 & 0 & 0 & 0 & 0 & 0 \\ 0 & -1 & 0 & 0 & 0 & 0 \\ 1 & 0 & 0 & 0 & 0 & 0 \\ 0 & 0 & 0 & 0 & 0 & 0 \end{pmatrix},$$

we may rewrite (11.192), (11.193) as

$$A^0(U)\partial_t U + \sum_{j=1}^{3} A^j \partial_j U = 0. \tag{11.194}$$

The initial condition for U is given by

$$U(t=0) = (\varepsilon^{-1}(D^0),\ \mu^{-1}(H^0)). \tag{11.195}$$

Theorem 5.8 can be applied to the initial value problem (11.194), (11.195).
Using the assumption that (E,H) and (D,B) are mapped into each other by smooth bijections we conclude that there is a unique local solution

$$V := \ (D,B) \in C^0([0,T), W^{s,2}) \cap C^1([0,T], W^{s-1,2})$$

for some $T > 0$, if $s > 3/2 + 1$, i.e. if $s \geq 3$, and $V^0 \in W^{s,2}$. The relation

$$\nabla' D(t,x) = 0, \quad \nabla' B(t,x) = 0$$

follows for all $t \in [0,T]$, $x \in \mathbb{R}^3$ from the differential equations for D and B which yield

$$\partial_t \nabla' D = \partial_t \nabla' B = 0$$

(provided $\nabla' D^0 = \nabla' B^0 = 0$, which is of course assumed to assure compatibility).
In analogy to the considerations made for nonlinear wave equations we obtain successively:

C: High energy estimates:

$$\|V(t)\|_{s,2} \leq c\|V^0\|_{s,2} \exp\{c \int_0^t \|V(r)\|_{1,\infty}^2\, dr\}, \quad t \in [0,T], c = c(s),$$

where

$$V^0 := V(t=0) = (D^0, B^0).$$

D: Weighted a priori estimates:

$$\sup_{0\le t\le t}(1+t)^{2/3}\|V(t)\|_{s_1,6}\le M_0<\infty,$$

where M_0 is independent of T, provided s_1 is sufficiently large and

$$\|V^0\|_{s_0,2}+\|V^0\|_{s_0,6/5}$$

is sufficiently small ($s_0>s_1$ being sufficiently large).

E: Final energy estimate:

$$\|V(t)\|_{s,2}\le K\|V^0\|_{s,2},\quad 0\le t\le T,$$

s being sufficiently large, V^0 being sufficiently small, K being independent of T (and V^0).

Remarks: In the Steps **D**, **E** a representation of V of the type

$$V(t)=W(t)V^0+\int_0^t W(t-r)F(V)(r)dr$$

is used (cf. the beginning of Chapter 11 and Chapters 7,8). The estimates for the linear part, (11.191), can be used for the integrand since the nonlinearity $F=(\nabla\times F_1,\nabla\times F_2)$ has divergence zero too: $\nabla'(\nabla\times F_j)=0, j=1,2$. Details may be found in the paper of F. Klaus [95].
Summarizing we have obtained the following global existence theorem.

Theorem 11.15 *We assume (11.182), (11.183). Then there exist an integer $s_0\ge 3$ and a $\delta>0$ such that the following holds:*
If (D^0,B^0) belongs to $W^{s,2}\cap W^{s,6/5}$ with $s\ge s_0$ and

$$\|(D^0,B^0)\|_{s,2}+\|(D^0,B^0)\|_{s,6/5}<\delta,$$

and if

$$\nabla' D_0=\nabla' B_0=0,$$

then there is a unique solution of the initial value problem to the nonlinear Maxwell equations (11.178) – (11.181) with

$$(D,B)\in C^0([0,\infty),W^{s,2})\cap C^1([0,\infty),W^{s-1,2}).$$

Moreover, we have

$$\begin{aligned}\|(D,B)(t)\|_\infty+\|(D,B)(t)\|_6 &= \mathcal{O}(t^{-2/3}),\\ \|(D,B)(t)\|_{s,2} &= \mathcal{O}(1)\quad \text{as}\quad t\to\infty.\end{aligned}$$

Remark: Initially anisotropic models, replacing $\varepsilon_0 E$ in (11.182) by $\varepsilon(0)E$, where $\varepsilon(0)$ is a diagonal matrix but not necessarily a multiple of the identity, have been studied by O. Liess [107, 108] in the context of crystal optics. There phenomena appear similar to those discovered for the equations of elasticity in the initially cubic case (in $\mathbb{R}^2$), cf. Subsection 11.1.2; the decay rate which could be proved up to now is weaker than that in the initially isotropic case.

11.7 Plate equations

In this section we are concerned with perturbations of the linear plate equation

$$y_{tt} + \Delta^2 y = 0$$

of the type

$$y_{tt} + \Delta^2 y = f(y_t, \nabla^2 y) + \sum_{i=1}^{n} b_i(y_t, \nabla^2 y)\partial_i y_t \tag{11.196}$$

with smooth nonlinear functions f and b_i, $i = 1, \ldots, n$.
$y = y(t, x)$ is a real function of $t > 0$ and $x \in \mathbb{R}^n$ with prescribed initial values

$$y(t = 0) = y_0, \quad y_t(t = 0) = y_1. \tag{11.197}$$

The assumption on the nonlinearities near zero will be

$$f(W) = \mathcal{O}(|W|^{\alpha+1}),\ b_i(W) = \mathcal{O}(|W|^{\alpha}),\quad i = 1, \ldots, n \text{ as } |W| \longrightarrow 0, \tag{11.198}$$

$$\text{for some} \quad \alpha \in \mathbb{N}.$$

The obvious difference in the equation (11.196) in comparison to those in the Sections 11.1 – 11.3, 11.5, 11.6 is that the nonlinearity does not contain the highest-order derivatives. This is a similar situation to that in Section 11.4 where we discussed Schrödinger equations. (Actually there is a close relation between the linear Schrödinger equation and the linear plate equation, see below.) The reason for only admitting semilinear nonlinearities is that the proof of the energy estimates given in [198] requires that the nonlinear terms on the right-hand side of equation (11.196) have to be controlled by those on the left-hand side. For example, a term like

$$\sum_{i,j,k=1}^{n} a_{ijk}(y_t, \nabla^2 y)\partial_i\partial_j\partial_k y$$

cannot be controlled there. Partial integrations in the typical L^2-inner products with this term generate derivatives of y_t because the term is not symmetric; but derivatives of y_t cannot be handled with the information one gets from the left-hand side.

Remarks: This difficulty with third-order terms already arises in proving a local existence theorem. We mention that there is a *local* existence theorem by P. Lesky which admits certain symmetric fourth-order nonlinearities, see [100] and also one by W. v. Wahl for the nonlinear clamped plate [188, 189]. The system (11.196), (11.197) has been discussed by F. Willems in [198] where those details which are omitted in the sequel may be found.

The transformation to a first-order system is given as follows. Let

$$V := (\partial_t y, \nabla^2 y).$$

With the notation

$$\delta_{ij,k\ell} := \left\{ \begin{array}{ll} 1 & \text{if} \quad (i,j) = (k,\ell) \\ 0 & \text{if} \quad (i,j) \neq (k,\ell) \end{array} \right\}, \qquad i,j,k,\ell = 1,\dots,n,$$

and $m := n^2 + 1$ we define the $m \times m$-matrices A_{ij} and $B_i = B_i(V)$ by

$$A_{ij} := \begin{pmatrix} 0 & (-\delta_{ij,k\ell})_{k\ell} \\ (\delta_{ij,k\ell})_{k\ell} & 0 \end{pmatrix}$$

and

$$B_i := \begin{pmatrix} b_i & 0 & \cdots & 0 \\ 0 & & & \\ \vdots & & 0 & \\ 0 & & & \end{pmatrix}$$

for $i,j = 1,\dots,n$.

Then V satisfies

$$V_t = \sum_{i,j=1}^{n} A_{ij}\partial_i\partial_j V + \sum_{i=1}^{n} B_i(V)\partial_i V + \tilde{f}(V), \tag{11.199}$$

and

$$V(t=0) = V^0 := (y_1, \nabla^2 y_0),$$

where

$$\tilde{f}(V) := (f(V), 0).$$

Defining

$$A(V) := -\sum_{i,j=1}^{n} A_{ij}\partial_i\partial_j - \sum_{i=1}^{n} B_i(V)\partial_i$$

and

$$A := A(0)$$

we obtain from (11.199) that V satisfies

$$V_t + AV = F(V), \quad V(t=0) = V^0, \tag{11.200}$$

where

$$F(V) := \tilde{f}(V) + (A(0) - A(V))\,V,$$

and the assumption (11.198) implies that F satisfies

$$F(V) = \mathcal{O}(|V|^{\alpha+1}) \quad \text{as} \quad |V| \longrightarrow 0. \tag{11.201}$$

Hence we have the desired general structure again and we shall continue going through the Steps **A**–**E** of the general scheme.

A: Decay for $F \equiv 0$:
Let y solve

$$y_{tt} + \Delta^2 y = 0, \quad y(t=0) = y_0, \quad y_t(t=0) = y_1.$$

We shall first obtain decay rates for y_t and Δy.
Let

$$w := y_t + i\Delta y.$$

Then w satisfies

$$w_t - i\Delta w = 0$$

with initial value

$$w(t=0) = y_1 + i\Delta y_0,$$

i.e. w is a solution of the linear Schrödinger equation, which we discussed in Section 11.4. The decay of w expressed in formula (11.143) yields

$$\|(y_t, \Delta y)(t)\|_q \leq c\,(1+t)^{-\frac{n}{2}(1-\frac{2}{q})}\|(y_1, \Delta y_0)\|_{N_p,p}, \tag{11.202}$$

$t \geq 0$, $c = c(q)$, where $2 \leq q \leq \infty$, $1/p + 1/q = 1$, and $N_p > n(1 - 2/q)$.
It is also possible to use a Fourier representation in order to obtain decay rates for $\|(y_t, \Delta y)(t)\|_q$, $2 \leq q < \infty$, similarly to the procedure in Section 11.1.2. Let $y_0 = 0$, $h := y_1$. Then

$$\begin{aligned} y_t(t,x) &= \mathcal{F}^{-1}\left(\cos(|\cdot|^2 t)\hat{h}(\cdot)\right)(x), \\ \Delta y(t,x) &= \mathcal{F}^{-1}\left(\sin(|\cdot|^2 t)\hat{h}(\cdot)\right)(x), \end{aligned}$$

i.e. it is sufficient to study

$$u(t,x) := \mathcal{F}^{-1}\left(e^{i|\cdot|^2 t}\hat{h}(\cdot)\right)(x). \tag{11.203}$$

We have the following version of Theorem 2.2 from H. Pecher's paper [138] (see also Lemma 11.3 in Section 11.1.2).

Lemma 11.16 *Let $\gamma \geq 0$, $m \in \mathbb{N}$ and*

$$\varrho := \begin{cases} n-1 & \text{if} \quad m = 1, \\ n & \text{if} \quad m \geq 2. \end{cases}$$

Then there is a constant $c > 0$ such that for all $v \in C_0^\infty(\mathbb{R}^n)$ and all $t > 0$ the estimate

$$\|\mathcal{F}^{-1}\left(\frac{e^{it|\cdot|^m}}{|\cdot|^{2m\gamma}}(\mathcal{F}v)(\cdot)\right)\|_q \leq ct^{-\frac{n}{m}(\frac{1}{p}-\frac{1}{q})+2\gamma}\|v\|_p$$

holds, provided

$$1 < p \leq 2 \leq q < \infty,\ 1/p + 1/q = 1,\ 1/p - 1/q \geq 2m\gamma/n,$$

$$(1/p - 1/2)(2n - m\varrho) \leq 2m\gamma.$$

(c only depends on q, n, m and γ.)

An application of Lemma 11.16 to n defined in (11.203), with $\gamma = s$ (to be defined below) and $m = 2$, leads to the following estimates:

$$\begin{aligned} \|u(t,\cdot)\|_q &= \|\mathcal{F}^{-1}\left(\frac{e^{it|\cdot|^2}}{|\cdot|^{4s}}|\cdot|^{4s}\hat{h}(\cdot)\right)\|_q \\ &= \|\mathcal{F}^{-1}\left(\frac{e^{it|\cdot|^2}}{|\cdot|^{4s}}\mathcal{F}(\Delta^{2s}h)(\cdot)\right)\|_q \\ &\leq ct^{-\frac{n}{2}(\frac{1}{p}-\frac{1}{q})+2s}\|\Delta^{2s}h\|_p \\ &\leq ct^{-\frac{n}{2}(\frac{1}{p}-\frac{1}{q})+2s}\|h\|_{N,p}, \end{aligned}$$

where

$$4s \leq N < 4s + 1.$$

The application of the last Lemma is possible if s satisfies

$$\frac{1}{4}\left(\frac{1}{p} - \frac{1}{q}\right)(2n - 2\varrho) \leq s \leq \frac{n}{4}\left(\frac{1}{p} - \frac{1}{q}\right).$$

The left-hand side of the last inequality equals zero by the definition of ϱ. For $t \geq 1$ we choose the smallest s which is possible, $s := 0$. This implies

$$\|u(t)\|_q \leq ct^{-\frac{n}{2}(1-\frac{2}{q})}\|h\|_p. \tag{11.204}$$

For $0 \leq t \leq 1$ we choose the largest s which is possible,

$$s := \frac{n}{4}\left(\frac{1}{p} - \frac{1}{q}\right),$$

which leads to

$$\|u(t)\|_q \leq c\|h\|_{N,p} \tag{11.205}$$

with

$$n\left(1 - \frac{2}{q}\right) \leq N < n\left(1 - \frac{2}{q}\right) + 1.$$

Combining (11.204) and (11.205) we obtain

$$\|(y_t, \Delta y)(t)\|_q \leq c(1+t)^{-\frac{n}{2}(1-\frac{2}{q})}\|h\|_{N,p}. \tag{11.206}$$

($y(t=0)$ is assumed to be zero, $h = y_t(t=0)$.)

Remarks: Observe that (11.206) has been obtained now only for q which is strictly less than infinity. Using Lemma 11.16 it would also be possible to estimate $\|y(t)\|_q$, with further restrictions on q.

There is also a representation of y in terms of Fresnel integrals which is obtained from the Fourier representation

$$y(t,x) = \mathcal{F}^{-1}\left(\frac{\sin(|\cdot|^2 t)}{|\cdot|^2}(\mathcal{F}h)(\cdot)\right)(t,x).$$

With

$$f(t,\xi) := \frac{1}{\sqrt{2\pi}^n}\frac{\sin(|\xi|^2 t)}{|\xi|^2}$$

we may write y as

$$y(t,x) = \mathcal{F}^{-1}\left[f(t,\cdot)\right](h(x-\cdot)),$$

where $[\ldots]$ denotes the distribution generated by $\ldots$ (applied to $h(x-\cdot) \in C_0^\infty(\mathbb{R}^n)$).

Let $n = 3$.

With the Fresnel integral W defined by

$$W(z) := \frac{1}{\sqrt{2\pi}^3}\int_0^z (\cos s^2 - \sin s^2)ds$$

we obtain for $t > 0$:

$$\mathcal{F}^{-1}\left[f(t,\cdot)\right] = \left[\frac{1}{|\cdot|}W\left(\frac{|\cdot|}{2\sqrt{t}}\right)\right]$$

(cf. R. Leis [98], F. Willems [198]); hence we have

$$y(t,x) = \frac{1}{\sqrt{2\pi}^3}\int_{\mathbb{R}^3} h(x-z)\frac{1}{|z|}\int_0^{\frac{|z|}{2\sqrt{t}}}(\cos s^2 - \sin s^2)ds\,dz. \tag{11.207}$$

Remark: The regularity of h should be asssumed appropriately such that the integrals in (11.207) converge, e.g. $h \in C_0^\infty$ as above.
From (11.207) we obtain

$$\Delta y(t,x) = -\frac{1}{4}\frac{1}{\sqrt{2\pi}^3} t^{-3/2} \int_{\mathbf{R}^3} h(x-z)\left(\cos(\frac{|z|^2}{4t}) + \sin(\frac{|z|^2}{4t})\right) dz$$

whence we conclude

$$\|\Delta y(t)\|_\infty \le \frac{1}{2}\frac{1}{\sqrt{2\pi}^3} t^{-3/2}\|h\|_1, \quad t > 0. \tag{11.208}$$

We also get from (11.207)

$$\begin{aligned}
\Delta y(t,x) &= \int_{\mathbf{R}^3} (\Delta h)(x-z)\frac{1}{|z|}W(\frac{|z|}{2\sqrt{t}})dz \\
&= 4t \int_{\mathbf{R}^3} (\Delta h)(x - 2\sqrt{t}z)\frac{1}{|z|}W(|z|)dz \\
&= 4t \int_0^\infty r\, W(r) \int_{S^2} (\Delta h)(x - 2\sqrt{t}rw)dw\, dr \\
&= 4t \int_0^\infty rW(r) \int_{S^2} \{-\int_{2\sqrt{t}r}^\infty \frac{d}{ds}(\Delta h)(x - sw)ds\}dw\, dr \\
&= 4t \int_0^\infty r\, W(r) \int_{2\sqrt{t}r}^\infty \int_{S^2} (\nabla\Delta h)(x - sw)w dw\, ds\, dr.
\end{aligned}$$

This implies

$$\begin{aligned}
\|\Delta y(t)\| &\le c\,\|h\|_{3,1} \int_0^\infty \frac{|W(r)|}{r}\, dr \qquad (11.209) \\
&\le c\,\|h\|_{3,1}, \quad (t \ge 0),
\end{aligned}$$

because there exists $c, r_0 > 0$ such that $|W(r)| \le c\,r$ if $r \le r_0$ and $W(r) \le cr^{-1}$ if $r \ge r_0$ (cf. [98, p. 226]).
With (11.208) and (11.209) we have an L^1–L^∞-estimate for Δy

$$\|\Delta y(t)\|_\infty \le c(1+t)^{-3/2}\|h\|_{3,1}, \quad t \ge 0. \tag{11.210}$$

From (11.207) we obtain

$$y_t(t,x) = -\frac{1}{4}\frac{1}{\sqrt{2\pi}^3}\, t^{-3/2} \int_{\mathbb{R}^3} h(x-y)\left(\cos\left(\frac{|z|^2}{4t}\right) - \sin\left(\frac{|z|^2}{4t}\right)\right) dz.$$

This implies

$$\|y_t(t)\|_\infty \le \frac{1}{2\sqrt{2\pi}^3}\, t^{-3/2}\|h\|_{3,1} \quad (t>0). \tag{11.211}$$

For $0 \le t \le 1$ we have the coarse estimate

$$\begin{aligned} \|y_t(t)\|_\infty &= \|\mathcal{F}^{-1}\left(\cos(|\cdot|^2 t)(\mathcal{F}h)(\cdot)\right)\|_\infty \qquad (11.212)\\ &\le \frac{1}{\sqrt{2\pi}^3}\|\cos(|\cdot|^2 t)(\mathcal{F}h)(\cdot)\|_1\\ &\le \frac{1}{\sqrt{2\pi}^3}\|(1+|\cdot|^2)^m(\mathcal{F}h)(\cdot)\|_\infty\|(1+|\cdot|^2)^{-m}\|_1\\ &\le \frac{c}{\sqrt{2\pi}^3}\|\mathcal{F}\left((1+\Delta)^m h\right)\|_\infty \quad \text{if} \quad m>3/2\\ &\le c\,\|h\|_{2m,1}\\ &\le c\,\|h\|_{4,1}. \end{aligned}$$

(The last estimate is called coarse because the known optimal order of derivatives (known from (11.202)) appearing on the right-hand side should be three instead of four.)
By (11.211) and (11.212) we have an L^1–L^∞-estimate for y_t

$$\|y_t(t)\|_\infty \le c(1+t)^{-3/2}\|h\|_{4,1}. \tag{11.213}$$

The two L^1–L^∞-estimates for Δy and y_t in (11.210) and (11.213) respectively can now be combined with the following L^2–L^2-estimates

$$\begin{aligned} \|y_t(t)\|_2 &= \|\mathcal{F}^{-1}\left(\cos(|\cdot|^2 t)(\mathcal{F}h)(\cdot)\right)\|_2 \qquad (11.214)\\ &= \|\cos(|\cdot|^2 t)(\mathcal{F}h)(\cdot)\|_2 \le \|h\|_2,\\ \|\Delta y(t,\cdot)\|_2 &= \|\mathcal{F}^{-1}\left(\sin(|\cdot|^2 t)\mathcal{F}h(\cdot)\right)\|_2 \qquad (11.215)\\ &= \|\sin(|\cdot|^2 t)(\mathcal{F}h)(\cdot)\|_2 \le \|h\|_2, \end{aligned}$$

to an L^p–L^q-decay estimate for $1 \le p \le q \le \infty$ as usual by interpolation.
Remark: The estimates (11.214), (11.215) follow without using the Fourier transform from the equations

$$y_{tt} + \Delta^2 y = 0, \quad y(t=0) = y_0(=0), \quad y_t(t=0) = y_1$$

by multiplication with y_t in L^2 which yields

$$\forall t \geq 0: \quad \|y_t(t)\|_2^2 + \|\Delta y(t)\|_2^2 = \|y_1\|_2^2 + \|\Delta y_0\|_2^2.$$

The inequality

$$\|\nabla^2 w\|_q \leq c\,\|\Delta w\|_q$$

which is valid for all $w \in W^{2,q}$, $2 \leq q < \infty$, with a constant $c = c(n,q) > 0$ (compare Dunford & Schwartz [29, pp. 1044ff] for the necessary tools), together with (11.206), finally yields the L^p–L^q-decay estimate for $V = (y_t, \nabla^2 y)$ (with $y(t=0) = 0$, hence $V^0 = (y_1, 0)$):

$$\|V(t)\|_q \leq c(1+t)^{-\frac{n}{2}(1-\frac{2}{q})}\|\,V^0\|_{N_p,p}, \quad t \geq 0,\, c = c(n,q),$$

where $2 \leq q < \infty$, $1/p + 1/q = 1$, and $n(1-2/q) \leq N_p < n(1-2/q)+1$.

B: Local existence and uniqueness:
The existence of a unique solution V to the nonlinear system (11.200) or, equivalently, to

$$V_t + A(V)V = \tilde{f}(V), \quad V(t=0) = V^0,$$

follows from the general results on nonlinear evolution equations given by T. Kato, see [82]. For fixed $w \in W^{s,2}$ the operator $-A(w)$, canonically defined on its domain $D\left(A(w)\right) \subset L^2$, is the generator of a so-called C_0-semigroups of type $(1,\beta)$, i.e. the operator norm of the corresponding semigroup can be bounded as follows:

$$\|e^{-tA(w)}\| \leq e^{\beta t},$$

for some $\beta \geq 0$, for all $t \geq 0$. To show that the general assumptions from [82] are satisfied will be omitted here, see [198] for details.
The result is the existence of a unique, local solution

$$V \in C^0([0,T], W^{s,2}) \cap C^1([0,T], W^{s-2,2})$$

for some $T > 0$, if $s > n/2 + 1$ and $V^0 \in W^{s,2}$.
Remark: There is also a local existence theorem contained in the paper [100] by P. Lesky. His approach works for a larger class of nonlinearities admitting certain fourth-order nonlinearities, but no third-order nonlinearities. W. von Wahl gives a local existence theorem for the nonlinear clamped plate in [188, 189].
The following Steps **C**–**E** now are obtained in the same way as the corresponding ones for the wave equation in the Chapters 6–8. We summarize:

C: High energy estimates:

$$\|V(t)\|_{s,2} \leq C\|V^0\|_{s,2} \exp\{C\int_0^t \|V(\tau)\|_{1,\infty}^{\alpha} d\tau\}, \quad t \in [0,T],\, C = C(s).$$

D: Weighted a priori estimates:

$$\sup_{0\leq t\leq T} (1+t)^{\frac{n}{2}(1-\frac{2}{q})}\|V(t)\|_{s_1,q} \leq M_0 < \infty,$$

where M_0 is independent of T, provided

$$q = 2\alpha + 2,$$

$$\frac{1}{\alpha}(1+\frac{1}{\alpha}) < \frac{n}{2},$$

s_1 is sufficiently large and

$$\|V^0\|_{s,2} + \|V^0\|_{s,\frac{2\alpha+2}{2\alpha+1}}$$

is sufficiently small ($s > s_1$ being sufficiently large).

E: Final energy estimate:

$$\|V(t)\|_{s,2} \leq K\|V^0\|_{s,2}, \quad 0 \leq t \leq T,$$

s being sufficiently large, V^0 being sufficiently small, K being independent of T (and V^0).
Summarizing we obtain the following global existence theorem.

Theorem 11.17 *We assume (11.198) with* $\frac{1}{\alpha}(1+\frac{1}{\alpha}) < \frac{n}{2}$. *Then there exist an integer* $s_0 > \frac{n}{2}+1$ *and a* $\delta > 0$ *such that the following holds:*
If $V^0 = (y_1, \nabla^2 y_0)$ *belongs to* $W^{s,2} \cap W^{s,p}$ *with* $s \geq s_0$ *and* $p = \frac{2\alpha+2}{2\alpha+1}$ *and*

$$\|V^0\|_{s,2} + \|V^0\|_{s,p} < \delta,$$

then there is a unique solution y *of the initial value problem to the nonlinear plate equation (11.196), (11.197) with*

$$(y_t, \nabla^2 y) \in C^0([0,\infty), W^{s,2}) \cap C^1([0,\infty), W^{s-2,2}).$$

Moreover, we have

$$\begin{aligned} \|(y_t,\nabla^2 y)(t,\cdot)\|_\infty + \|(y_t,\nabla^2 y)(t,\cdot)\|_{2\alpha+2} &= \mathcal{O}(t^{-\frac{n}{2}\frac{\alpha}{\alpha+1}}), \\ \|(y_t,\nabla^2 y)(t,\cdot)\|_{s,2} &= \mathcal{O}(1) \quad \text{as} \quad t\to\infty. \end{aligned}$$

Remark: Concerning the regularity assumptions on the coefficients f and $b_i, i = 1, \ldots, n$ in the equation (11.196) the same holds as for the systems discussed before. The smoothness assumption $f, b_i \in C^\infty$ can be weakened, namely in this case to the requirement

$$f \in C^{s+1}, \quad b_i \in C^s$$

(cf. [198]).

12 Further aspects and questions

We have considered a set of initial value problems of different kinds and we have found common structures and common starting points for proving global existence theorems for small data. It is natural to carry over the methods to other evolution equations from mathematical physics or from other branches of the applied sciences. We shall not pursue this here. It is also obvious that in view of the breadth of the subject of nonlinear evolution equations we have only dealt with a very specific part. In this section we wish to point out some related questions and current research problems.

We start with looking at initial-*boundary* value problems, where x varies in an exterior domain Ω or an interior domain Ω. Then boundary conditions have to be prescribed for the unknown function V on the boundary $\partial\Omega$. The difficulties arising from the presence of $\partial\Omega$ are enormous. This starts with the simple fact that differentiating the equation with respect to x is not compatible with the boundary conditions in general. Also we do not have explicit representation formulae at hand (unless Ω has special symmetries) preventing us from carrying over the linear part, Step **A** in the general scheme, directly. In order to illustrate these problems a little bit and to illustrate some of the methods which lead to related results we shall give a short outline of some ideas involved in discussing boundary value problems.

Exterior domains.

Let $\Omega \subset \mathbb{R}^n$ be an exterior domain, i.e. for $n \geq 2$, Ω is a domain with non-empty, bounded complement, and for $n = 1$, $\Omega := (0, \infty)$. Let $\partial\Omega$ be smooth.

There are results concerning special equations, e.g. for the equations of heat-conductive, compressible, viscous fluids in three space dimensions by Matsumura & Nishida [120] or for the incompressible Navier–Stokes equations by H. Iwashita [54], who also proves L^p–L^q-estimates for the linearized problem (*Claude Louis Marie Henri Navier,* 15.2.1785 – 23.8.1836; *George Gabriel Stokes,* 13.8.1819 – 1.2.1903). General results e.g. for fully nonlinear wave equations have been obtained by Y. Shibata [163] and by Shibata & Tsutsumi [167].

In order to obtain decay rates for solutions to the associated linearized problems we mention two methods.

<u>The approach using the Laplace transform.</u>

This method applies the Laplace transform with respect to t, discusses the resolvent of the resulting stationary equations in detail and then uses the information just obtained in the inverse transform. As an example we consider the system

$$\begin{aligned} u_{tt} - \Delta u &= 0 \quad \text{in} \quad \mathbb{R} \times \Omega, \\ u &= 0 \quad \text{in} \quad \mathbb{R} \times \partial\Omega, \end{aligned}$$

$$u(t=0) = 0,\ u_t(t=0) = u_1 \quad \text{in } \Omega.$$

If

$$\tilde{u}(k,\cdot) = \int_0^\infty e^{-ikt} u(t,\cdot)dt,$$

then

$$(\Delta + k^2)\tilde{u}(k,\cdot) = -u_1(\cdot),$$

hence

$$\tilde{u}(k,x) = -(R(k^2)u_1)(x),$$

where $R(k^2)$ denotes the resolvent $(\Delta + k^2)^{-1}$. At first $R(k^2)$ is only defined for $k^2 \in \mathbb{C} \setminus [0,\infty)$. u is then given by

$$u(t,x) = \frac{1}{2\pi} \int_{-\infty-ic}^{\infty-ic} e^{ikt} \tilde{u}(k,x)\, \mathrm{d}k, \quad c > 0 \text{ arbitrary},$$

or

$$u(t,x) = -\frac{1}{2\pi} \int_{-\infty-ic}^{\infty-ic} e^{ikt} (R(k^2)u_1)(x)\, \mathrm{d}k.$$

The asymptotic behavior of u as $t \longrightarrow \infty$ can be described if the behavior of $R(k^2)$ near $k = 0$ and for $|k| \longrightarrow \infty$ is known well enough. B.R. Vainberg [184, 185] proved that $R(k^2)$ can be holomorphically extended to $k^2 \in [0,\infty)$ as an operator from L^2-functions with compact support in $\mathbb{R}^n$ into $H^2_{loc}(\Omega)$ ($\equiv$ space of functions u which are locally in $W^{2,2}(\Omega)$ and for which $e^{|\cdot|}\nabla^\beta u(\cdot)$ is in $L^2(\Omega)$ for $|\beta| \leq 2$). Moreover, $R(\cdot)$ can be estimated as follows.

1. $|k| \longrightarrow \infty$:
   $|||R(k^2)||| \sim |k|^{-1}$ (operator norm in L^2_{loc}) if Ω is "non-trapping" (see the definition below).

2. $|k| \longrightarrow 0$:
   $R(k^2)$ can be developed in a Laurent series. (*Pierre Alphonse Laurent,* 18.7.1813 – 2.9.1854.)

This leads to a decay rate for the local L^2-norm of Du (local energy decay) and can be combined with cut-off techniques to a global L^p–L^q-estimate, see Shibata & Tsutsumi [167]. We remark that the minimal rate of decay is determined through the part 2: "$|k| \longrightarrow 0$" where the condition "Ω non-trapping" is not needed.
An (exterior) domain Ω is called *non-trapping* if the following holds.

$$\forall a > 0 \quad \exists T = T(\Omega, a) > 0 \quad \forall u_1 \in L^2_a(\Omega) : u \in C^\infty([T,\infty) \times \overline{\Omega}_a),$$

where

$$L_a^2(\Omega) := \left\{ f \in L^2(\Omega) \mid \operatorname{supp} f \subset \overline{\Omega}_a \, (= \overline{\Omega \cap B(0,a)}) \right\}.$$

Remark: The convexity of $\mathbb{R}^n \setminus \Omega$ implies that Ω is non-trapping, see R.B. Melrose [122], K. Yamamoto [201]. For further geometrical interpretations, like "all rays which hit $\partial\Omega$ and which propagate according to the laws of geometrical optics move away from $\partial\Omega$ in finite time; no ray is trapped, even not asymptotically", see Morawetz, Ralston & Strauss [131].

The advantage of the approach above consists of its great generality in applications, e.g. for damped or for undamped problems, for self-adjoint or non-self-adjoint problems; see [16, 30, 59, 167, 183]. For damped problems the assumption "Ω is non-trapping" is not needed. A small disadvantage consists in the complexity of the arguments used for the study of $R(k^2)$. Moreover one usually obtains a local energy decay, not directly global L^p–L^q-estimates.

For damped problems and star-shaped obstacles $\mathbb{R}^n \setminus \Omega$ we present a simple method using generalized eigenfunctions which is also directly applicable to operators with variable coefficients.

Ansatz via generalized eigenfunctions.

We observe that the decay of solutions to the heat equation $u_t - \Delta u = 0$ in $\mathbb{R}^n$ can easily be obtained by using the Fourier transform and exploiting the fact that the kernel of this transform, namely $(2\pi)^{-n/2} e^{ix\xi}$, is uniformly bounded with respect to x and ξ. This leads to the following ansatz.

Let $\Omega \subset \mathbb{R}^n$ be an exterior domain, $n \geq 3$, with smooth boundary $\partial\Omega$, and let

$$\begin{aligned} A &: D(A) \subset L^2(\Omega) \longrightarrow L^2(\Omega), \\ D(A) &:= \{ v \in W_0^{1,2}(\Omega) \mid \sum_{m,k=1}^{n} \partial_m a_{mk}(\cdot) \partial_k v(\cdot) \in L^2(\Omega) \}, \end{aligned}$$

$Av(\cdot) := -\sum_{m,k=1}^{n} \partial_m a_{mk}(\cdot)\partial_k v(\cdot)$ where $a_{mk} = a_{km}$ is a real-valued, smooth function of $x \in \overline{\Omega}, a_{mk}(x) = \delta_{mk}$ for $|x| > r_0$, for some fixed $r_0 > 0$, with $\partial\Omega \subset B(0, r_0), m, k = 1, \ldots, n$. We assume

$$\forall x \in \overline{\Omega} \quad \forall \xi \in \mathbb{R}^n : \quad \sum_{m,k=1}^{n} a_{mk}(x) \xi_m \xi_k \geq a_0 |\xi|^2,$$

with some fixed constant $a_0 > 0$.

It is well-known that there is a generalized eigenfunction expansion (also called generalized Fourier transform) $\mathcal{F}_+ : L^2(\Omega) \longrightarrow L^2(\mathbb{R}^n)$, $\mathcal{F}_+$ being unitary, with the property

$$\mathcal{F}_+(\varphi(A)w)(\xi) = \varphi(|\xi|^2)(\mathcal{F}_+ w)(\xi) \tag{12.1}$$

for functions $\varphi(A)$ of A defined by the spectral theorem for the self-adjoint operator A (see R. Leis [98], C.H. Wilcox [197]).
Moreover, we have

$$(\mathcal{F}_+ w)(\xi) = \int_\Omega \overline{\psi(x,\xi)} w(x) dx \equiv \hat{w}(\xi)$$

and

$$(\mathcal{F}_+^{-1} \hat{w})(x) = \int_{\mathbb{R}^n} \psi(x,\xi) \hat{w}(\xi) d\xi.$$

The kernel ψ is uniquely determined by the following conditions (12.2) – (12.6).
Let $j \in C^\infty(\mathbb{R}^n), j \geq 0,\ j(r) = 0$ for $r \leq r_1$, and $j(r) = (2\pi)^{-n/2}$ for $r \geq r_1 + 1$, where $r_1 > r_0$ is fixed. Then

$$\psi(x,\xi) = j(|x|)\, e^{ix\xi} + \psi'(x,\xi), \tag{12.2}$$

$$\forall \xi \in \mathbb{R}^n \,:\, (1 - j(|\cdot|))\, \psi(\cdot,\xi) \in D(A), \tag{12.3}$$

$$\forall \xi \in \mathbb{R}^n \quad \forall r > 0 \,:\, j(|\cdot|)\psi(\cdot,\xi) \in W^{2,2}(\Omega_r), \tag{12.4}$$

$$\forall \xi \in \mathbb{R}^n \,:\, \sum_{m,k=1}^{n} \left(\partial_m a_{mk}(\cdot)\partial_k + |\xi|^2\right) \psi'(\cdot,\xi) = -\left(\Delta + |\xi|^2\right)\left(j(|\cdot|)e^{i\cdot\xi}\right), \tag{12.5}$$

$$\forall \xi \in \mathbb{R}^n \,:\, \psi'(\cdot,\xi) \quad \text{satisfies the outgoing radiation condition:} \tag{12.6}$$

$$\begin{aligned} \frac{\partial \psi'(x,\xi)}{\partial |x|} - i|\xi|\psi'(x,\xi) &= \mathcal{O}(|x|^{-(n+1)/2}), \\ \psi'(x,\xi) &= \mathcal{O}(|x|^{-(n-1)/2}). \end{aligned}$$

Remarks: Such a generalized eigenfunction expansion was given first for the Schrödinger operator in $\mathbb{R}^3$ by T. Ikebe [52]. Later on this was extended to higher dimensions and to perturbations of the Laplace operator, also for exterior domains; see the papers of S. Agmon [3], Alsholm & Schmidt [5], A. Majda [113], K. Mochizuki [127], N.A. Shenk [160] and Shenk & Thoe [161].
For the proof of the existence of ψ' one can use the *principle of limiting absorption*, see [98, 197]. This principle holds for a larger class of operators e.g. for certain Maxwell operators. In view of (12.5) $\psi(\cdot,\xi)$ is called a generalized eigenfunction and the name generalized eigenfunction expansion is justified. In view of (12.1) and the remarks at the beginning we are interested in pointwise estimates on ψ and ψ' respectively.
If $Av = -\Delta v$, i.e. $a_{mk} = \delta_{mk}$, and $\mathbb{R}^n \setminus \Omega$ is star-shaped, $n \geq 3$, we have

$$\exists m \in \mathbb{N} \quad \exists c > 0 \quad \forall x \in \overline{\Omega} \quad \forall \xi \in \mathbb{R}^n \,:\, |\psi(x,\xi)| \leq c(1+|\xi|)^m, \tag{12.7}$$

see Morawetz & Ludwig [130] and [153]. Using (12.1), the factor $(1+|\xi|)^m$ appearing in (12.7) will finally turn into a differentiation of the initial data. As an application one can prove a global existence theorem for small data to the following fourth-order nonlinear parabolic initial-boundary value problem (see [153]).

$$\begin{aligned} u_t + \Delta^2 u &= f(u, \nabla u, \nabla^2 u, \nabla^3 u, \nabla^4 u) \quad \text{in} \quad [0,\infty) \times \Omega, \\ u = \Delta u &= 0 \quad \text{in} \quad [0,\infty) \times \partial\Omega, \\ u(t=0) &= u_0 \quad \text{in} \quad \Omega, \end{aligned}$$

where

$$f \quad \text{is smooth,} \quad f(w) = \mathcal{O}(|w|^2) \quad \text{near} \quad w = 0,$$

$$\mathbb{R}^n \setminus \Omega \quad \text{is star-shaped,} \quad n > 4.$$

In the case that the operator A really has variable coefficients one has the following. If $\mathbb{R}^3 \setminus \Omega$ is star-shaped and if

$$\min_{x \in \overline{\Omega}} \left(2 \min_{|\xi|=1} \sum_{m,k=1}^{3} a_{mk}(x)\xi_m\xi_k - 9 \max_{m,k=1,2,3} |\nabla a_{mk}(x) \cdot x| \right) > 0 \tag{12.8}$$

holds, then we have

$$\forall r > 0 \quad \exists c > 0 \quad \forall x \in \overline{\Omega}_r \quad \forall \xi \in \mathbb{R}^3 : |\psi(x,\xi)| \leq c(1 + |\xi|^2), \tag{12.9}$$

see C.O. Bloom [13] and [146]. The local character (with respect to x) of the estimate (12.9) still leads to global L^p–L^q-estimates since it allows one to prove a local energy decay result which is sufficiently strong to combine it with the corresponding initial value problem (here $A = -\Delta$ in $\mathbb{R}^3$) using cut-off functions.

Remark: The approach of Vainberg should directly lead to a removal of the star-shapedness assumption.

As an application of (12.9) one can prove a global existence theorem for small data to the following non-homogeneous second-order damped wave equation (see [147]).

$$\begin{aligned} u_{tt} - \sum_{m,k=1}^{3} \partial_m a_{mk} \partial_k u \ + \ u_t &= h(u, u_t, \nabla u, \nabla u_t, \nabla^2 u) \quad \text{in } [0,\infty) \times \Omega, \\ u &= 0 \quad \text{in} \quad [0,\infty) \times \partial\Omega, \\ u(t=0) &= u_0, \ u_t(t=0) = u_1 \quad \text{in} \quad \Omega, \end{aligned}$$

where

$$h \quad \text{is smooth,} \quad h(w) = \mathcal{O}(|w|^3) \quad \text{near } w = 0,$$

$$\mathbb{R}^3 \setminus \Omega \quad \text{is star-shaped,}$$

$$(a_{mk})_{mk} \quad \text{satisfies} \quad (12.8).$$

We remark that the method of Vainberg also works for problems with variable coefficients; see Iwashita & Shibata [55] for systems of second-order.

Interior domains.

We notice that the decay of solutions to the linearized system was essential in the proofs of the global existence theorems in exterior domains including $\mathbb{R}^n$. This is not always given in bounded domains. For heat equations the decay is even stronger, namely exponentially, but for wave equations with Dirichlet or Neumann boundary conditions there is no decay at all, but oscillations appear. Therefore we shall concentrate in considering the following nonlinear wave equation for a real-valued function $u = u(t,x)$:

$$u_{tt} - \Delta u = f(u, u_t, \nabla u, \nabla u_t, \nabla^2 u) \quad \text{in } \mathbb{R} \times \Omega,$$
$$u(t=0) = u_0, \; u_t(t=0) = u_1 \qquad \text{in } \Omega,$$
$$u|_{\partial\Omega} = 0 \quad \text{or} \quad \left.\frac{\partial u}{\partial \nu}\right|_{\partial\Omega} = 0 \quad (\nu : \text{ outer normal}),$$
$$f(w) = \mathcal{O}(|w|^{\alpha+1}) \quad \text{near } w = 0, \quad \text{for some } \alpha \in \mathbb{N}.$$

$\Omega \subset \mathbb{R}^n$ is bounded with smooth boundary $\partial\Omega$ and f is assumed to be smooth.
One known result on the formation of singularities for bounded domains is that of Klainerman & Majda [93] in one space dimension for the initial-boundary value problem

$$u_{tt} = (K(u_x))_x, \tag{12.10}$$

$$u(t=0) = \varepsilon\phi, \quad u_t(t=0) = \varepsilon\psi, \tag{12.11}$$

x varying in the bounded interval $[0, L]$, $L > 0$, with boundary conditions

$$u(t,0) = u(t,L) = 0, \tag{12.12}$$

or

$$u_x(t,0) = u_x(t,L) = 0. \tag{12.13}$$

Let $K'(0) = 1$ and let $\alpha \in \mathbb{N}$ be the first integer with

$$K^{(\alpha+1)}(0) \neq 0.$$

Then there is a constant $C = C(\phi, \psi)$ and an $\varepsilon_0 > 0$ such that a C^2-solution of (12.10) – (12.12) develops singularities at the time $T = C\varepsilon^{-\alpha}$, provided $\varepsilon < \varepsilon_0$. If K is an odd function the same conclusion holds in case that we replace the Dirichlet condition (12.12) by the Neumann condition (12.13).
This includes the equation for a nonlinear vibrating string where

$$K(u_x) = \frac{u_x}{\sqrt{1+u_x^2}}.$$

On the other hand, if the boundary conditions are of dissipative type, namely

$$K(u_x(t,0)) - \tau u_t(t,0) = 0, \quad u_t(t,L) = 0, \tag{12.14}$$

$0 < \tau < \infty$ fixed, then there is again a global solution for small data as was shown by Greenberg & Li [40]; see also Alber & Cooper [4], and Shibata & Zheng [169] for a corresponding result in higher dimensions.
We remark that the existence of global small solutions to nonlinear wave equations in bounded domains is also known if there is a damping term appearing in the equations, i.e. if a term "cu_t", with a positive constant $c > 0$, is added to the left-hand side of the differential equation for u, cf. Y. Shibata [162, 163].
In order to illustrate the effect that boundary conditions a priori might have we recall the fact that solutions to the nonlinear parabolic equation

$$u_t - \Delta u = u^2 \tag{12.15}$$

in a bounded domain $\Omega \subset \mathbb{R}^n$ will exist globally for a small initial value $u(t=0) = u_0$ in the case of the Dirichlet boundary condition, cf. Zheng [205], but the solution in general blows up in finite time in case of the Neumann boundary condition which can easily be seen from studying

$$v(t) := \int_\Omega u(t,x)\,dx.$$

Namely, v satisfies

$$\frac{d}{dt}v \geq Mv^2, \quad \text{with } M = (\text{volume}(\Omega))^{-1}.$$

Hence there is a blow-up if $v_0 := v(0) > 0$ as t approaches $(v_0 M)^{-1}$, cf. Zheng [204].
The formation of singularities can also be seen directly from the example

$$u(t,x) := (T_0 - t)^{-1}, \quad T_0 > 0 \text{ given.}$$

u solves (12.15) with Neumann boundary condition in any space dimension and blows up as t approaches T_0 although $u(t=0)$ is small if T_0 is large. This example is of course very special and connected with the fact that constants are solutions of the (linear) stationary Neumann problem while the first blow-up observation holds for rather arbitrary data (but satisfying $v(0) > 0$, a requirement also being related to the constant function).
The last two examples can be carried over to the wave equation too, namely for $\alpha \in \mathbb{N}$ the function

$$u(t,x) := \frac{\sqrt[\alpha]{\frac{4}{\alpha^2} + \frac{2}{\alpha}}}{\sqrt[\alpha]{(t-T_0)^2}}, \quad T_0 > 0 \text{ given,}$$

(x varying in a bounded domain $\Omega \subset \mathbb{R}^n$, $n \in \mathbb{N}$ arbitrary), solves

$$u_{tt} - \Delta u = u^{\alpha+1}$$

with Neumann boundary condition and with data $u(t=0)$, $u_t(t=0)$ which are small for large T_0. u is smooth as long as t is less than T_0, and u tends to infinity as t approaches T_0.

Also the former example can be carried over to nonlinear wave equations with Neumann boundary condition. For $\alpha \in \mathbb{N}$, let u be a C^2-solution to

$$u_{tt} - \Delta u = u^{\alpha+1}$$

with initial values

$$u(t=0) = u_0, \quad u_t(t=0) = u_1,$$

satisfying the Neumann boundary condition

$$\left.\frac{\partial u}{\partial \nu}\right|_{\partial\Omega} = 0.$$

We assume that the following holds:

$$\beta := \int_\Omega u_0 > 0, \quad \gamma := \int_\Omega u_1 > 0. \tag{12.16}$$

Then v, with

$$v(t) := \int_\Omega u(t,x)\,dx,$$

satisfies

$$v'' \geq M v^{\alpha+1}, \quad \text{with } M = M(\text{volume}(\Omega), \alpha). \tag{12.17}$$

From the assumption (12.16) it follows that

$$v' > 0 \quad \text{and} \quad v > 0.$$

Now we conclude from (12.17) that

$$\frac{1}{2}v'(t)^2 - \frac{M}{\alpha+2}v(t)^{\alpha+2} \geq \frac{1}{2}v'(0)^2 - \frac{M}{\alpha+2}v(0)^{\alpha+2} \equiv P.$$

Thus we conclude that

$$v'(t) \geq \sqrt{2\frac{M}{\alpha+2}v(t)^{\alpha+2} + 2P}.$$

Hence if u is a solution on $[0,T]$ then necessarily

$$T = \int_0^{v(T)} \frac{dw}{\sqrt{2\frac{M}{\alpha+2}w^{\alpha+2} + 2P}} \leq \int_0^\infty \frac{dw}{\sqrt{2\frac{M}{\alpha+2}w^{\alpha+2} + 2P}} < \infty.$$

(See R.T. Glassey [39], Payne & Sattinger [137] and Zheng & Chen [207] for further examples of blow-up for semilinear wave equations.)

The last two examples show that neither the magnitude of n nor the magnitude of the degree of vanishing of the nonlinearity near zero ($= \alpha + 1$) in connection with the smallness of the data may prevent a local smooth solution from developing singularities in finite time.

In contrast to the parabolic problem the study of the Dirichlet problem for the wave equation is expected to lead to similar blow-up phenomena. To show this we shall present fully nonlinear blow-up examples for radially symmetric solutions both for Dirichlet and Neumann boundary conditions. The radial symmetry will reduce the space dimension to one and we shall apply the above mentioned result of Klainerman & Majda [93]. This is not an immediate consequence for arbitrary wave equations since the equation

$$u_{tt} - \Delta u = 0$$

turns into

$$\bar{u}_{tt} - \bar{u}_{rr} - \frac{n-1}{r}\bar{u}_r = 0, \tag{12.18}$$

and this is not in conservation form that would be needed to carry over the arguments of the proof in [93]. Moreover the trick of using a periodic extension (being antisymmetric for the Dirichlet case resp. symmetric for the Neumann case) of a solution in $\Omega = (0, L)$ to $\Omega = \mathbb{R}^1$ does not work because of the term $\bar{u}_r/r$ appearing in (12.18). Hence we shall apply an appropriate transformation first.
Here is the precise statement of the result. Set $\Omega_3 := \{x \in \mathbb{R}^3 \mid 1 < |x| < 2\}$.

Theorem 12.1 *For every $\alpha \in \mathbb{N}$ there are (smooth) nonlinearities $f = f(x, u, \nabla u, \nabla^2 u)$, $f(x, w) = \mathcal{O}(|w|^{\alpha+1})$ near $w = 0$, uniformly in x, such that there is no global C^2-solution to the initial-boundary value problem*

$$u_{tt} - \Delta u = f(\cdot, u, \nabla u, \nabla^2 u) \quad \textit{in } \mathbb{R} \times \Omega_3, \tag{12.19}$$

$$u(t=0) = \varepsilon\phi, \quad u_t(t=0) = \varepsilon\psi \quad \textit{(radially symmetric) in } \Omega_3, \tag{12.20}$$

$$u|_{\partial\Omega_3} = 0, \tag{12.21}$$

$\varepsilon > 0$ small, no matter how smooth the data ϕ and ψ are or how small ε is or how large α is. Namely, there is a constant $C = C(\phi, \psi) > 0$ and an $\varepsilon_0 = \varepsilon_0(\phi, \psi) > 0$ such that the solution develops a singularity as t approaches $C\varepsilon^{-\alpha}$, provided $\varepsilon < \varepsilon_0$.
The same conclusion holds under the boundary condition

$$\left.\frac{\partial u}{\partial \nu}\right|_{\partial\Omega_3} = 0 \quad (\nu: \textit{ outer normal}). \tag{12.22}$$

The nonlinearities are given by

$$f(x, u, \nabla u, \nabla^2 u) := \left(\frac{4x\nabla u}{|x|^2} + \Delta u\right)\left(K'(u + x\nabla u) - 1\right), \tag{12.23}$$

where K is an arbitrary smooth function (being odd in the case of the Neumann boundary condition (12.22)) which satisfies

$$K'(w) = 1 + \mathcal{O}(|w|^\alpha) \quad \textit{near } w = 0.$$

Remarks: A local solution exists and is necessarily radially symmetric (cf. the remarks following the proof of Theorem 12.1). Global solutions exist for small data if the Dirichlet (resp. Neumann) boundary condition (12.21) (resp. (12.22)) is replaced by the dissipative boundary conditions (cf. (12.14))

$$K(\bar{u}(t,1)+\bar{u}_r(t,1))-\tau\bar{u}_t(t,1)=0, \quad \bar{u}_t(t,2)=0, \tag{12.24}$$

where $0<\tau<\infty$ is a fixed parameter, $\bar{u}(t,r)=u(t,x)$, $r=|x|$ and $\bar{u}_r$ denotes the radial derivative.

PROOF of Theorem 12.1:
We are looking for transformations $v(t,r)=\Phi\left(t,r,\bar{u}(t,r)\right)$ which carry over the differential equation for $\bar{u}$, namely

$$\bar{u}_{tt}-\bar{u}_{rr}-\frac{n-1}{r}\bar{u}_r=f(\cdots)$$

into a differential equation for v of the type

$$v_{tt}-v_{rr}=\tilde{f}(\cdots),$$

then admitting the application of [93].
Let us start with a general nonlinearity g, and we consider $\Omega\subset\mathbb{R}^n$ instead of Ω_3 for a moment, n not necessarily being equal to 3,

$$u_{tt}-\Delta u=g(\cdot,u,\nabla u,\nabla^2 u).$$

We assume that there is a local, smooth, radially symmetric solution $u(t,x)=\bar{u}(t,r)$, $r=|x|$. Making the ansatz

$$v(t,r):=p(r)\bar{u}(t,r).$$

we have

$$v_{tt}=p\bar{u}_{tt},\ v_r=p'\bar{u}+p\bar{u}_r,\ v_{rr}=p''\bar{u}+2p'\bar{u}_r+p\bar{u}_{rr},$$

hence

$$v_{tt}-v_{rr}=p\left(\bar{u}_{tt}-\bar{u}_{rr}-\frac{n-1}{r}\bar{u}_r\right)+\left(p\frac{n-1}{r}-2p'\right)\bar{u}_r+p''\bar{u}.$$

With the requirement in mind that we wish to end up with a nonlinearity that is at least quadratic, we require $p''=0$, i.e. $p(r)=ar+b$ with some $a,b\in\mathbb{R}$ and also $p\frac{n-1}{r}-2p'=0$, i.e. $(ar+b)(n-1)-2ar=0$, whence necessarily $b=0$ and $n=3$ follows, $a\in\mathbb{R}$ arbitrary, (without loss of generality $a=1$).
That is, $p(r)=r$ and v with $v(t,r)=r\bar{u}(t,r)$ satisfies (in Ω_3 now)

$$v_{tt}-v_{rr}=r\left(\bar{u}_{tt}-\bar{u}_{rr}-\frac{2}{r}\bar{u}_r\right)=rg.$$

We would like to have that v satisfies

$$v_{tt}-(K(v_r))_r=0,$$

or, equivalently,

$$v_{tt} - v_{rr} = v_{rr}(K'(v_r) - 1).$$

Since $v_r = \bar{u} + r\bar{u}_r$, $v_{rr} = 2\bar{u}_r + r\bar{u}_{rr}$, the right-hand side reads in terms of $\bar{u}$

$$(2\bar{u}_r + r\bar{u}_{rr})(K'(\bar{u} + r\bar{u}_r) - 1).$$

Hence we have to choose g as

$$g := \left(\frac{2}{r}\bar{u}_r + \bar{u}_{rr}\right)(K'(\bar{u} + r\bar{u}_r) - 1).$$

Since

$$\bar{u}(t,r) = u(t,x),\ r = |x|,\ \bar{u}_r = \frac{x}{|x|}\nabla u;\ \bar{u}_{rr} = \frac{n-1}{|x|}\left(\frac{x}{|x|}\nabla u\right) + \Delta u = \frac{2}{|x|}\left(\frac{x}{|x|}\nabla u\right) + \Delta u,$$

we see that g in terms of u equals f as defined by (12.23) in Theorem 12.1. Consequently, if u satisfies (12.19), (12.20), (12.21) (resp. (12.22)), then v satisfies — as long as it exists and for $1 \le r \le 2$ — the relations

$$v_{tt} - (K(v_r))_r = 0, \tag{12.25}$$

$$v(0,r) = r\varepsilon\phi(x), \quad v_t(0,r) = r\varepsilon\psi(x), \quad |x| = r, \tag{12.26}$$

$$v(t,1) = v(t,2) = 0, \tag{12.27}$$

$$(\text{resp.} \quad v_r(t,1) = v_r(t,2) = 0). \tag{12.28}$$

It follows from Klainerman & Majda [93] that v develops a singularity in the second derivatives at time $T = C\varepsilon^{-\alpha}$ which gives the desired result for u.

Q.E.D.

The simple ansatz $v(t,r) = p(r)\bar{u}(t,r)$ only works in three space dimensions, while a more general dependence $v(t,r) = \Phi(r, \bar{u}(t,r))$ leads to difficulties for $n \neq 3$.

The reason why the polynomial ansatz $v = r\bar{u}$ works in three space dimensions is that the fundamental solution $f_n = f_n(r)$ to the equation

$$f_n''(r) + \frac{n-1}{r}f_n'(r) = 0$$

has the property that the following recursive formula holds (up to constants which are not essential in the sequel):

$$f_{n+2}(r) = \frac{f_n'(r)}{r}.$$

This implies that

$$f_1'(r) = rf_3(r)$$

and we observe that f_1' satisfies the same differential equation as f_1.

Of course one could recursively derive similar formulae for the higher order fundamental solutions in terms of the derivatives of f_1, for example

$$n = 5: \quad f_1'''(r) = 3r f_5(r) + r^2 f_5'(r)$$

or

$$n = 7: \quad {f_1}^{(5)}(r) = 15r f_7(r) + 9r^2 f_7'(r) + r^3 f_7''(r),$$

but these formulae are not appropriate for our problem because there appear derivatives of f_5 and f_7 respectively on the right-hand side. This would imply, e.g. for $n = 5$, that, if $\bar{u}$ satisfies

$$\bar{u}_{tt} - \bar{u}_{rr} - \frac{4}{r}\bar{u}_r = g,$$

then v, defined by

$$v := 3r\bar{u} + r^2\bar{u}_r$$

satisfies

$$v_{tt} - v_{rr} = 3rg + r^2 g_r,$$

hence

$$3rg + r^2 g_r = v_{rr}(K'(v_r) - 1)$$

should hold, while it is not clear how to define g depending on at most second derivatives of u. This becomes even worse for $n = 7, 8, \ldots$ due to the appearance of higher derivatives of g.

The existence of a local solution to (12.19), (12.20), (12.21) (resp. (12.22)), is obvious by construction since a local solution to (12.25), (12.26), (12.27) (resp. (12.28)), exists. Independently, the existence of a unique local solution to (12.19), (12.20), (12.21) would follow from the general local existence theorem by Shibata & Tsutsumi [168], observing that the solution to radially symmetric data necessarily must be radially symmetric for all times. The latter follows from the uniqueness of the solution and the fact that if u is the local solution, then w with $w(t,x) := u(t,Px)$, $P \in O(3) \equiv$ orthogonal group in $\mathbb{R}^3$, satisfies the same differential equation, initial conditions and boundary conditions, and hence w must coincide with u which means that u is radially symmetric.

The remark following the statement of Theorem 12.1 concerning dissipative boundary conditions is now obvious since

"u satisfies the boundary conditions (12.24)"

is equivalent to

"v satisfies the boundary conditions (12.14)",

and then v globally exists for small data according to Greenberg & Li [40].

We would like now to close the section by listing a few related questions and to point out some open problems.

- Necessary conditions for the global existence of small, smooth solutions:

  Most of the theorems in Chapter 1 and Chapter 11 only provide sufficient conditions. Here sharp results are required, which means an investigation of possible blow-up situations.

- Non-homogeneous media, variable coefficients:

  In the previous Chapters the operators A appearing in the linear main part had constant coefficients, corresponding for example to homogeneous media in elasticity. The fact that constant coefficients were considered was essential for the derivation of decay rates. The simple reason is the availability of appropriate representation formulae for solutions to the linearized system. Here are many open questions, cf. the discussion above where a non-homogeneous example was treated in an exterior domain.

- Weak solutions:

  In the case that there are no global smooth solutions it is natural to ask whether there are global weak solutions. This question has been answered only in rather specific situations, e.g. in one space dimension, in general.

- Arbitrary domains:

  As far as we have studied boundary value problems in this section, the boundary was assumed to be smooth. This is important for the regularity theory which plays an important role in proving the global existence theorems for smooth solutions. Besides the typical interior and exterior boundary value problems mentioned above there are domains with other geometries of interest, for example domains with an infinite boundary like half planes, waveguides or unbounded cylinders, the latter two categories being of the type

  $$\Omega = \mathbb{R}^{n-m} \times \Omega', \quad \Omega' \subset \mathbb{R}^m \text{ bounded}, \ 1 \leq m < n.$$

  Here there are already new phenomena arising in the linear theory, for example the *principle of limiting amplitude* being not valid in certain cases (see P. Werner [195]). This principle allows statements on the asymptotic behavior of solutions to linear wave equations assuming a time-periodic force as $t \to \infty$ and is always satisfied in exterior domains if $n \geq 3$.

  We shall consider waveguides in Chapter 13.

- Boundary conditions:

  Often it is necessary to investigate initial-boundary value problems for each set of boundary conditions separately. In contrast to this it turns out that for linear and for nonlinear wave equations one obtains corresponding results for both the Dirichlet and the Neumann boundary condition. But this is not self-evident. For example, looking at the homogeneous, isotropic equations of elasticity outside a ball in $\mathbb{R}^3$, one finds that for the Dirichlet boundary condition the local energy (local L^2-norms of derivatives of the displacement vector) decays exponentially, but it does not decay with a rate for the corresponding Neumann boundary condition due to the presence of surface waves. The local energy even grows if one assumes a suitable mixed boundary condition (see Ikehata & Nakamura [53]). These problems hence require new ansätze.

- Individual equations:

  It was mentioned at several places throughout this book that the common structure which was found for all the systems discussed here does not mean putting all the features under one cover. Specific properties of specific equations lead to sharp results. Here further research is required for each individual system.

- Numerical investigations:

  Last, but not least, we wish to emphasize that numerical investigations deserve a great interest, e.g. for the shock wave analysis for hyperbolic problems. This is important not only with regard to the applications, e.g. in gas dynamics, but also because interesting hints for further theoretical, analytical research are expected.

13 Evolution equations in waveguides

Now we extend the considerations to initial-*boundary* value problems for *waveguides* $\Omega \subset \mathbb{R}^n$. In the main part, we consider nonlinear wave equations and Schrödinger equations as well as step **A** for the equations of elasticity and the Maxwell equations in *flat* or *classical waveguides*. These are domains Ω of the type

$$\Omega = \mathbb{R}^l \times \mathcal{B} \ni (x', x''), \quad \mathcal{B} \subset \mathbb{R}^{n-l} \text{ bounded}, \tag{13.1}$$

where $1 \leq l \leq n-1$. Typical examples are provided in the following Figures 13.1–3:

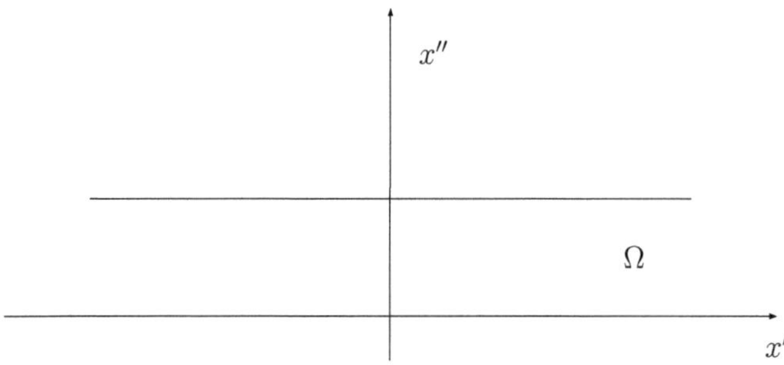

Figure 13.1: $n = 2, l = 1$: infinite strip

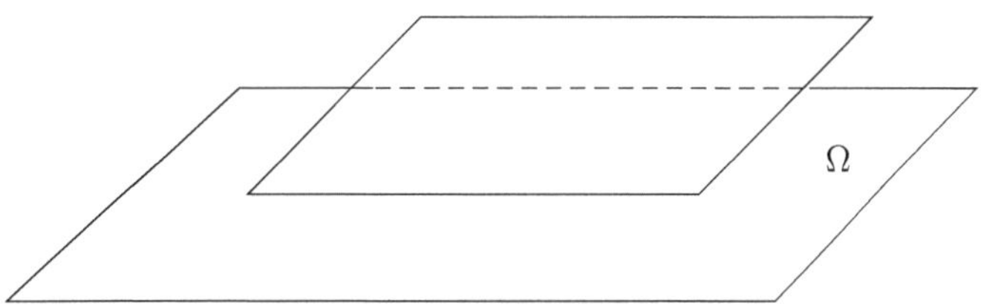

Figure 13.2: $n = 3, l = 2$: infinite plate

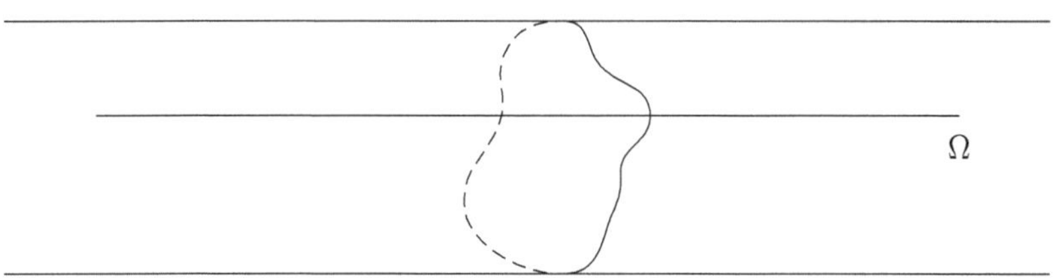

Figure 13.3: $n = 3, l = 1$: infinite cylinder

In the last part, we shall also look at generalized waveguides, see Figure 13.4.
The combination of bounded and unbounded parts makes the situation in waveguides different from the Cauchy problem, but also different from bounded or exterior domains.

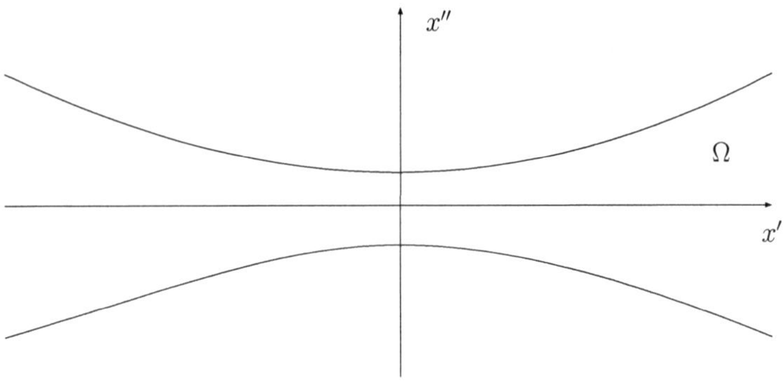

Figure 13.4: $n = 2, l = 1$: generalized waveguide

13.1 Nonlinear wave equations

We consider the fully nonlinear equations

$$u_{tt} - \Delta u + mu = f(u, u_t, \nabla u, \nabla u_t, \nabla^2 u) \tag{13.2}$$

for a function $u = u(t,x), t \geq 0, x \in \Omega \subset \mathbb{R}^n, m \geq 0$ being a constant, with initial conditions

$$u(t=0) = u_0, \quad u_t(t=0) = u_1, \tag{13.3}$$

and Dirichlet boundary conditions

$$u(t,\cdot) = 0 \quad \text{on } \partial\Omega. \tag{13.4}$$

Ω is a flat waveguide as in (13.1) with smooth boundary $\partial\Omega$. For $m = 0$ we have wave equations, whereas $m > 0$ corresponds to Klein-Gordon equations. In contrast to Cauchy problems we also consider a dependence of the smooth nonlinear function f on u, for the case $m = 0$.

13.1.1 Linear part

Following the general steps **A**–**E**, we first characterize the asymptotic behavior of solutions to the linearized equations

$$u_{tt} - \Delta u + mu = 0 \quad \text{in } [0,\infty) \times \Omega, \tag{13.5}$$

$$u = 0 \quad \text{in } [0,\infty) \times \partial\Omega, \tag{13.6}$$

$$u(0,\cdot) = u_0,\ u_t(0,\cdot) = u_1 \quad \text{in } \Omega. \tag{13.7}$$

The idea will be to use an eigenfunction expansion in the bounded part $\mathcal{B}$ of $\Omega = \mathbb{R}^l \times \mathcal{B}$, and exploit the known behavior of functions defined in $[0,\infty) \times \mathbb{R}^l$.
For $x \in \Omega$ we write

$$x = (x', x'') \quad \text{with } x' \in \mathbb{R}^l,\ x'' \in \mathcal{B},$$

$$\Delta = \sum_{j=1}^{n} \partial_j^2, \qquad \Delta' = \sum_{j=1}^{l} \partial_j^2, \qquad \Delta'' = \sum_{j=l+1}^{n} \partial_j^2,$$

and, analogously,

$$\nabla, \nabla', \nabla''.$$

Let

$$A : D(A) \subset L^2(\Omega) \to L^2(\Omega),$$

$$D(A) := W^{2,2}(\Omega) \cap W_0^{1,2}(\Omega), \quad A\varphi := -\Delta\varphi,$$

and

$$A' : D(A') \subset L^2(\mathbb{R}^l) \to L^2(\mathbb{R}^l),$$

$$D(A') := W^{2,2}(\mathbb{R}^l), \quad A'\varphi := -\Delta'\varphi,$$

$$A'' : D(A'') \subset L^2(\mathcal{B}) \to L^2(\mathcal{B}),$$

$$D(A'') := W^{2,2}(\mathcal{B}) \cap W_0^{1,2}(\mathcal{B}), \quad A''\varphi := -\Delta''\varphi.$$

A'' is self-adjoint, having a complete orthonormal set $(w_j)_{j\in\mathrm{N}}$ of eigenfunctions corresponding to eigenvalues $(\lambda_j)_{j\in\mathrm{N}}$ satisfying

$$0 < \lambda_1 \leq \lambda_2 \leq \ldots \leq \lambda_j \to \infty \quad \text{as } j \to \infty.$$

The spectra of A' and A, respectively, are

$$\sigma(A') = [0, \infty), \quad \sigma(A) = [\lambda_1, \infty)$$

and are purely continuous, cp. [99].
The L^1-L^∞-decay of solutions to (13.5) – (13.7) is described as follows.

Theorem 13.1 *Let*

$$K_2 := \left[\frac{n}{2}\right] + \left[\frac{n-l}{2}\right] + 4, \quad K_3 := \left[\frac{l+3}{2}\right] + n - l + 1,$$

and

$$u_0 \in D(A^{K_2/2}) \cap W^{K_2+K_3+1,1}(\Omega),$$

$$u_1 \in D(A^{(K_2-1)/2}) \cap W^{K_2+K_3-1,1}(\Omega).$$

Then the unique solution u to (13.5) – (13.7) satisfies

$$\|\,(u(t), u_t(t), \nabla u(t))\,\|_{L^\infty(\Omega)} \leq \frac{c}{(1+t)^{l/2}} \left(\|u_0\|_{W^{K_2+K_3+1,1}(\Omega)} + \|u_1\|_{W^{K_2+K_3-1,1}(\Omega)}\right),$$

where the positive constant c depends at most on m.

Remarks:

1. Of course, c may also depend on the fixed n and $\mathcal{B}$.
2. The decay rate $l/2$ is the same for wave equations ($m = 0$) and for Klein-Gordon equations ($m > 0$), wheras it is different for the Cauchy problem in $\mathbb{R}^n$, where we have $(n-1)/2$ for $m = 0$ and $n/2$ for $m > 0$. For the case $m = 0$ one notices the interesting behavior for $l = n - 1$ where one has the same decay as for the Cauchy problem. This holds, for example, for an infinite strip in $\mathbb{R}^2$ or for the region between two parallel planes in $\mathbb{R}^3$.
3. In general, for fixed n, the decay becomes weaker as the number of bounded dimensions increases.

PROOF of Theorem 13.1:

Case 1: $t \geq \frac{1}{\sqrt{\lambda_1}}$.

Expanding $u = u(t, x', x'')$ for fixed (t, x'), one has

$$u(t, x', x'') = \sum_{j=1}^{\infty} v_j(t, x') w_j(x'')$$

with

$$v_j(t, x') = \langle u(t, x', \cdot), w_j \rangle_{L^2(\mathcal{B})}.$$

By (13.5) we obtain

$$0 = u_{tt} - \Delta u + mu = \sum_{j=1}^{\infty} (v_{j,tt} - \Delta' v_j + \lambda_j v_j + m v_j) w_j.$$

Denoting

$$v_{j,0}(x') := \langle u_0(x', \cdot), w_j \rangle_{L^2(\mathcal{B})}, \quad v_{j,1} := \langle u_1(x', \cdot), w_j \rangle_{L^2(\mathcal{B})}$$

we conclude that v_j satisfies

$$v_{j,tt} - \Delta' v_j + (m + \lambda_j) v_j = 0 \qquad \text{in } [0, \infty) \times \mathbb{R}^l, \tag{13.8}$$

$$v_j(0, \cdot) = v_{j,0}, \quad v_{j,t}(0, \cdot) = v_{j,1} \qquad \text{in } \mathbb{R}^l, \tag{13.9}$$

that is, both for $m = 0$ and for $m > 0$, v_j satisfies a Klein-Gordon equation in all of $\mathbb{R}^l$. For v_j we can exploit the decay of solutions to a Cauchy problem for the Klein-Gordon equation, but we need a version describing the behavior of the mass which is depending on λ_j in our situation.

Lemma 13.2 *Let*

$$K_1 := \left[\frac{l+3}{2}\right]$$

and let $M \geq M_0 > 0$, $v_0 \in W^{K_1,1}(\mathbb{R}^l)$, $v_1 \in W^{K_1-1,1}(\mathbb{R}^l)$. Then the unique solution v to

$$v_{tt} - \Delta' v + Mv = 0 \qquad \textit{in } [0, \infty) \times \mathbb{R}^l,$$

$$v(0, \cdot) = v_0, \quad v_t(0, \cdot) = v_1 \quad \textit{in}$$

$$R^l,$$

satisfies for $t \geq \frac{1}{\sqrt{M}}$

$$\|v(t)\|_{L^\infty(\mathbf{R}^l)} \leq \frac{c}{t^{l/2}} \left(M^{\frac{1}{4}} \|v_0\|_{W^{K_1,1}(\mathbf{R}^l)} + M^{\frac{l-2}{4}} \|v_1\|_{W^{K_1-1,1}(\mathbf{R}^l)} \right),$$

$c > 0$ *being a constant depending at most on* M_0.

PROOF: Let

$$\tilde{v}(t,x') := v\left(\frac{t}{\sqrt{M}}, \frac{x'}{\sqrt{M}}\right).$$

Then $\tilde{v}$ satisfies

$$\tilde{v}_{tt} - \Delta' \tilde{v} + \tilde{v} = 0,$$

$$\tilde{v}(0,x') = v_0\left(\frac{x'}{\sqrt{M}}\right) =: \tilde{v}_0(x'), \quad \tilde{v}_t(0,x') = \frac{1}{\sqrt{M}} v_1\left(\frac{x'}{\sqrt{M}}\right) =: \tilde{v}_1(x').$$

By the known estimates for Klein-Gordon equations from Chapter 11.5, we have for $t \geq 1$

$$|\tilde{v}(t,x')| \leq \frac{c}{t^{l/2}} \left(\|\tilde{v}_0\|_{W^{K_1,1}(\mathbf{R}^l)} + \|\tilde{v}_1\|_{W^{K_1-1,1}(\mathbf{R}^l)} \right). \tag{13.10}$$

A substitution $y' = \frac{x'}{\sqrt{M}}$, $dx' = (\sqrt{M})^l dy'$, yields

$$\|\tilde{v}_0\|_{L^1(\mathbf{R}^l)} = M^{l/2} \|v_0\|_{L^1(\mathbf{R}^l)}.$$

Since

$$\partial_j \tilde{v}_0(x') = \frac{1}{\sqrt{M}} (\partial_j v_0)\left(\frac{x'}{\sqrt{M}}\right), \quad \frac{1}{\sqrt{M}} \leq \frac{1}{\sqrt{M_0}},$$

we obtain

$$\|\tilde{v}_0\|_{W^{K_1,1}(\mathbf{R}^l)} \leq c M^{l/2} \|v_0\|_{W^{K_1,1}(\mathbf{R}^l)}, \tag{13.11}$$

and

$$\|\tilde{v}_1\|_{W^{K_1-1,1}(\mathbf{R}^l)} \leq c M^{(l-1)/2} \|v_1\|_{W^{K_1-1,1}(\mathbf{R}^l)}, \tag{13.12}$$

where $c = c(M_0)$. Combining (13.10) – (13.12) we get

$$\begin{aligned} |v(t,x')| &= |\tilde{v}(\sqrt{M}t, \sqrt{M}x')| \\ &\leq \frac{c}{(\sqrt{M}\,t)^{l/2}} \left(M^{l/2} \|v_0\|_{W^{K_1,1}(\mathbf{R}^l)} + M^{(l-1)/2} \|v_1\|_{W^{K_1-1}(\mathbf{R}^l)} \right). \end{aligned}$$

Q.E.D.

Corollary 13.3 *Let*

$$v_0 \in W^{K_1+1,1}(\mathbb{R}^l), \quad v_1 \in W^{K_1,1}(\mathbb{R}^l).$$

Then we have for $t \geq \frac{1}{\sqrt{M}}$

$$\begin{aligned}\|\nabla' v(t)\|_{L^\infty(\mathbf{R}^l)} &\leq \frac{c}{t^{l/2}}\left(M^{l/4}\|v_0\|_{W^{K_1+1,1}(\mathbf{R}^l)} + M^{(l-2)/4}\|v_1\|_{W^{K_1,1}(\mathbf{R}^l)}\right),\\ \|v_t(t)\|_{L^\infty(\mathbf{R}^l)} &\leq \frac{c}{t^{l/2}}\left(M^{(l+2)/4}\|v_0\|_{W^{K_1+1,1}(\mathbf{R}^l)} + M^{l/4}\|v_1\|_{W^{K_1,1}(\mathbf{R}^l)}\right),\end{aligned}$$

where $c > 0$ *depends at most on* M_0.

PROOF: $h_j := \partial_j v$, $j \in \{1, \ldots, l\}$, satisfies the same Klein-Gordon equation as v, now with initial data

$$h_j(0) = \partial_j v_0, \quad h_{j,t}(0) = \partial_j v_1$$

yielding the first claimed estimate by using Lemma 13.2.
$\tilde{h} := v_t$ also satisfies the same differential equation, but with initial data

$$\tilde{h}(0) = v_1, \quad \tilde{h}_t(0) = (\Delta' - M)v_0.$$

Again using Lemma 13.2, we have

$$\begin{aligned}|\tilde{h}(t, x')| &\leq \frac{c}{t^{l/2}}\left(M^{l/4}\|v_1\|_{W^{K_1,1}(\mathbf{R}^l)} + M^{(l-2)/4}\|(\Delta' - M)v_0\|_{W^{K_1-1,1}(\mathbf{R}^l)}\right),\\ &\leq \frac{c}{t^{l/2}}\left(M^{l/4}\|v_1\|_{W^{K_1,1}(\mathbf{R}^l)} + M^{(l+2)/4}\|v_0\|_{W^{K_1+1,1}(\mathbf{R}^l)}\right).\end{aligned}$$

Q.E.D.

Now we return to the proof of Theorem 13.1. By Sobolev's embedding theorem and elliptic regularity, we have

$$\begin{aligned}|u(t, x', x'')|^2 + |\nabla'' u(t, x', x'')|^2 &\leq c\|u(t, x', \cdot)\|^2_{W^{K_5,2}(\mathcal{B})} \\ &\leq c\|(A'')^{K_5/2} u(t, x', \cdot)\|^2_{L^2(\mathcal{B})},\end{aligned} \tag{13.13}$$

where

$$K_5 := \left[\frac{n-l}{2}\right] + 3.$$

Concerning elliptic regularity we have for $A_1 \in \{A, A''\}$, with $\langle \cdot, \cdot \rangle_{L^2}$ and $\| \cdot \|_{L^2}$ denoting the inner product in $L^2(\Omega)$ respectively the norm in $L^2(\mathcal{B})$,

$$\lambda_1 \|\varphi\|^2_{L^2} \leq \langle A_1 \varphi, \varphi \rangle_{L^2} = \|\nabla \varphi\|^2_{L^2} \tag{13.14}$$

implying

$$\|\varphi\|_{L^2} \leq \frac{1}{\sqrt{\lambda_1}} \|A_1^{1/2} \varphi\|_{L^2}$$

and

$$\|\varphi\|_{L^2}^2 + \|\nabla\varphi\|_{L^2}^2 \leq c\|A_1^{1/2}\varphi\|_{L^2}^2, \quad \varphi \in D(A_1^{1/2}). \tag{13.15}$$

Now we use the elliptic regularity expressed in

Lemma 13.4 *Let $j \in \mathbb{N}, \varphi \in D(A_1)$, $1 < p < \infty$, $\varphi \in L^p(\Omega)$ resp. $L^p(\mathcal{B}), A_1\varphi \in W^{j,p}(\Omega)$ resp. $W^{j,p}(\mathcal{B})$. Then $\varphi \in W^{j+2,p}(\Omega)$ resp. $W^{j+2,p}(\mathcal{B})$ and*

$$\|\varphi\|_{W^{2+j,p}} \leq c\|A_1\varphi\|_{W^{j,p}}$$

holds, where c is a positive constant at most depending on j and p.

The proof of Lemma 13.4 is given for bounded domains in [174, Chapter II, Theorem 9.1.]. It carries over to a flat waveguide Ω since it has a bounded cross section, and hence the Poincaré inequality can be used.
Applying Lemma 13.4 successively, we obtain

Corollary 13.5 *Let $j \in \mathbb{N}, \varphi \in D(A_1^j)$, $1 < p < \infty$, and $\varphi, A_1\varphi, \cdots, A_1^j\varphi \in L^p(\Omega)$ resp. $L^p(\mathcal{B})$. Then we have*

$$\|\varphi\|_{W^{2j,p}} \leq c\|A_1^j\varphi\|_{L^p},$$

c at most depending on j and p.

Combining Lemma 13.4 and (13.15) we conclude for $j \in \mathbb{N}$ and $\varphi \in D(A_1^{(2j+1)/2})$

$$\begin{aligned} \|\varphi\|_{W^{2j+1,2}} &= \|\varphi\|_{W^{2j-1+2,2}} \leq c\|A_1\varphi\|_{W^{2j-1,2}} \leq \ldots \\ &\leq c\|A_1^j\varphi\|_{W^{1,2}} \leq c\|A_1^{j+\frac{1}{2}}\varphi\|_{L^2}. \end{aligned} \tag{13.16}$$

Corollary 13.5 and (13.16) yield for any $j \in \mathbb{N}, \varphi \in D(A^{j/2})$

$$\|\varphi\|_{W^{j,2}} \leq c(j)\|A_1^{j/2}\varphi\|_{L^2}. \tag{13.17}$$

This last estimate has been used in the inequality in (13.13). We proceed in the pointwise estimate (13.13) and obtain, using Lemma 13.2,

$$\begin{aligned} |u(t,x',x'')|^2 + |\nabla''u(t,x',x'')|^2 &\leq c\sum_{j=1}^{\infty} \lambda_j^{K_5}|v_j(t,x')|^2 \\ &\leq \frac{c}{t^l}\sum_{j=1}^{\infty} \lambda_j^{K_5}\Big((m+\lambda_j)^{l/2}\|\langle v_{j,0}\|_{W^{K_1,1}(\mathbf{R}^l)}^2 \\ &\quad + (m+\lambda_j)^{(l-2)/2}\|\langle v_{j,1}\|_{W^{K_1-1}(\mathbf{R}^l)}^2\Big) \\ &\leq \frac{c}{t^l}\sum_{j=1}^{\infty} \lambda_j^{K_5+l/2}\|\langle u_0, w_j\rangle_{L^2(\mathcal{B})}\|_{W^{K_1,1}(\mathbf{R}^l)}^2 \\ &\quad +\lambda_j^{K_5+l/2-1}\|\langle u_1, w_j\rangle_{L^2(\mathcal{B})}\|_{W^{K_1-1,1}(\mathbf{R}^l)}^2. \end{aligned}$$

Observing

$$\langle u_0, w_j\rangle_{L^2(\mathcal{B})} = \frac{1}{\lambda_j^{K_2/2}} \left\langle (A'')^{K_2/2} u_0, w_j \right\rangle_{L^2(\mathcal{B})},$$
$$\langle u_1, w_j\rangle_{L^2(\mathcal{B})} = \frac{1}{\lambda_j^{(K_2-1)/2}} \left\langle (A'')^{(K_2-1)/2} u_1, w_j \right\rangle_{L^2(\mathcal{B})},$$

we conclude

$$\begin{aligned} &|u'(t,x',x'')|^2 + |\nabla'' u(t,x',x'')|^2 \\ &\le \tfrac{c}{t^l} \sum_{j=1}^{\infty} \tfrac{1}{\lambda_j^{K_2-K_5-l/2}} \left(\|\langle (A'')^{K_2/2} u_0, w_j\rangle_{L^2(\mathcal{B})}\|^2_{W^{K_1,1}(\mathbf{R}^l)} \right. \\ &\quad \left. + \|\langle (A'')^{(K_2-1)/2} u_1, w_j\rangle_{L^2(\mathcal{B})}\|^2_{W^{K_1-1,1}(\mathbf{R}^l)} \right). \end{aligned} \tag{13.18}$$

To obtain the convergence of $\sum_{j=1}^{\infty} \frac{1}{\lambda_j^{K_2-K_5-l/2}}$ we need some information on the asymptotic behavior of the eigenvalues λ_j as $j \to \infty$.

Lemma 13.6 *The eigenvalues $(\lambda_j)_j$ of A'' satisfy*

$$\lambda_j \ge c j^{\frac{2}{n-l}}, \quad j \in \mathbb{N},$$

where $c > 0$ is independent of j (but depends on $\mathcal{B}$).

PROOF: This kind of estimate goes back to early work of Weyl[1]. We use [2, Theorem 14.6] yielding for the number $N(\lambda)$ of eigenvalues satisfying $\lambda_j \le \lambda$,

$$N(\lambda) = c\lambda^{\frac{n-l}{2}} + \mathcal{O}(\lambda^{\frac{n-l}{2}}), \text{ as } \lambda \to \infty,$$

with $c = c(\mathcal{B}) > 0$. Hence

$$N(\lambda) \le c\lambda^{\frac{n-l}{2}} \quad \text{for} \quad \lambda \ge \lambda_1,$$

implying the assertion since $N(\lambda_j) \ge j$.

Q.E.D.

Lemma 13.6 implies

$$\lambda_j^{K_2-K_5-\frac{l}{2}} \ge c j^{(K_2-K_5-l/2)\frac{2}{n-l}}.$$

By the choices of K_2 and K_5 we have

$$(K_2 - K_5 - \frac{l}{2})\frac{2}{n-l} > 1,$$

hence,

$$\sum_{j=1}^{\infty} \frac{1}{\lambda_j^{K_2-K_5-l/2}} < \infty. \tag{13.19}$$

[1] Hermann Weyl, 9.11.1885 – 9.12.1955

Moreover,

$$\begin{aligned}|\langle (A'')^{K_2/2}u_0, w_j\rangle_{L^2(\mathcal{B})}| &\leq \|(A'')^{K_2/2}u_0\|_{L^2(\mathcal{B})} \\ &\leq c\|u_0\|_{W^{K_2,2}(\mathcal{B})} \\ &\leq c\|u_0\|_{W^{K_2+n-l+1,1}(\mathcal{B})},\end{aligned} \tag{13.20}$$

similarly

$$|\langle (A'')^{(K_2-1)/2}u_1, w_j\rangle_{L^2(\mathcal{B})}| \leq c\|u_1\|_{W^{K_2+n-l,1}(\mathcal{B})}. \tag{13.21}$$

Combining (13.18) – (13.21) and observing the definition of K_3, we obtain

$$|u(t,x',x'')|^2 + |\nabla'' u(t,x,x'')|^2 \leq \frac{c}{t^l}\left(\|u_0\|^2_{W^{K_2+K_3,1}(\Omega)} + \|u_1\|^2_{W^{K_2+K_3-2,1}(\Omega)}\right). \tag{13.22}$$

Analogously, we get

$$\begin{aligned}|\nabla' u(t,x',x'')|^2 &\leq c\|(A'')^{(K_5-1)/2}\nabla' u(t,x',\cdot)\|^2_{L^2(\mathcal{B})} \\ &= c\sum_{j=1}^{\infty}\lambda_j^{K_5-1}|\langle \nabla' u(t,x',\cdot), \omega_j\rangle_{L^2(\mathcal{B})}|^2 \\ &\leq \text{r.h.s. of (13.22)}.\end{aligned} \tag{13.23}$$

The time derivative is estimated by

$$\begin{aligned}|u_t(t,x',x'')|^2 &\leq c\|u_t(t,x',\cdot)\|^2_{W^{K_5-1,2}(\mathcal{B})} \\ &\leq \frac{c}{t^l}\left(\|u_0\|^2_{W^{K_2+K_3+1,1}(\Omega)} + \|u_1\|^2_{W^{K_2+K_3-1,1}(\Omega)}\right).\end{aligned} \tag{13.24}$$

Case 2: $\quad 0 \leq t \leq \frac{1}{\sqrt{\lambda_1}}$.
Let

$$K_4 := [\frac{n}{2}] + 2.$$

Then $u_0 \in D(A^{K_4/2}), u_1 \in D(A^{(K_4-1)/2})$ since $K_4 \leq K_2$. Let $(P_\lambda)_{\lambda\in\mathbb{R}}$ denote the spectral family of A. Then

$$u(t) = \int_{\lambda_1}^{\infty} \cos(\sqrt{m+\lambda}\,t)dP_\lambda u_0 + \int_{\lambda_1}^{\infty} \frac{\sin(\sqrt{m+\lambda}\,t)}{\sqrt{m+\lambda}}dP_\lambda u_1.$$

This implies for $t \geq 0, x \in \Omega$, using (13.17),

$$\begin{aligned}|u(t,x)| + |\nabla u(t,x)| &\leq c\|u(t,\cdot)\|_{W^{K_4,2}(\Omega)} \leq c\|A^{K_4/2}u(t,\cdot)\|_{L^2(\Omega)} \\ &\leq c\left(\left(\int_{\lambda_1}^{\infty}\lambda^{K_4}\cos^2(\sqrt{m+\lambda}\,t)d\|P_\lambda u_0\|^2_{L^2(\Omega)}\right)^{1/2}\right. \\ &\quad \left. + \left(\int_{\lambda_1}^{\infty}\frac{\lambda^{K_4}}{m+\lambda}\sin^2(\sqrt{m+\lambda}\,t)d\|P_\lambda u_1\|^2_{L^2(\Omega)}\right)^{1/2}\right)\end{aligned} \tag{13.25}$$

$$
\begin{aligned}
&\leq c\left(\|A^{K_4/2}u_0\|_{L^2(\Omega)} + \|A^{(K_4-1)/2}u_1\|_{L^2(\Omega)}\right)\\
&\leq c\left(\|u_0\|_{W^{K_4,2}(\Omega)} + \|u_1\|_{W^{K_4-1,2}(\Omega)}\right)\\
&\leq \frac{c}{(1+t)^{l/2}}\left(\|u_0\|_{W^{K_2+K_3+1,1}(\Omega)} + \|u_1\|_{W^{K_2+K_3-1,1}(\Omega)}\right),
\end{aligned}
$$

observing $0 \leq t \leq \frac{1}{\sqrt{\lambda_1}}$. Analogously, we have

$$|u_t(t,x)| \leq \text{right-hand side of (13.25)}. \tag{13.26}$$

Combining (13.25), (13.26) with (13.22) – (13.24) completes the proof of Theorem 13.1.

Q.E.D.

We obtain the L^2–L^2-estimate from multiplying the differential equation (13.5) by u_t in $L^2(\Omega)$, yielding

$$\frac{d}{dt}\left(\|u_t(t)\|^2_{L^2(\Omega)} + \|\nabla u(t)\|^2_{L^2(\Omega)} + m\|u(t)\|^2_{L^2(\Omega)}\right) = 0.$$

Integrating in time t and using Poincaré's inequality (cp. (13.14))

$$\|u(t)\|_{L^2(\Omega)} \leq c\|\nabla u(t)\|_{L^2(\Omega)},$$

we have proved

Theorem 13.7 *Let $u_0 \in D(A), u_1 \in D(A^{1/2})$. Then the solution u to (13.5) – (13.7) satisfies for $t \geq 0$*

$$\|u(t)\|_{W^{1,2}(\Omega)} + \|u_t(t)\|_{L^2(\Omega)} \leq c\left(\|u_0\|_{W^{1,2}(\Omega)} + \|u_1\|_{L^2(\Omega)}\right),$$

where c depends at most on m (and $\mathcal{B}$).

The L^p–L^q-decay we conclude from Theorems 13.1 and 13.7 by interpolation.

Theorem 13.8 *Let the assumptions of Theorem 13.1 be satisfied, let u be the unique solution to (13.5) – (13.7), and let $2 \leq q \leq \infty, 1/p + 1/q = 1$.*

(i) If $u_0 = 0$, then u satisfies for $t \geq 0$

$$\|(u, u_t, \nabla u)(t)\|_{L^q(\Omega)} \leq \frac{c}{(1+t)^{(1-\frac{2}{q})\frac{l}{2}}}\|u_1\|_{W^{N_p,p}(\Omega)},$$

where

$$N_p := \begin{cases} (1-\frac{2}{q})(K_2+K_3-1), & \text{if } q \in \{2,\infty\} \\ [(1-\frac{2}{q})(K_2+K_3-1)]+1, & \text{if } 2 < q < \infty, \end{cases}$$

and c depends at most on q and m.

(ii) If additionally $u_1 \in D(A^{(K_2-1)/2}) \cap W^{K_2+K_3,1}(\Omega)$, *then* u *satisfies for* $t \geq 0$

$$\|(u, u_t, \nabla u)(t)\|_{L^q(\Omega)} \leq \frac{c}{(1+t)^{(1-\frac{2}{q})\frac{l}{2}}} \|(u_0, u_1, \nabla u_0)\|_{W^{\widetilde{N}_p,p}(\Omega)},$$

where

$$\widetilde{N}_p := \begin{cases} (1-\frac{2}{q})(K_2+K_3), & \text{if } q \in \{2, \infty\} \\ [(1-\frac{2}{q})(K_2+K_3)] + 1, & \text{if } 2 < q < \infty, \end{cases}$$

and c *depends at most on* q *and* m.

One can also treat other boundary conditions than the Dirichlet one (13.6), e.g. Neumann boundary conditions

$$\frac{\partial u}{\partial \nu} = 0 \quad \text{in} \quad [0,\infty) \times \partial\Omega, \tag{13.27}$$

where $\nu = \nu(x)$ denotes the exterior normal in $x \in \partial\Omega$. Of course, there is the eigenfunction of $-\Delta''$, say w_0, corresponding to the eigenvalue $\lambda_0 = 0$. Then the coefficient

$$v_0(t, x') := \langle u(t, x', \cdot), w_0 \rangle_{L^2(\mathcal{B})}$$

only satisfies a wave equation for $m = 0$, not a Klein-Gordon equation,

$$v_{0,tt}(t, x') - \Delta' v_0(t, x') = 0.$$

Hence, the L^∞-decay is of order $t^{-\frac{l-1}{2}}$ (instead of $t^{-\frac{l}{2}}$). Indeed, if $\tilde{u} = \tilde{u}(t, x')$ is a solution to

$$\begin{aligned} \tilde{u}_{tt} - \Delta' \tilde{u} &= 0 \quad \text{in } [0,\infty) \times \mathbb{R}^l, \\ \tilde{u}(0, \cdot) &= \varphi \quad \text{in } \mathbb{R}^l, \\ \tilde{u}_t(0, \cdot) &= \psi \quad \text{in } \mathbb{R}^l, \end{aligned}$$

with $\varphi, \psi \neq 0$, then

$$u(t, x', x'') := \tilde{u}(t, x') w_0(x'')$$

solves the linear wave equation ((13.5) with $m = 0$) in $[0,\infty) \times \Omega$ and satisfies the Neumann boundary condition (13.27), together with nonvanishing data. The decay in L^∞ is only of order $t^{-\frac{l-1}{2}}$. On the other hand, if the initial data u_0, u_1 are orthogonal to the eigenfunctions w_0 (= constant),

$$\int_{\mathcal{B}} u_0(x', x'') dx'' = 0 = \int_{\mathcal{B}} u_1(x', x'') dx'',$$

for $x' \in \mathbb{R}^l$, then the better decay of oder $t^{-\frac{l}{2}}$ follows again.
In the next step we consider right-hand sides $f = f(t,x)$ in

$$u_{tt} - \Delta u + mu = f \quad \text{in} \quad [0,\infty) \times \Omega, \tag{13.28}$$

$$u = 0 \quad \text{in} \quad [0,\infty) \times \partial\Omega, \tag{13.29}$$

$$u(0,\cdot) = u_0,\ u_t(0,\cdot) = u_1 \quad \text{in} \quad \Omega. \tag{13.30}$$

A simple use of a variation of constants formula would require a set of boundary conditions for $f = f(t,\cdot)$ for fixed t; cp. the conditions on u_0, u_1 in Theorem 13.1. In view of the application to the nonlinear problem (13.2) – (13.4) we take a different approach. Let $B := A + m$, i.e.,

$$u_{tt} + Bu = f,$$

$$u(0) = u_0,\ u_t(0) = u_1.$$

Then B^{-1} exists since $\sigma(B) = [\lambda_1 + m, \infty)$ with $\lambda_1 > 0, m \geq 0$, and

$$v := B^{-1} u_{tt}$$

satisfies

$$v_{tt} + Bv = B^{-1} f_{tt} =: g,$$

$$v(0) = B^{-1} u_{tt}(0), \quad v_t(0) = B^{-1} u_{ttt}(0),$$

i.e., v satisfies the same differential equation, but with a right-hand side g that belongs to $D(B)$, hence it satisfies certain boundary conditions.
In this way, a higher regularity of f in t is needed and replaces the generally missing boundary conditions. The desired estimates for u will then be obtained from those for v via

$$u = B^{-1} f - v.$$

Let $T \in (0,\infty]$ and $K \in \mathbb{N}_0$ be arbitrary but fixed. We look for solutions u to

$$u_{tt}(t) + Bu(t) = f(t), \quad t \geq 0, \tag{13.31}$$

$$u(t) \in D(B), \quad t \geq 0, \tag{13.32}$$

$$u(0) = u_0, \quad u_t(0) = u_1, \tag{13.33}$$

$$u \in \bigcap_{j=2K}^{2K+2} C^j\big([0,T], W^{2K+2-j,2}(\Omega)\big). \tag{13.34}$$

Let

$$u_j := \left(\left(\frac{d}{dt}\right)^j u\right)(0) \equiv d_t^j u(0)$$

for $j = 0, 1, \ldots, 2K+2$. Then

$$u_j = \begin{cases} \sum_{k=0}^{\frac{j}{2}-1} (\Delta - m)^k f^{(j-2-2k)}(0) + (\Delta - m)^{\frac{j}{2}} u_0, & j \geq 2 \text{ even}, \\ \sum_{k=0}^{\frac{j-1}{2}-1} (\Delta - m)^k f^{(j-2-2k)}(0) + (\Delta - m)^{\frac{j-1}{2}} u_1, & j \geq 3 \text{ odd}, \end{cases} \tag{13.35}$$

where $j = 2, 3, \ldots, 2K+2$, and $f^{(m)} := d_t^m f$, $m \in \mathbb{N}_0$.
Exemplarily,

$$\begin{aligned} j = 2: \quad u_{tt}(0) &= (\Delta - m)u_0 + f(0), \\ j = 3: \quad u_{ttt}(0) &= (\Delta - m)u_1 + f_t(0), \\ j = 4: \quad d_t^4 u(0) &= (\Delta - m)u_{tt}(0) + f_{tt}(0) \\ &= (\Delta - m)^2 u_0 + (\Delta - m)f(0) + f_{tt}(0). \end{aligned}$$

Theorem 13.9 *Let $f \in C^{2K}([0,T], L^2(\Omega))$. If u is the solution to (13.31) – (13.34), then*

$$v := u + \sum_{j=0}^{K-1} (-B)^{-(j+1)} f^{(2j)} \tag{13.36}$$

is the solution to

$$v_{tt} + Bv = (-B)^{-K} f^{(2K)}, \quad t \geq 0, \tag{13.37}$$

$$v(t) \in D(B), \quad t \geq 0, \tag{13.38}$$

$$v(0) = (-B)^{-K} u_{2K}, \quad v_t(0) = (-B)^{-K} u_{2K+1}, \tag{13.39}$$

$$v \in \bigcap_{j=0}^{2} C^j\Big([0,T], W^{2-j,2}(\Omega)\Big), \tag{13.40}$$

and

$$v = (-B)^{-K} d_t^{2K} u. \tag{13.41}$$

Proof: Let

$$\tilde{v} := (-B)^{-K} d_t^{2K} u$$

then

$$\begin{aligned} (d_t^2 + B)\tilde{v} &= (-B)^{-K} d_t^{2K} (d_t^2 + B) u \\ &= (-B)^{-K} f^{(2K)}, \end{aligned}$$

hence $\tilde{v}$ solves (13.37) – (13.40). Defining

$$\tilde{u} := \tilde{v} - \sum_{j=0}^{K-1} (-B)^{-(j+1)} f^{(2j)},$$

we have

$$\begin{aligned}(d_t^2+B)\tilde{u} &= (d_t^2+B)\tilde{v} - \sum_{j=0}^{K-1}(d_t^2+B)(-B)^{-(j+1)}f^{(2j)}\\ &= (-B)^{-K}f^{(2K)} - \sum_{j=0}^{K-1}(-B)^{-(j+1)}f^{(2j+2)} - \sum_{j=0}^{K-1}(-B)^{-j}f^{(2j)}\\ &= f,\end{aligned}$$

$$\begin{aligned}\tilde{u}(0) &= \tilde{v}(0) - \sum_{j=0}^{K-1}(-B)^{-(j+1)}f^{(2j)}(0)\\ &= (-B)^{-K}u_{2K} - \sum_{j=0}^{K-1}(-B)^{-(j+1)}f^{(2j)}(0)\\ &= (-B)^{-K}\left(\sum_{k=0}^{K-1}(\Delta-m)^k f^{(2K-2-2k)}(0) + (\Delta-m)^K u_0\right)\\ &\quad -\sum_{j=0}^{K-1}(-B)^{-(j+1)}f^{(2j)}(0)\\ &= u_0 + \sum_{k=0}^{K-1}(-B)^{-(K-k)}f^{(2(K-k-1))}(0) - \sum_{j=0}^{K-1}(-B)^{-(j+1)}f^{(2j)}(0)\\ &= u_0,\end{aligned}$$

and, analogously,

$$\tilde{u}_t(0) = u_1.$$

Hence $\tilde{u}$ satisfies (13.31) – (13.33), so by uniqueness we get $\tilde{u} = u$, hence

$$\begin{aligned}\tilde{v} &= \tilde{u} + \sum_{j=0}^{K-1}(-B)^{-(j+1)}f^{(2j)}\\ &= u + \sum_{j=0}^{K-1}(-B)^{-(j+1)}f^{(2j)}\\ &= v,\end{aligned}$$

i.e.,

$$v = (-B)^K\, d_t^{2K} u$$

solves (13.37) – (13.40).

Q.E.D.

Before stating a general result on the L^p-L^q-decay, we formulate necessary compatibility conditions for the data.

Definition 13.10 *Let $K \in \mathbb{N}_0$. Then (u_0, u_1, f) satisfies the "compatibility condition of order 2K" if, for $j = 0, 1, \ldots, 2K+1$,*

$$u_j \in W_0^{1,2}(\Omega) \cap W^{2K+2-j,2}(\Omega) \cap W^{2K+2-j,1}(\Omega), \tag{13.42}$$

and

$$u_{2K+2} \in L^2(\Omega). \tag{13.43}$$

Theorem 13.11 *Let*

$$K \geq \frac{K_2+K_3-1}{2}, \qquad f \in \bigcap_{j=0}^{2K} C^j\left([0,T], W^{2K-j,2}(\Omega)\cap W^{2K-j,1}(\Omega)\right),$$

and let (u_0, u_1, f) satisfy the compatibility condition of order $2K$. Let $2 \leq q < \infty$ and $1/p + 1/q = 1$. Let u be the solution to (13.28) – (13.30) satisfying (13.34). Then we have

$$\begin{aligned}
&\|u(t), u_t(t), \nabla u(t)\|_{L^q(\Omega)} \\
&\quad \leq \frac{c}{(1+t)^{(1-\frac{2}{q})\frac{l}{2}}}\left(\|(u_0,u_1,\nabla u_0)\|_{W^{2K+1,p}(\Omega)} + \sum_{j=0}^{2K-2}\|f^{(j)}(0)\|_{W^{2K-j,p}(\Omega)}\right) \\
&\quad\quad + c\int_0^t \frac{1}{(1+t-\tau)^{(1-\frac{2}{q})\frac{l}{2}}}\|f^{(2K)}(\tau)\|_{L^p(\Omega)}d\tau \\
&\quad\quad + c\sum_{j=0}^{2K-1}\|f^{(j)}(t)\|_{W^{2K-1-j,p}(\Omega)},
\end{aligned}$$

where the constant $c > 0$ depends at most on m (and q).

Proof: 1. According to Theorem 13.9, the function v defined by (13.36) solves (13.37) – (13.41). We observe

$$u_{2K} \in W_0^{1,2}(\Omega)\cap W^{2,2}(\Omega) = D(B), \quad u_{2K+1} \in W_0^{1,2}(\Omega) = D(B^{1/2}),$$

and

$$\begin{aligned}
v_0 &:= v(0) = (-B)^{-K}u_{2K} \in D(B^{K+1}) \subset W^{2K+2,2}(\Omega), \\
v_1 &:= v_t(0) = (-B)^{-K}u_{2K+1} \in D(B^{K+1/2}) \subset W^{2K+1,2}(\Omega), \\
g(t) &:= (-B)^{-K}f^{2K}(t) \in D(B^K) \subset W^{2K,2}(\Omega), \quad t \geq 0.
\end{aligned}$$

We conclude for the solution $\hat{v}$ to

$$\hat{v}_{tt} + B\hat{v} = 0, \quad \hat{v}(0) = v_0, \quad \hat{v}_t(0) = v_1$$

from Theorem 13.8 (ii) that

$$\|\left(\hat{v}(t), \hat{v}_t(t), \nabla\hat{v}(t)\right)\|_{L^q(\Omega)} \leq \frac{c}{(1+t)^{(1-\frac{2}{q})\frac{l}{2}}}\|(v_0,v_1,\nabla v_0)\|_{W^{\tilde{N}_p,p}(\Omega)} \tag{13.44}$$

holds. The solution $\check{v}$ to

$$\check{v}_{tt} + B\check{v} = g, \quad \check{v}(0) = 0, \quad \check{v}_t(0) = 0$$

is given by

$$\check{v}(t) = \int_0^t v^\tau(t-\tau)d\tau,$$

where v^τ solves, for fixed $0 \le \tau \le t$,

$$v_{tt}^\tau + Bv^\tau = 0, \quad v^\tau(0) = 0, \quad v_t^\tau(0) = g(\tau).$$

Using Theorem 13.8 (i) we conclude, since $g(\tau) \in D(B^K)$,

$$\| \big(\check{v}(t), \check{v}_t(t), \nabla\check{v}(t)\big) \|_{L^q(\Omega)} \le c \int_0^t \frac{1}{(1+t-\tau)^{(1-\frac{2}{q})\frac{l}{2}}} \|g(\tau)\|_{W^{\widetilde{N}_p,p}(\Omega)} d\tau. \tag{13.45}$$

Combining (13.44), (13.45) we get for $v = \hat{v} + \check{v}$

$$\begin{aligned} \|\big(v(t), v_t(t), \nabla v(t)\big)\|_{L^q(\Omega)} \le\ & \frac{c}{(1+t)^{(1-\frac{2}{q})\frac{l}{2}}} \|(v_0, v_1, \nabla v_0)\|_{W^{N_p,p}(\Omega)} \\ & + c\int_0^t \frac{1}{(1+t-\tau)^{(1-\frac{2}{q})\frac{l}{2}}} \|g(\tau)\|_{W^{N_p,p}(\Omega)} d\tau. \end{aligned} \tag{13.46}$$

2. Since $p > 1$ we can apply Corollary 13.5 to conclude, using $N_p \le K_2 + K_3 - 1 \le 2K$,

$$\|g(\tau)\|_{W^{N_p,p}(\Omega)} \le c\|B^K g(\tau)\|_{L^p(\Omega)} = c\|f^{(2K)}(\tau)\|_{L^p(\Omega)}.$$

We also have $\widetilde{N}_p \le 2K+1$ and get

$$\|(v_0, v_1, \nabla v_0)\|_{W^{\widetilde{N}_p,p}(\Omega)} \le c\Big(\|u_{2K}\|_{W^{2,p}(\Omega)} + \|u_{2K+1}\|_{W^{1,p}(\Omega)}\Big).$$

Hence (13.46) turns into

$$\begin{aligned} \|\big(v(t), v_t(t), \nabla v(t)\big)\|_{L^q(\Omega)} \le\ & \frac{c}{(1+t)^{(1-\frac{2}{q})\frac{l}{2}}} \Big(\|u_{2K}\|_{W^{2,p}(\Omega)} + \|u_{2K+1}\|_{W^{1,p}(\Omega)}\Big) \\ & + c\int_0^t \frac{1}{(1+t-\tau)^{(1-\frac{2}{q})\frac{l}{2}}} \|f^{(2K)}(\tau)\|_{L^p(\Omega)} d\tau. \end{aligned} \tag{13.47}$$

3. By (13.36), i.e., $v = u + \sum_{j=0}^{K-1}(-B)^{-(j+1)} f^{(2j)}$, we have

$$\begin{aligned} \|\big(u(t), u_t(t), \nabla u(t)\big)\|_{L^q(\Omega)} \le\ & \|\big(v(t), v_t(t), \nabla v(t)\big)\|_{L^q(\Omega)} \\ & + \sum_{j=0}^{K-1} \Big(\|B^{-(j+1)} f^{(2j)}(t)\|_{L^q(\Omega)} + \|B^{-(j+1)} f^{(2j+1)}(t)\|_{L^q(\Omega)} \\ & + \|\nabla B^{-(j+1)} f^{(2j)}(t)\|_{L^q(\Omega)}\Big). \end{aligned} \tag{13.48}$$

Since $n \le 2K$ we conclude

$$\begin{aligned} \|B^{-(j+1)} f^{(2j)}(t)\|_{W^{1,q}(\Omega)} \le\ & c\|B^{-(j+1)} f^{(2j)}(t)\|_{W^{2K+1,p}(\Omega)} \\ \le\ & c\|f^{(2j)}(t)\|_{W^{2K-2j-1,p}(\Omega)}, \end{aligned} \tag{13.49}$$

and, analogously,

$$\|B^{-(j+1)} f^{(2j+1)}(t)\|_{L^q(\Omega)} \leq c\|f^{(2j+1)}(t)\|_{W^{2K-2j-2,p}(\Omega)}. \tag{13.50}$$

Combining (13.48) – (13.50) we obtain

$$\begin{aligned}\|\big(u(t), u_t(t), \nabla u(t)\big)\|_{L^q(\Omega)} &\leq \|\big(v(t), v_t(t), \nabla v(t)\big)\|_{L^q(\Omega)} \\ &\quad + c\sum_{j=0}^{2K-1} \|f^{(j)}(t)\|_{W^{2K-1-j,p}}.\end{aligned} \tag{13.51}$$

4. Finally, see (13.35), we have

$$\begin{aligned}\|u_{2K}\|_{W^{2,p}(\Omega)} &\leq \sum_{k=0}^{K-1} \|f^{(2K-2-2k)}(0)\|_{W^{2k+2,p}(\Omega)} + \|u_0\|_{W^{2K+2,p}}(\Omega) \\ &= \sum_{j=0}^{K-1} \|f^{(2j)}(0)\|_{W^{2K-2j,p}(\Omega)} + \|u_0\|_{W^{2k+2,p}}(\Omega),\end{aligned} \tag{13.52}$$

$$\begin{aligned}\|u_{2K+1}\|_{W^{1,p}(\Omega)} &\leq \sum_{k=0}^{K-1} \|f^{(2K-1-2k)}(0)\|_{W^{2k+1,p}(\Omega)} + \|u_1\|_{W^{2K+1,p}}(\Omega) \\ &= \sum_{j=0}^{K-1} \|f^{(2j)}(0)\|_{W^{2K-2j-1,p}(\Omega)} + \|u_1\|_{W^{2K+1,p}}(\Omega).\end{aligned} \tag{13.53}$$

Combining (13.47), (13.51) – (13.53), the assertion of the Theorem follows.

Q.E.D.

We conclude the linear part with optimality considerations. The following example will show that the decay rates proved above are optimal. This will be done in looking for a solution to the linear problem

$$u_{tt}(t,x) - \Delta u(t,x) + m\, u(t,x) = 0 \quad \text{in } [0,\infty) \times \Omega, \tag{13.54}$$

$$u(t,x) = 0 \quad \text{in } [0,\infty) \times \partial\Omega, \tag{13.55}$$

$$u(0,x) = u_0(x),\ u_t(0,x) = u_1(x) \quad \text{in } \Omega, \tag{13.56}$$

$m = const. \geq 0$, which has, as $t \to \infty$, exactly the L^∞-decay $\mathcal{O}\ (t^{-l/2})$. For this purpose let $(w_j)_{j\in\mathbb{N}}$ denote again the orthonormal system of eigenfunctions to the eigenvalues $(\lambda_j)_{j\in\mathbb{N}}$ of $-\Delta''$ in $\mathcal{B}$. Let $v_0 \in C_0^\infty(\mathbb{R}^l)$ and

$$\begin{aligned} u_0(x) = u_0(x', x'') &:= v_0(x')\, w_1(x'') \qquad \text{for } x \in \Omega, \\ u_1 &:= 0. \end{aligned}$$

For u with these data to be a solution to (13.54) – (13.56) it is sufficient and necessary to have

$$u(t,x) = v(t,x')\ w_1(x'') \tag{13.57}$$

with v satisfying

$$v_{tt}(t,x') - \Delta' v(t,x') + (m+\lambda_1)v(t,x') = 0 \quad \text{in } [0,\infty)\times \mathbb{R}^l, \tag{13.58}$$

$$v(0,x') = v_0(x'), \quad v_t(0,x') = 0 \quad \text{in } \mathbb{R}^l. \tag{13.59}$$

For this we observe for u satisfying (13.57)

$$-\Delta u(t,x) = (-\Delta' - \Delta'')\,(v(t,x')\; w_1(x'')) = (-\Delta' + \lambda_1)\; v(t,x')\; w_1(x'').$$

For the solution v to the Klein-Gordon system (13.58), (13.59), the asymptotic behavior is known to be (cp. [101])

$$v(t,x') = \frac{(m+\lambda_1)^{l/4}}{(2\pi)} \frac{\cos(t\sqrt{m+\lambda_1}+\frac{l\pi}{4})}{t^{l/2}} \int_{\mathbb{R}^l} v_0(x')\,dx' + \mathcal{O}\Big(\frac{1}{t^{(l+1)/2}}\Big) \tag{13.60}$$

as $t\to\infty$, for any fixed $x'\in\mathbb{R}^l$. So we get the sharp L^∞-decay $\mathcal{O}(t^{-l/2})$ for $u(t,x) = v(t,x')\; w_1(x'')$.

13.1.2 Nonlinear part

We turn to the fully nonlinear system, i.e., we look for the existence and for the asymptotic behavior of solutions u to

$$u_{tt} - \Delta u + mu = f(u,u_t,\nabla u,\nabla u_t,\nabla^2 u) \quad \text{in } [0,\infty)\times\Omega, \tag{13.61}$$

$$u = 0 \quad \text{in } [0,\infty)\times\partial\Omega, \tag{13.62}$$

$$u(0,\cdot) = u_0,\; u_t(0,\cdot) = u_1 \quad \text{in } \Omega, \tag{13.63}$$

where $m \geq 0$ as before.

In [167] nonlinear wave equations ($m=0$) were studied in *exterior* domains $\tilde\Omega$ (i.e. $\mathbb{R}^n\backslash\tilde\Omega$ is bounded), which are *non-trapping*. Nonlinearities of the type $f_1(t,x,u_t,\nabla u,\nabla u,\nabla^2 u)$ not involving u were considered there. The dependence on t and x could also be dealt with here but is just replaced by f as in (13.61) for simplicitiy. The methods from [167] also apply to the case $m>0$ in exterior domains. Here, we can also treat $f=f(u,\ldots)$ depending on u because of Poincaré's estimate which allows to estimate in each place norms of u by the corresponding norms of ∇u. The strategy in [167] consists in

(i) having a local existence theorem available,

(ii) proving L^p–L^q-estimates for the linearized system, and

(iii) proving a priori estimates for the local solution exploiting (ii).

This coarse description (i) – (iii), of course, reminds of the general scheme **A** – **E** from the Cauchy problem but part (ii) in section 4.1 in [167] uses the local energy decay property for non-trapping domains which is not available in our case. But we have already proved the general L^p–L^q-decay result in Theorem 13.11, and we can proceed getting a priori estimates as in sections 6 and 7 in [167].
A local existence theorem can be taken from Theorem 1.1 of [100].
Let

$$K, M \in \mathbb{N}, \quad M \geq K_2 + K_3 + 1, \quad K \geq 2M, \tag{13.64}$$

and suppose that

$$f \in C^{2K-1}(\mathbb{R}^{n^2+2n+2}), \quad |f(W)| = O(|W|^{\alpha+1}) \text{ as } |W| \to 0, \tag{13.65}$$

where

$$\alpha \geq \alpha(l) := \begin{cases} 3, & \text{if } l = 1, \\ 2, & \text{if } 2 \leq l \leq 4, \\ 1, & \text{if } l \geq 5. \end{cases} \tag{13.66}$$

Let

$$q(l) := 2\alpha(l) + 2, \qquad p(l) := \frac{2\alpha(l)+2}{2\alpha(l)+1} \tag{13.67}$$

be associated Hölder exponents, and let

$$d(l) := \left(1 - \frac{2}{q(l)}\right)\frac{l}{2} = \frac{\alpha(l)}{\alpha(l)+1}\frac{l}{2} = \begin{cases} 3/8, & \text{if } l = 1, \\ 2/3, & \text{if } l = 2, \\ 1, & \text{if } l = 3, \\ 4/3, & \text{if } l = 4, \\ l/4, & \text{if } l \geq 5. \end{cases} \tag{13.68}$$

The number $d(l)$ is the decay rate of the $L^{q(l)}$-norm $\|u(t)\|_{L^{q(l)}}$ for the linearized problem, $\alpha = \alpha(l)$ is determined by the condition

$$\frac{1}{\alpha}\left(1 + \frac{1}{\alpha}\right) < \frac{l}{2}, \tag{13.69}$$

and then q resp. p by

$$\frac{1}{q} + \frac{1}{p} = 1 \quad \text{and} \quad \frac{\alpha}{q} + \frac{1}{2} = \frac{1}{p}, \tag{13.70}$$

as in the Cauchy problem. Denoting by $\overline{D}^k v$ all derivatives of v in t and x up to order k, we have

Theorem 13.12 *Suppose that $\Omega = \mathbb{R}^l \times \mathcal{B}$, where $\mathcal{B} \subset \mathbb{R}^{n-l}$ is bounded with smooth boundary $\partial\mathcal{B}$, and that (13.64) – (13.68) hold. Then there is $0 < \varepsilon < 1$ such that if*

$u_0 \in W^{2K,2}(\Omega) \cap W^{2K-1,p(l)}(\Omega)$, $u_1 \in W^{2K-1,2}(\Omega) \cap W^{2K-2,p(l)}(\Omega)$ and (u_0, u_1, f) satisfies a compatibility condition corresponding to (13.42), (13.43) of order $2K$, and

$$\|u_0\|_{W^{2K,2}(\Omega)} + \|u_0\|_{W^{2K-1,p(l)}(\Omega)} + \|u_1\|_{W^{2K-1,2}(\Omega)} + \|u_1\|_{W^{2K-2,p(l)}(\Omega)} < \varepsilon,$$

then there exists a unique solution

$$u \in \bigcap_{j=0}^{2K} C^j\left([0,\infty), W^{2K-j,2}(\Omega)\right) \subset C^2\left([0,\infty) \times \overline{\Omega}\right)$$

to (13.61) – (13.63) satisfying

$$\sup_{t\geq 0}\left(\|\overline{D}^K u(t)\|_{L^2(\Omega)} + (1+t)^{d(l)}\|\overline{D}^{K-M} u(t)\|_{L^{q(l)}(\Omega)}\right) \leq c_1, \tag{13.71}$$

where the constant c_1 depends at most on l, m and Ω.

PROOF: Since the local existence theorem guarantees that the local solution

$$u \in \bigcap_{j=0}^{2K} C^j\ \left([0,T), W^{2K-j,2}(\Omega)\right) \qquad \text{for some } T > 0,$$

can be continued with respect to $t \geq T$ as long as u is sufficiently small, it suffices to prove (13.71). The (lengthy) details in the spirit of our general scheme can be found in [101] or [12].

Q.E.D.

Looking at the Cauchy problem or at the case of exterior domains, one might expect sharper results for a few (unbounded) spatial dimensions $l = 1, 2, 3, 4$ – sharper with respect to the admissible $\alpha = \alpha(l)$, cp. (13.66). This cases have been investigated in part by Metcalfe, Sogge, Stewart and Perry in [123, 124].

13.2 Schrödinger equations

For Schrödinger equations

$$u_t - i\Delta u = f(u, \nabla u), \tag{13.72}$$

where $u = u(t,x) \in \mathbb{C}, t \geq 0, x \in \Omega \subset \mathbb{R}^n, \Omega$ a waveguide as before, with initial condition

$$u(0,\cdot) = u_0, \tag{13.73}$$

and Dirichlet boundary condition

$$u(t,\cdot) = 0 \text{ on } \partial\Omega, \tag{13.74}$$

we can argue as for nonlinear wave equations, cp. [14]. We assume for the smooth nonlinear function f, that

$$\frac{\partial f(W)}{\partial(\partial_j u)} \text{ is real}, \quad 1 \leq j \leq n, \tag{13.75}$$

and

$$|f(W)| = \mathcal{O}(|W|^{\alpha+1}) \quad \text{as } |W| \to 0 \tag{13.76}$$

hold for some $\alpha \in \mathbb{N}$.

The decay of solutions to the linearized problem, where $f = f(t,x)$, is first given for $f = 0$.

Theorem 13.13 *Let*

$$K_1 := 2(n-l)+4, \quad K_2 := l+1+\left[\frac{n-l+1}{2}\right],$$

and

$$u_0 \in D(A^{K_1/2}) \cap W^{K_1+K_2,1}(\Omega).$$

Then the unique solution u to (13.72) – (13.74), with $f=0$, satisfies for $2 \le q \le \infty$, $1/p+1/q=1$, and for $t \ge 0$:

$$\|u(t)\|_{L^q(\Omega)} \le \frac{c}{(1+t)^{(1-\frac{2}{q})\frac{l}{2}}} \|u_0\|_{W^{N_p,p}(\Omega)},$$

where

$$N_p := \begin{cases} (1-\frac{2}{q})(K_1+K_2)-1, & \text{if } q \in \{2,\infty\}, \\ [(1-\frac{2}{q})(K_1+K_2)], & \text{if } 2<q<\infty. \end{cases}$$

The positive constant c depends at most on q.

Considering linearized equations with $f = f(t,x)$, the compatibility condition of order $K \in \mathbb{N}_0$ for (u_0, f) is given by requiring

$$u_j \in W^{2K+2-2j,2}(\Omega) \cap W_0^{1,2}(\Omega), \quad j = 0,1,\dots,K, \tag{13.77}$$

$$u_{K+1} \in L^2(\Omega), \tag{13.78}$$

where

$$u_j := \left(d_t^j u\right)(0,\cdot) = (i\Delta)^j u_0 + \sum_{k=0}^{j-1} (i\Delta)^k f^{(j-1-k)}(0,\cdot).$$

The L^p-L^q-decay for the linearized equations with right-hand side $f = f(t,x)$ is then given by

Theorem 13.14 *Let K_1, K_2 be given as in Theorem 13.13,*

$$K \ge \frac{K_1+K_2}{2},$$

$$f \in \bigcap_{j=0}^{K-1} C^j\Big([0,\infty),\, W^{2K-2-2j,2}(\Omega) \cap W^{2K-2-2j,1}(\Omega)\Big),$$

$$f \in C^K\Big([0,\infty),\, L^2(\Omega)\cap L^1(\Omega)\Big),$$

and let (u_0, f) *satisfy the compatibility condition (13.77), (13.78) of order* K. *Let* $2 \le q < \infty$ *and* $1/p + 1/q = 1$. *Then the unique solution* u *to (13.72) – (13.74), with* $f = f(t,x)$, *satisfies*

$$\|u(t)\|_{L^q(\Omega)} \le \frac{c}{(1+t)^{(1-\frac{2}{q})\frac{l}{2}}}\Big(\|u_0\|_{W^{2K,p}(\Omega)} + \sum_{j=0}^{K-1}\|f^{(j)}(0)\|_{W^{2K-2-2j,\,p}}\Big)$$

$$+ c\int_0^t \frac{1}{(1+t-\tau)^{(1-\frac{2}{q})\frac{l}{2}}}\|f^{(K)}(\tau)\|_{L^p(\Omega)}d\tau$$

$$+ c\sum_{j=0}^{K-1}\|f^{(j)}(t)\|_{W^{2K-2-2j,\,p}}(\Omega),$$

where the constant $c > 0$ *depends at most on* q.

For the nonlinear system (13.72) – (13.76) we assume

$$\alpha \ge \alpha(l) := \begin{cases} 3, & \text{if} \quad l = 1, \\ 2, & \text{if} \quad 2 \le l \le 4, \\ 1, & \text{if} \quad l \ge 5, \end{cases} \tag{13.79}$$

$$q(l) := 2\alpha(l) + 2, \quad p(l) := \frac{2\alpha(l)+2}{2\alpha(l)+1}, \tag{13.80}$$

and

$$d(l) := \left(1 - \frac{1}{2q(l)}\right)\frac{l}{2}, \tag{13.81}$$

cp. (13.66) – (13.68). Let

$$K, M \in \mathbb{N}, \quad M \ge 2\left[\frac{K_1 + K_2 + 1}{2}\right] + 1, \quad K \ge 2M, \tag{13.82}$$

where K_1, K_2 are given in Theorem 13.13.

Theorem 13.15 *Suppose that* $\Omega = \mathbb{R}^l \times \mathcal{B}$, *where* $\mathcal{B} \subset \mathbb{R}^{n-l}$ *is bounded with smooth boundary* $\partial\mathcal{B}$, *and that (13.79) – (13.82) hold. Then there is* $0 < \varepsilon < 1$ *such that if* $u_0 \in W^{2K,2}(\Omega) \cap W^{2K,p(l)}(\Omega)$ *and* (u_0, f) *satisfy a compatibility condition corresponding to (13.77), (13.78) of order* K, *and*

$$\|u_0\|_{W^{2K,2}(\Omega)} + \|u_0\|_{W^{2K,p\,(l)}(\Omega)} < \varepsilon,$$

then there exists a unique solution

$$u \in \bigcap_{j=0}^{K} C^j\Big([0,\infty),\, W^{2K-2j,2}(\Omega)\Big),$$

to (13.72) – (13.76) satisfying

$$\sup_{t\ge 0}\Big(\|\bar{D}^K u(t)\|_{L^2(\Omega)} + (1+t)^{d(l)}\|\bar{D}^{K-M}u(t)\|_{L^{q(l)}(\Omega)} \le c_1\Big),$$

where the constant c_1 *depends at most on* l *and* Ω.

We recall that for Cauchy problems we have local existence theorems for more general, quasilinear Schrödinger equations, see [115, 116]. For Theorem 13.15 we may use [83].

13.3 Equations of elasticity and Maxwell equations

We consider, following [102], the equations of (homogeneous, isotropic) elasticity for the displacement vector u,

$$u_{tt} - \mu\Delta u - (\mu+\lambda)\nabla \operatorname{div} u = f, \tag{13.83}$$

where $u : [0,\infty) \times \Omega \to \mathbb{R}^n$, $n = 2, 3$, with the Lamé constants λ, μ satisfying $\mu > 0$, $2\mu + n\lambda > 0$, cp. [63], and where

$$\Omega = \mathbb{R}^l \times \mathcal{B}, \quad \mathcal{B} \subset \mathbb{R}^{n-l} \text{ bounded},$$

is a waveguide as before. $f : [0,\infty) \times \Omega \to \mathbb{R}^n$ is a given function, and Δu is to be read in each component. In contrast to the situation with the Laplace operator appearing in wave equations,

$$\Delta = \Delta' + \Delta'',$$

the operator of elasticity

$$E = \mu\Delta + (\mu+\lambda)\nabla \operatorname{div}$$

does not split up into $E = E' + E''$, where E' and. E'' only acts on x' and x'', respectively, since, for example for $n = 2$,

$$E = \begin{pmatrix} \mu\Delta + (\mu+\lambda)\partial_1^2 & (\mu+\lambda)\partial_1\partial_2 \\ & \\ (\mu+\lambda)\partial_2\partial_1 & \mu\Delta + (\mu+\lambda)\partial_2^2 \end{pmatrix}.$$

We shall consider the equations (13.83) together with initial conditions

$$u(0,\cdot) = u^0, \quad u_t(0,\cdot) = u^1, \tag{13.84}$$

and with Maxwell type boundary conditions

$$\nu \times u(t,\cdot) = 0, \quad \operatorname{div} u(t,\cdot) = 0 \quad \text{on } \partial\Omega, \tag{13.85}$$

or

$$\nu \cdot u(t,\cdot) = 0, \quad \nu \times (\nabla \times u(t,\cdot)) = 0 \quad \text{on } \partial\Omega. \tag{13.86}$$

Below, we shall see that the boundary conditions (13.85) resp. (13.86), which are typical for Maxwell equations, can be read for waveguides as common Dirichlet or Neumann type boundary conditions. The reason for choosing these boundary conditions is connected with the problem of non-splitting E into $E' + E''$ mentioned above and the idea of overcoming this difficulty by projection techniques. To understand this we think for a moment of the Cauchy problem with $\Omega = \mathbb{R}^3$, where we have the orthogonal decomposition of L^2-vector fields,

$$\left(L^2(\mathbb{R}^3)\right)^3 = \overline{\nabla W^{1,2}(\mathbb{R}^3)} \oplus D_0(\mathbb{R}^3), \tag{13.87}$$

with $D_0(\mathbb{R}^3)$ denoting the fields with divergence zero; and $\overline{\nabla W^{1,2}(\mathbb{R}^3)}$ is the L^2-closure of gradients of functions in $W^{1,2}(\mathbb{R}^3)$. Decomposing the displacement vector

$$u = u^{po} + u^s$$

correspondingly, we obtain a decomposition of (13.83) into

$$u^{po}_{tt} - (2\mu + \lambda)\Delta u^{po} = f^{po}, \quad u^s_{tt} - \mu\Delta u^s = f^s. \tag{13.88}$$

For this we use

$$\nabla \times u^{po} = 0, \quad \operatorname{div} u^s = 0,$$

and the formula

$$\Delta = \nabla \operatorname{div} - \nabla \times \nabla\times \tag{13.89}$$

turning (13.83) into

$$u_{tt} - (2\mu + \lambda)\nabla \operatorname{div} u + \mu\nabla \times \nabla \times u = f$$

yielding

$$\left(u^{po}_{tt} - (2\mu + \lambda)\nabla \operatorname{div} u^{po}\right) + \left(u^s_{tt} + \mu\nabla \times \nabla \times u^s\right) = f^{po} + f^s,$$

or (13.88).

Unfortunately, the decomposition (13.87) and its variants in domains with boundaries are not compatible with the usual Dirichlet boundary conditions (similar for the corresponding elastic Neumann boundary conditions). But it turns out that the Maxwell boundary conditions (13.85) or (13.86) are compatible with the following variants of (13.87).

Before specifying these decompositions, we remark that we also have corresponding decompositions in two space dimensions, as well as formula (13.89), if we define for a vector field $H : \mathbb{R}^2 \to \mathbb{R}^2$ and a scalar function $h : \mathbb{R}^2 \to \mathbb{R}$

$$\nabla \times H := \partial_1 H_2 - \partial_2 H_1, \quad \nabla \times h := \begin{pmatrix} \partial_2 h \\ -\partial_1 h \end{pmatrix}.$$

Now Ω being a waveguide again, we use the following orthogonal decomposition in case we consider the boundary conditions (13.85):

$$\left(L^2(\Omega)\right)^n = \overline{\nabla W^{1,2}_0(\Omega)} \oplus D_0(\Omega), \tag{13.90}$$

which follows from the projection theorem. Decomposing u correspondingly, we have

$$u = u^{po} + u^s, \quad u^{po}(t,\cdot) \in \overline{\nabla W_0^{1,2}(\Omega)}, \quad u^s(t,\cdot) \in D_0(\Omega).$$

The compatibility of the boundary conditions (13.85) with the decomposition (13.90) is reflected in a decomposition of the boundary conditions as follows.
u^s satisfies

$$u^s_{tt} + \mu\nabla \times \nabla \times u^s = f^s, \quad \operatorname{div} u^s = 0,$$

$$u^s(0,\cdot) = u^{0,s}, \quad u^s_t(0,\cdot) = u^{1,s},$$

$$\nu \times u^s(t,\cdot) = 0 \quad \text{on } \partial\Omega. \tag{13.91}$$

The boundary condition (13.91) will be satisfied in the weak sense,

$$u^s(t,\cdot) \in R^0(\Omega),$$

where $R^0(\Omega)$ generalizes the classical boundary condition as usual, for $n = 3$:

$$R^0(\Omega) := \Big\{ v \in \big(L^2(\Omega)\big)^n \,|\, \nabla \times v \in \big(L^2(\Omega)\big)^n, \text{ and}$$

$$\forall F \in \big(L^2(\Omega)\big)^n, \nabla \times F \in \big(L^2(\Omega)\big)^n : \int_\Omega v(\nabla \times F) = \int_\Omega (\nabla \times v) F \Big\},$$

for $n = 2$:

$$R^0(\Omega) := W_0^{1,2}(\Omega).$$

We remark that $R^0(\Omega)$ equals the completion of C_0^∞-fields with respect to the norm (cf. [98])

$$\|\cdot\|_R := \big(\|\cdot\|^2_{L^2} + \|\nabla \times \cdot\|^2_{L^2}\big)^{1/2}.$$

u^{po} satisfies

$$u^{po}_{tt} - (2\mu + \lambda)\nabla \operatorname{div} u^{po} = f^{po}, \quad u^{po} \in \overline{\nabla W_0^{1,2}(\Omega)},$$

$$u^{po}(0,\cdot) = u^{0,po}, \quad u^{po}_t(0,\cdot) = u^{1,po},$$

$$\nu \times u^{po}(t,\cdot) = 0, \quad \operatorname{div} u^{po}(t,\cdot) = 0 \ \text{ on } \partial\Omega.$$

The boundary condition $\nu \times u^{po}(t,\cdot) = 0$ is automatically satisfied since $u^{po} \in \overline{\nabla W_0^{1,2}(\Omega)} \subset R^0(\Omega)$. Thus, we obtain both for $\beta = po$ and $\beta = s$ that

$$u^\beta_{tt} - \tau_\beta \Delta u^\beta = f^\beta, \tag{13.92}$$

$$u^\beta(0,\cdot) = u^{0,\beta}, \quad u^\beta_t(0,\cdot) = u^{1,\beta}, \tag{13.93}$$

$$\nu \times u^\beta(t,\cdot) = 0, \quad \operatorname{div} u^\beta(t,\cdot) = 0 \ \text{ on } \partial\Omega, \tag{13.94}$$

with

$$\tau_\beta := \begin{cases} 2\mu + \lambda, & \text{for } \beta = po, \\ \mu, & \text{for } \beta = s \end{cases}$$

being positive.

The initial-boundary value problem (13.92) – (13.94) is of Maxwell type corresponding to the second-order equation for the electric field with so-called electric boundary conditions, see [98, 190, 191, 192]. The existence theory is well-known.

Turning to the boundary condition (13.86), we use the orthogonal decomposition

$$\left(L^2(\Omega)\right)^n = R_0(\Omega) \oplus \overline{\nabla \times R^0(\Omega)}, \tag{13.95}$$

where $R_0(\Omega)$ denotes the fields with vanishing rotation. u is now decomposed into

$$u = u^{po} + u^s, \quad u^{po}(t,\cdot) \in R_0(\Omega), \quad u^s(t,\cdot) \in \overline{\nabla \times R^0(\Omega)}.$$

Similar agreements as above yield (13.92), (13.93) and now, for $\beta \in \{po, s\}$,

$$\nu \times (\nabla \times u^\beta(t,\cdot)) = 0, \quad \nu \cdot u^\beta(t,\cdot) = 0 \ \text{ on } \partial\Omega. \tag{13.96}$$

The first part of the boundary conditions is interpreted in the sense $\nabla \times u^\beta(t,\cdot) \in R^0(\Omega)$. The second part is formulated weakly by saying $u^\beta \in D^0(\Omega)$ where

$$D^0(\Omega) := \left\{ v \in \left(L^2(\Omega)\right)^n \,\middle|\, \operatorname{div} v \in L^2(\Omega) \text{ and} : \forall g \in W^{1,2}(\Omega) : \int_\Omega v(\nabla g) = -\int_\Omega (\operatorname{div} v) g \right\}.$$

The space $D^0(\Omega)$ equals the completion of C_0^∞-fields with respect to the norm (cf. [98])

$$\|\cdot\|_D := \left(\|\cdot\|_{L^2}^2 + \|\operatorname{div}\cdot\|_{L^2}^2\right)^{1/2}.$$

The initial-boundary value problem (13.92), (13.93), (13.96) is of Maxwell type corresponding to so-called magnetic boundary conditions, and the existence theory is well-known too.

Consequently, in order to finally obtain decay rates for the displacement vector, we will look at the Maxwell equations under electric and magnetic boundary conditions, respectively.

Before proceeding in this direction, we examine the electric boundary condition (13.85) and the magnetic boundary condition (13.86) for the displacement vector u in a waveguide. It turns out that, in many cases, these boundary conditions in waveguides have a meaning in terms of natural Dirichlet or Neumann type boundary conditions for the components of the displacement vector.

First, we consider the two-dimensional case, where we have essentially only one situation, namely Ω being a strip with cross section $(0,1)$ without loss of generality,

$$\Omega = \mathbb{R} \times (0,1).$$

The first boundary conditions are (13.85), which are in two space dimensions equivalent to

$$-\nu_2 u_1 + \nu_1 u_2 = 0, \quad \partial_1 u_1 + \partial_2 u_2 = 0 \ \text{ on } \partial\Omega.$$

Since $\nu = (0, \pm 1)'$ and $\partial/\partial\nu = \pm\partial_2$, this is equivalent to

$$u_1 = \frac{\partial}{\partial\nu} u_2 = 0 \quad \text{ on } \partial\Omega.$$

Hence it represents a free movement in the normal direction and no shear movement. The second boundary conditions are (13.86) or, equivalently,

$$\nu_1 u_1 + \nu_2 u_2 = 0, \quad \nu_2(-\partial_2 u_1 + \partial_1 u_2) = 0, \quad \nu_1(\partial_2 u_1 - \partial_1 u_2) = 0,$$

which is, in view of $\nu = (0, \pm 1)'$, equivalent to

$$\frac{\partial}{\partial\nu} u_1 = u_2 = 0 \quad \text{ on } \partial\Omega,$$

representing a free shear movement.

Second, we consider the three-dimensional case $n = 3$ with $l = 2$, where Ω, essentially, represents the region between two planes, i.e. without loss of generality

$$\Omega = \mathbb{R}^2 \times (0, 1).$$

The first boundary conditions (13.85) are now equivalent to

$$\begin{aligned} \nu_2 u_3 - \nu_3 u_2 &= 0, \\ \nu_3 u_1 - \nu_1 u_3 &= 0, \\ \nu_1 u_2 - \nu_2 u_1 &= 0, \\ \partial_1 u_1 + \partial_2 u_2 + \partial_3 u_3 &= 0. \end{aligned} \tag{13.97}$$

Observing $\nu = (0, 0, \pm 1)$ and $\partial/\partial\nu = \pm\partial_3$, we obtain

$$u_1 = u_2 = \frac{\partial}{\partial\nu} u_3 = 0 \quad \text{ on } \partial\Omega, \tag{13.98}$$

with a, now, obvious interpretation. The second boundary conditions (13.86) are

$$\begin{aligned} \nu_1 u_1 + \nu_2 u_2 + \nu_3 u_3 &= 0, \\ \nu_2(\partial_1 u_2 - \partial_2 u_1) - \nu_3(\partial_3 u_1 - \partial_1 u_3) &= 0, \\ \nu_3(\partial_2 u_3 - \partial_3 u_2) - \nu_1(\partial_1 u_2 - \partial_2 u_1) &= 0, \\ \nu_1(\partial_3 u_1 - \partial_1 u_3) - \nu_2(\partial_2 u_3 - \partial_3 u_2) &= 0, \end{aligned} \tag{13.99}$$

or, equivalently,

$$\frac{\partial}{\partial\nu} u_1 = \frac{\partial}{\partial\nu} u_2 = u_3 = 0 \quad \text{ on } \partial\Omega. \tag{13.100}$$

Third, we have the three-dimensional infinite cylinder, $n = 3, l = 1$,

$$\Omega = \mathbb{R} \times \mathcal{B},$$

where $\mathcal{B} \subset \mathbb{R}^2$ is a bounded domain.
Observing $\nu = (0, \nu_2, \nu_3)'$, we obtain for the first boundary conditions (13.85) from (13.97)

$$u_1 = 0, \quad \nu_2 u_3 - \nu_3 u_2 = 0, \quad \partial_2 u_2 + \partial_3 u_3 = 0. \tag{13.101}$$

For the second boundary conditions (13.86) we get from (13.99)

$$\begin{aligned} \nu_2 u_2 + \nu_3 u_3 &= 0, \\ \nu_2(\partial_1 u_2 - \partial_2 u_1) - \nu_3(\partial_3 u_1 - \partial_1 u_3) &= 0, \\ \partial_2 u_3 - \partial_3 u_2 &= 0. \end{aligned} \tag{13.102}$$

The boundary conditions (13.101) and (13.102), respectively, become more transparent for cylindrically symmetrical domains $\mathcal{B}$, where Ω is a classical cylinder, and this means for $\mathcal{B}$ that

$$x'' \in \mathcal{B} \Rightarrow Rx'' \in \mathcal{B}$$

for all $R \in O(2)$, the set of orthogonal real 2×2 matrices. Typical examples for $\mathcal{B}$ are balls or annular domains.
We call a vector field $u : \Omega \to \mathbb{R}^3$ *cylindrically symmetrical* if we have for all $x_1 \in \mathbb{R}$, $x'' = (x_2, x_3) \in \mathcal{B}$ and $R \in O(2)$:

$$u_1(x_1, Rx'') = u_1(x_1, x''), \quad (u_2, u_3)'(x_1, Rx'') = R(u_2, u_3)'(x_1, x'').$$

That is, u is cylindrically symmetrical if, for fixed x_1, the first component $u_1(x_1, \cdot)$ as a scalar field, and the second and third components together as a vector field $(u_2, u_3)'(x_1, \cdot)$ are radially symmetrical in $\mathcal{B}$. Therefore, we have the following characterizations, cp. [63],

Lemma 13.16 *$u : \Omega \to \mathbb{R}^3$ is cylindrically symmetrical $\Leftrightarrow$ there exist functions*

$$h, \phi : \mathbb{R} \times \mathbb{R}_0^+ \to \mathbb{R}$$

such that for all $(x_1, x'') \in \Omega$

$$u_1(x_1, x'') = h(x_1, r), \quad (u_2, u_3)'(x_1, x'') = \phi(x_1, r)x''$$

holds, where $r := |x''| = \sqrt{x_2^2 + x_3^2}$.

Our initial-boundary value problem (13.83) – (13.85), resp. (13.83) – (13.86), turns out to be cylindrically invariant, i.e. we have

Lemma 13.17 *Let Ω be cylindrically symmetrical. If the data u^0, u^1 are cylindrically symmetrical, then the solution $u(t,\cdot)$ to (13.83) – (13.85), resp. (13.83) – (13.86), with $f=0$, is cylindrically symmetrical for all $t \geq 0$.*

PROOF: Let $R = (r_{jk})_{1\leq j,k\leq 2} \in O(2)$, and let

$$\widetilde{R} := \begin{pmatrix} 1 & 0 & 0 \\ 0 & r_{11} & r_{12} \\ 0 & r_{21} & r_{22} \end{pmatrix}.$$

Then $\widetilde{R}$ is an orthogonal 3×3 matrix. For $t\geq 0$ and $x=(x_1,x'')\in\Omega$ let

$$v(t,x) := \widetilde{R}' u(t,\widetilde{R}x).$$

Since

$$v_{tt}(t,x) = \widetilde{R}'u_{tt}(t,\widetilde{R}x), \quad \Delta v(t,x) = \widetilde{R}'(\Delta u)(t,\widetilde{R}x), \quad \nabla \operatorname{div} v(t,x) = \widetilde{R}'(\nabla \operatorname{div} u)(t,\widetilde{R}x),$$

we conclude that v satisfies the same differential equation as u and has the same initial values. By uniqueness of solutions it only remains to show that v satisfies the same boundary conditions as u, that is, the invariance of the boundary conditions under cylindrical symmetry.

For the first boundary conditions (13.85) this can be seen as follows. First note that

$$\begin{pmatrix} \nu_2(x_1,Rx'') \\ \nu_3(x_1,Rx'') \end{pmatrix} = R \begin{pmatrix} \nu_2(x_1,x'') \\ \nu_3(x_1,x'') \end{pmatrix},$$

and thus, using $\nu_1 = 0$,

$$\nu(\widetilde{R}x) = \widetilde{R}\nu(x).$$

This implies

$$\begin{aligned} \nu(x)\times \widetilde{R}'u(t,\widetilde{R}x) &= \left(\widetilde{R}'\widetilde{R}\nu(x)\right)\times \widetilde{R}'u(t,\widetilde{R}x) \\ &= \det(R)\widetilde{R}'\left(\widetilde{R}\nu(x)\times u(t,\widetilde{R}x)\right) \\ &= \det(R)\widetilde{R}'\left(\nu(\widetilde{R}x)\times u(t,\widetilde{R}x)\right) \\ &= 0 \quad \text{on } \partial\Omega \end{aligned}$$

since $\nu\times u(t,\cdot)=0$ on $\partial\Omega$. Hence

$$\nu \times v(t,\cdot) = 0 \quad \text{on } \partial\Omega. \tag{13.103}$$

A short calculation shows

$$\operatorname{div}\left(\widetilde{R}'\,u(t,\widetilde{R}x)\right) = (\operatorname{div} u)(t,\widetilde{R}x).$$

Therefore, we have

$$\operatorname{div} v(t, \cdot) = 0 \quad \text{on } \partial\Omega, \tag{13.104}$$

since $\operatorname{div} u(t, \cdot) = 0$ on $\partial\Omega$. By (13.103), (13.104), v satisfies the same boundary conditions (13.85) as u. For the second boundary conditions (13.86) one has on $\partial\Omega$

$$\nu(x)\, v(t,x) = \left(\widetilde{R}\,\nu(x)\right) u(t, \widetilde{R}x) = \nu(\widetilde{R}x)\, u(t, \widetilde{R}x) = 0. \tag{13.105}$$

Using

$$\nabla \times \left(\widetilde{R}' u(t, \widetilde{R}x)\right) = (\det R)\widetilde{R}'(\nabla \times u)(t, \widetilde{R}x)$$

we obtain in the same way

$$\nu(x) \times \left(\nabla \times v(t,x)\right) = \widetilde{R}'\left(\nu(\widetilde{R}x) \times (\nabla \times u)(t, \widetilde{R}x)\right) = 0 \tag{13.106}$$

on $\partial\Omega$. By (13.105), (13.106), v satisfies the same boundary conditions (13.86) as u.

Q.E.D.

For a cylindrically symmetrical solution

$$u(t, x_1, x'') = \begin{pmatrix} h(t, x_1, r) \\ \phi(t, x_1, r)x'' \end{pmatrix},$$

according to Lemma 13.16, we can now write the first boundary conditions (13.85) as

$$h = 0, \quad 2\phi + r\phi_r = 0 \quad \text{on } \partial\Omega,$$

cp. (13.101).

The second boundary conditions (13.86), resp. (13.102), can be rewritten as

$$h_r = 0, \quad \phi = 0 \quad \text{on } \partial\Omega, \tag{13.107}$$

since

$$\nu_2 u_2 + \nu_3 u_3 = r\phi$$

and

$$\nu_2(\partial_1 u_2 - \partial_2 u_1) - \nu_3(\partial_3 u_1 - \partial_1 u_3) = r(\phi_{x_1} - h_r).$$

In terms of u we have from (13.107)

$$\frac{\partial}{\partial \nu} u_1 = u_2 = u_3 = 0 \quad \text{on } \partial\Omega,$$

cp. (13.98).

Concerning the boundary conditions we remark that the case (13.85) was already studied for elasticity by Weyl[2] [196].

[2]Weyl gave a motivation as follows: *"Sie* [the boundary condition (13.85)] *wird für uns dadurch wesentlich, dass sie nach dem Schema ≪ Elastischer Körper ⟶* FRESNELS *elastischer Aether ⟶ elektromagnetischer Aether ≫ den Übergang von der Elastizitätstheorie zur Potentialtheorie zu Wege bringt."*

Now, coming back to the Maxwell initial-boundary value problems (13.92), (13.93), (13.94) resp. (13.96), we define the two Maxwell operators M_1, M_2 with

$$M_j : D(M_j) \subset \left(L^2(\Omega)\right)^n \longrightarrow \left(L^2(\Omega)\right)^n, \quad j = 1, 2,$$

by

$$D(M_1) := \left\{u \in \left(L^2(\Omega)\right)^n \mid u \in R^0(\Omega), \operatorname{div} u \in W_0^{1,2}(\Omega), \Delta u \in \left(L^2(\Omega)\right)^n\right\},$$

$$D(M_2) := \left\{u \in \left(L^2(\Omega)\right)^n \mid u \in D^0(\Omega), \nabla \times u \in R^0(\Omega), \Delta u \in \left(L^2(\Omega)\right)^n\right\},$$

and

$$M_j u := -\tau \Delta u,$$

where $\tau = \mu$ resp. $\tau = 2\mu + \lambda$. It is known ([193, 194]) that M_j is a positive self-adjoint operator with purely continuous spectrum

$$\sigma(M_j) = [\delta, \infty), \quad j = 1, 2,$$

where δ satisfies

$$\delta \left\{ \begin{array}{c} > \\ = \end{array} \right\} 0 \quad \text{if} \left\{ \begin{array}{l} j = 1 \text{ and } \mathcal{B} \text{ is simply connected} \\ j = 2 \text{ or } \mathcal{B} \text{ is multiply connected} \end{array} \right\}.$$

The following assertion on L^p-regularity for the Maxwell operators is an extension from [96], where the case of a bounded domain is studied.

Lemma 13.18 *Let $m \in \mathbb{N}_0$, $u \in D(M_j)$, $j = 1$ or $j = 2$, $1 < p < \infty$, $u \in \left(L^p(\Omega)\right)^n$, $M_j u \in \left(W^{m,p}(\Omega)\right)^n$. Then $u \in \left(W^{m+2,p}(\Omega)\right)^n$ and*

$$\|u\|_{W^{m+2,p}(\Omega)} \leq c\|(M_j + 1)\|_{W^{m,p}(\Omega)},$$

where $c > 0$ is a constant at most depending on m and p (and j).

To apply again the methods used for the classical wave equation with Dirichlet boundary conditions in Chapter 13.1, we need knowledge of the eigenvalue distribution for the different operators acting on the bounded cross section $\mathcal{B}$.
We have the following six cases:

I–IV: $n = 2, 3$ and $l = n - 1$, boundary conditions (13.85) or (13.86),

V: $n = 3, l = 1$, boundary conditions (13.85),

VI: $n = 3, l = 1$, boundary conditions (13.86), cylindrical symmetry.

Lemma 13.19 *The eigenvalues $(\varrho_m)_m$ for the Laplace operator studied in the cross section $\mathcal{B}$ under the different boundary conditions arising in the cases I – VI satisfy*

$$\varrho_m \geq c m^{\frac{2}{n-l}},$$

where $c > 0$ is independent of $m \in \mathbb{N}$.

PROOF: For the cases I – IV and VI, the boundary conditions reduce to Dirichlet or Neumann type boundary conditions, and we refer to Chapter 13.1. For case V we refer to [121].

Q.E.D.

There arise zero eigenvalues of the operators, for Neumann type boundary conditions in cases I – IV, VI, but also in case V, here with one-dimensional eigenspace arising from $R_0(\mathcal{B}) \cap R^0(\mathcal{B}) \cap D_0(\mathcal{B})$. Initial values living in these eigenspaces lead to smaller decay rates. In the following we consider the part in the orthogonal complement of these eigenspaces.

Definition 13.20 *w^0 satisfies condition (N) if the projection of $w^0(x', \cdot)$, for every fixed $x' \in \mathbb{R}^l$, onto the null space of the operator's part in the cross section vanishes.*

Now, we can carry over the methods from Chapter 13.1 to the Maxwell systems for $z = z(t, x) \in \mathbb{R}^n$,

$$z_{tt} - \tau \Delta z = f, \tag{13.108}$$

$$z(0, \cdot) = z^0, \quad z_t(0, \cdot) = z^1, \tag{13.109}$$

with either, on $\partial\Omega$,

$$\nu \times z(t, \cdot) = 0, \quad \operatorname{div} z(t, \cdot) = 0, \tag{13.110}$$

or

$$\nu \cdot z(t, \cdot) = 0, \quad \nu \times (\nabla \times z(t, \cdot)) = 0, \tag{13.111}$$

where $\tau > 0$.

Theorem 13.21 *Let*

$$K \geq \frac{1}{2}\left(\left[\frac{n}{2}\right] + \left[\frac{l+1}{2}\right] + n\right)$$

, and let $z^0, z^1, f(t, \cdot)$ satisfy, for $t \geq 0$, condition (N), and let (z^0, z^1, f) satisfy the compatibility condition of order $2K \in \mathbb{N}$, i.e.

$$\frac{d^j}{dt^j} z(0, \cdot) \in D(M) \text{ for } j = 0, 1, \ldots, 2K, \quad z^{2K+1} \in D(M^{1/2}),$$

where M is either M_1 or M_2. Moreover assume

$$z^0 \in W^{2K+2,2}(\Omega) \cap W^{2K+2,1}(\Omega), \quad z^1 \in W^{2K+1,2}(\Omega) \cap W^{2K+1,1}(\Omega),$$

$$f \in \bigcap_{j=0}^{2K} C^j\Big([0,\infty), W^{2K-j,2}(\Omega) \cap W^{2K-j,1}(\Omega)\Big).$$

Then the unique solution z to (13.108), (13.109) and (13.110) resp. (13.111) satisfies for $2 \leq q < \infty$ and $1/p + 1/q = 1$

$$\begin{aligned}
&\|(z(t,\cdot),\, z_t(t,\cdot),\, \nabla z(t,\cdot))\|_{L^q(\Omega)} \\
&\quad \leq \frac{c}{(1+t)^{(1-\frac{2}{q})\frac{l}{2}}} \Big(\|(z^0, z^1, \nabla z^0)\|_{W^{2K,p}(\Omega)} + \sum_{j=0}^{2K-1} \|f^{(j)}(0)\|_{W^{2K-1-j,p}(\Omega)}\Big) \\
&\quad + c\int_0^t \frac{1}{(1+t-s)^{(1-\frac{2}{q})\frac{l}{2}}} \sum_{j=0}^{2K} \|f^{(j)}(s)\|_{L^p(\Omega)}\, ds + c \sum_{j=0}^{2K-1} \|f^{(j)}(t)\|_{W^{2K-1-j,p}(\Omega)},
\end{aligned}$$

where the constant $c > 0$ does not depend on z^0, z^1, f or t.

Finally, we obtain the L^p-L^q-decay results for the elastic system (13.83), (13.84), (13.85) resp. (13.86) in decomposing u and f into

$$u = u^{po} + u^s, \qquad f = f^{po} + f^s,$$

with corresponding decomposition

$$\big(L^2(\Omega)\big)^3 \equiv \mathcal{H}^{po} \oplus \mathcal{H}^s.$$

Let $P^\beta, \beta \in \{po, s\}$ denote the projection operator

$$P^\beta : \big(L^2(\Omega)\big)^n \to \mathcal{H}^\beta. \tag{13.112}$$

Theorem 13.22 *Assume that the projections $P^{po}u^0, P^{po}u^1, P^{po}f(t,\cdot)$, and $P^s u^0, P^s u^1, P^s f(t,\cdot)$, for $t \geq 0$, satisfy condition (N). Let*

$$K \geq \frac{1}{2}\left(\left[\frac{n}{2}\right] + \left[\frac{l+1}{2}\right] + n\right),$$

$$u^0 \in W^{2K+2,2}(\Omega) \cap W^{2K+2,1}(\Omega), \quad u^1 \in W^{2K+1,2}(\Omega) \cap W^{2K+1,1}(\Omega),$$
$$f \in \bigcap_{j=0}^{2K} C^j\Big([0,\infty), W^{2K-j,2}(\Omega) \cap W^{2K-j,1}(\Omega)\Big).$$

Suppose that (u^0, u^1, f) satisfy the (corresponding) compatibility condition of order $2K$. Then, for $2 \leq q < \infty$, $1/p + 1/q = 1$, the unique solution u to (13.83), (13.84), (13.85) resp. (13.86) satisfies

$$\|(u(t,\cdot),\, u_t(t,\cdot),\, \nabla u(t,\cdot))\|_{L^q(\Omega)}$$

$$\begin{aligned}
&\le \frac{c}{(1+t)^{(1-\frac{2}{q})\frac{l}{2}}} \Big(\|(P^{po}u^0, P^{po}u^1, \nabla P^{po}u^0, P^s u^0, P^s u^1, \nabla P^s u^0)\|_{W^{2K,p}(\Omega)} \\
&\quad + \sum_{j=0}^{2K-1} \|(P^{po} f^{(j)}(0),\; P^s f^{(j)}(0)\|_{W^{2K-1-j,p}(\Omega)} \Big) \\
&\quad + c \int_0^t \frac{1}{(1+t-s)^{(1-\frac{2}{q})\frac{l}{2}}} \sum_{j=0}^{2K} \|(P^{po} f^{(j)}(s),\; P^s f^{(j)}(s))\|_{L^p(\Omega)}\, ds \\
&\quad + c \sum_{j=0}^{2K-1} \|(P^{po} f^{(j)}(t),\; P^s f^{(j)}(t))\|_{W^{2K-1-j,p}(\Omega)},
\end{aligned}$$

where the constant $c > 0$ does not depend on u^0, u^1, f or t.

In order to remove the projection operator P^β, for $\beta \in \{po, s\}$, in the estimates, one has to know the continuity of P^β given in (13.112) in Sobolev spaces $W^{N,p}(\Omega)$, i.e.

$$\|P^\beta v\|_{W^{N,p}(\Omega)} \le c_1 \|v\|_{W^{N,p}}(\Omega),$$

where c_1 is independent of v. For bounded domains, we could refer to [96] where the case $N = 0$ is discussed in detail. For the waveguides considered here, we refer to [102] for the decomposition (13.90) used for the boundary conditions (13.85), both for $n = 2$ and $n = 3$, and also for the decomposition (13.95) used for the boundary conditions (13.86) if $n = 2$ (the case $n = 3$ remains open here).

13.4 General waveguides

The flat waveguides of the type

$$\Omega = \mathbb{R}^l \times \mathcal{B}$$

which we studied in the previous chapters, will now be generalized to *repulsive* waveguides, typically domains like in Figure 13.5, where $(x', x'') \in \mathbb{R}^n \times \mathbb{R}^{\bar{m}}$.

Definition 13.23 *Let Ω be an open subset of $\mathbb{R}^n \times \mathbb{R}^{\bar{m}}$ with Lipschitz boundary $\partial\Omega$ and $n, \bar{m} \ge 1$. Let ν denote the exterior normal.*
Then Ω is called repulsive with respect to the x'-variables if we have for all $(x', x'') \in \partial\Omega$

$$\nu(x', x'') \cdot (x', 0) \le 0.$$

A non-repulsive domain is shown in Figure 13.6.
We remark that Ω being a repulsive domain implies that the cross sections of Ω for fixed $x'' \in \mathbb{R}^{\bar{m}}$,

$$\{x' \in \mathbb{R}^n \mid (x', x'') \in \Omega\},$$

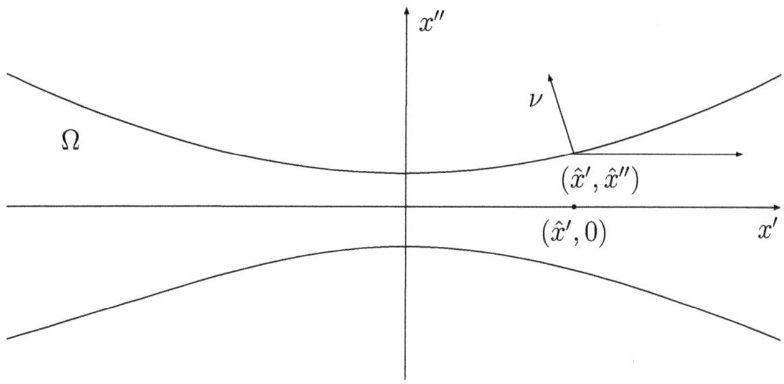

Figure 13.5: repulsive generalized waveguide

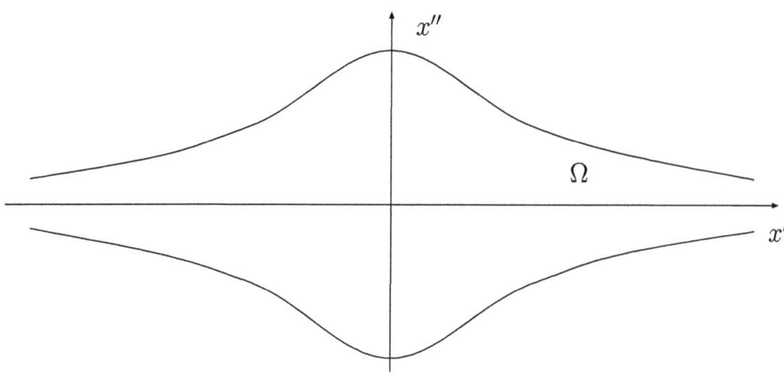

Figure 13.6: non-repulsive generalized waveguide

are *non-trapping* exterior domains.
The L^p-L^q-decay estimates proved for different equations in flat waveguides in the previous chapters are often called *dispersive* estimates. This kind of estimates has also been obtained by Dreher [27] for wave equations in unbounded conical sets. We shall not obtain analogous estimates in the non-flat case below. However, Schrödinger equations in all of $\mathbb{R}^{n+\bar{m}}$ satisfy weaker but more general estimates called *Strichartz estimates* (*Robert S. Strichartz*, *: 14.10.1943), which can be extended to our situation of repulsive waveguides.
In addition, we prove *smoothing estimates* for Schrödinger equations as well as for wave and for Klein-Gordon equations. It will be possible to allow an additional real-valued potential

$$V = V(x', x'').$$

If

$$H := -\Delta' - \Delta'' + V$$

denotes the associated Schrödinger operator, then, for example, smoothing for a solution $u = u(t, \cdot) \in D(H)$ to

$$i\, u_t + Hu = 0, \quad u(0, \cdot) = u_0,$$

will mean that for $\varepsilon > 0$ and some constant $c > 0$ the estimate

$$\|\langle x' \rangle^{-\frac{1}{2}-\varepsilon}(-\Delta')^{\frac{1}{4}}\, u\|_{L_t^2 L^2(\Omega)} \le c\|u_0\|_{L^2(\Omega)},$$

holds, by which we gain half a derivative. Here we use the notation

$$\langle x' \rangle := (1 + |x'|^2)^{\frac{1}{2}},$$

and

$$L_t^2 L^2(\Omega) := L^2\big((0,\infty), L^2(\Omega)\big).$$

Of course, the potential V will have to satisfy certain conditions:

$$V \ge 0, \quad -x'\nabla_{x'}\Big(|x'|V(x', x'')\Big) \ge 0. \tag{13.113}$$

Condition (13.113) is, for example, satisfied for some potentials $0 \le V = \widetilde{V}(|x'|, x'')$ decaying at least as $1/|x'|$ as $|x'| \to \infty$, because (13.113) turns into

$$\widetilde{V}(r', \cdot) + r'\frac{\partial}{\partial r'}\widetilde{V}(r', \cdot) \le 0$$

implying

$$\widetilde{V}(r', \cdot) \le \frac{\widetilde{V}(1, \cdot)}{r'}.$$

Examples are given by

$$V(x', x'') = \frac{g(x'')}{|x'|^{m_1}},$$

for $m_1 \ge 1$ and functions $g \ge 0$.

As an immediate consequence of the smoothing estimates we will deduce that there are no eigenvalues of H, since the presence of bound states would contradict the L^2-integrability in time of the solution.

For flat waveguides we have a purely continuous spectrum. This is also true for certain locally perturbed waveguides, in particular for any local perturbation Ω of $(0,1) \times \mathbb{R}^{n-1}$, for which $\nu(x) \cdot (x', 0) \le 0$ holds for any $x \in \partial\Omega$, see [132]. On the other hand, going back to [155, 76] one can easily construct local perturbations where the Dirichlet Laplacian has eigenvalues *below* its essential spectrum. But there may also exist eigenvalues embedded into the essential spectrum, see for example [200], where the following example is given. Let $D \subset \mathbb{R}^2$ be bounded, star-shaped with respect to the origin and invariant under the orthogonal group. Let $\varrho \in C^0(\mathbb{R}^k)$ be positive, $\varrho(x) = 1$ for large $|x|$ and $\max \varrho > 1$. Then the perturbed waveguide

$$\Omega_1 := \bigcap_{x \in \mathbb{R}^k} \Big(\{x\} \times \varrho(x) D\Big)$$

has an unbounded sequence of multiple eigenvalues embedded into the continuous spectrum.

Thus we see that suitable conditions on the shape of the domain, like repulsivity, are essential in order to exclude eigenvalues and to ensure dispersion. Conversely, in the presence of bumps in the wrong direction, we expect, in general, concentration of energy and disruption of dispersion.

The method to obtain the smoothing estimates will be to prove estimates for the resolvent operator

$$\mathcal{R}(z) := (H-z)^{-1}$$

for $z \notin \mathbb{R}$ of the type

$$\|\nabla_{x'}\mathcal{R}(z)f\|_{X_1}^2 + \|\mathcal{R}(z)f\|_{X_3}^2 + |z|\,\|\mathcal{R}(z)f\|_{X_1}^2 \le 5000\,n^2\|f\|_{X^*}^2, \tag{13.114}$$

and then to apply the concept of *H-smoothing* introduced by Kato [78] (*Tosio Kato*, 25.8.1917 – 2.10.1999) in the context of scattering theory. The norms used in (13.114) are called Morrey-Campanato norms (*Charles Bradfield Morrey Jr.*, 23.7.1907 – 29.4.1984, *Sergio Campanato*, 17.2.1930 – 1.3.2005) and are given by

$$\|f\|_{X_1} := \sup_{R>0} R^{-\frac{1}{2}}\|f\|_{L^2(\{|x'|\le R\})}, \qquad \|f\|_{X_3} := \sup_{R>0} R^{-\frac{3}{2}}\|f\|_{L^2(\{|x'|\le R\})},$$

$$\|f\|_{X^*} := \sum_{j\in\mathbb{Z}} 2^{\frac{j}{2}}\|f\|_{L^2(\{2^{j-1}\le|x'|\le 2^j\})},$$

see [37]. We also need

$$\|f\|_{X_2} := \sup_{R>0} R^{-1}\|f\|_{L^2(\{|x'|=R\})}.$$

Lemma 13.24 *(i)* $\|fg\|_{L^1(\Omega)} \le \|f\|_{X_1}\|g\|_{X^*}$,

(ii) $\|fg\|_{L^1(\Omega\cap\{R\le|x'|\le 2R\})} \le 4R^2\|f\|_{X_1}\|g\|_{X_3}$,

(iii) $\|fgh\|_{L^1(\Omega)} \le 2\|f\|_{X_3}\|g\|_{X^*}\||x'|h\|_{L^\infty(\Omega)}$,

(iv) $\|fg\|_{L^1(\Omega\cap\{|x'|\le R\})} \le 2R\|f\|_{X_3}\|g\|_{X^*}$,

(v) $\|f\|_{X_3} \le \|f\|_{X_2}$.

Proof: Let $\Omega_j := \Omega\cap\{2^{j-1}\le|x'|\le 2^j\}$.

$$\begin{aligned}
(i):\quad \|fg\|_{L^1(\Omega)} &= \sum_{j\in\mathbb{Z}}\|fg\|_{L^1(\Omega_j)} \le \sum_{j\in\mathbb{Z}}\|f\|_{L^2(\Omega_j)}\,\|g\|_{L^2(\Omega_j)}\\
&\le \sum_{j\in\mathbb{Z}}(2^j)^{-\frac{1}{2}}\|f\|_{L^2(\{|x'|\le 2^j\})}2^{\frac{j}{2}}\|g\|_{L^2(\Omega_j)}\\
&\le \|f\|_{X_1}\sum_{j\in\mathbb{Z}}2^{\frac{j}{2}}\|g\|_{L^2(\Omega_j)}\\
&\le \|f\|_{X_1}\|g\|_{X^*}.
\end{aligned}$$

$$
\begin{aligned}
(ii): \quad & \|fg\|_{L^1(\Omega\cap\{R\le|x'|\le 2R\})} \\
& \le (2R)^{-\frac{1}{2}}\|f\|_{L^2(\{|x'|\le 2R\})}(2R)^{\frac{1}{2}}(2R)^{\frac{3}{2}}(2R)^{-\frac{3}{2}}\|g\|_{L^2(\{|x'|\le 2R\})} \\
& \le (2R)^2\|f\|_{X_1}\|g\|_{X_3}.
\end{aligned}
$$

$$
\begin{aligned}
(iii): \quad \|fgh\|_{L^1(\Omega)} &= \|\frac{f}{|x'|}g|x'|h\|_{L^1(\Omega)} \\
&\le \sum_{j\in\mathbf{Z}} 2^{-\frac{j}{2}}\|\frac{f}{|x'|}\|_{L^2(\Omega_j)}2^{\frac{j}{2}}\|g\|_{L^2(\Omega_j)}\|\,|x'|h\|_{L^\infty(\Omega)} \\
&\le \sum_{j\in\mathbf{Z}} 2^{-\frac{j}{2}}\frac{1}{2^{j-1}}\|f\|_{L^2(\{|x'|\le 2^j\})}2^{\frac{j}{2}}\|g\|_{L^2(\Omega_j)}\||x'|h\|_{L^\infty(\Omega)} \\
&\le 2\|f\|_{X_3}\sum_{j\in\mathbf{Z}} 2^{\frac{j}{2}}\|g\|_{L^2(\Omega_j)}\|\,|x'|h\|_{L^\infty(\Omega)} \\
&= 2\|f\|_{X_3}\|g\|_{X^*}\|\,|x'|h\|_{L^\infty(\Omega)}.
\end{aligned}
$$

(iv): Follows from (iii) using $h := \chi_{\{|x'|\le R\}}$ (characteristic function).

$$
\begin{aligned}
(v): \quad \|f\|_{X_3}^2 &= \sup_{R>0} R^{-3}\|f\|^2_{L^2(\{|x'|\le R\})} = \sup_{R>0} R^{-3}\int_0^R \|f\|^2_{L^2(\{|x'|=\varrho\})}d\varrho \\
&\le \sup_{R>0} R^{-3}R^2\int_0^R \varrho^{-2}\|f\|^2_{L^2(\{|x'|=\varrho\})}d\varrho \\
&\le \sup_{R>0} R^{-1}\sup_{0\le\varrho\le R}\varrho^{-2}\|f\|^2_{L^2(\{|x'|=\varrho\})}\int_0^R 1d\varrho \\
&= \|f\|^2_{X_2}.
\end{aligned}
$$

Q.E.D.

A comparison to standard weighted L^2-norms will be useful, with weights of the form

$$
\langle x'\rangle_R := (R+\frac{|x'|^2}{R})^{\frac{1}{2}}, \qquad \text{for } R>0. \tag{13.115}
$$

Observing

$$
R+\frac{|x'|^2}{R} \ge \max\left\{R, \frac{|x'|^2}{R}\right\},
$$

we obtain for all $s>0$, after extending u as zero outside Ω,

$$
\begin{aligned}
\int_{\mathbf{R}^{n+\tilde{m}}} \left(R+\frac{|x'|^2}{R}\right)^{-s}|u|^2dx'dx'' &\le R^{-s}\int_{|x'|\le R}|u|^2 + R^s\int_{|x'|>R}|x'|^{-2s}|u|^2 \qquad (13.116)\\
&\le R^{-s}\int_{|x'|\le R}|u|^2 + R^s\sum_{j\ge j_R}\int_{\{2^{j-1}\le|x'|<2^j\}}|x'|^{-2s}|u|^2 \\
&\equiv I+II,
\end{aligned}
$$

where $j_R := [\log_2 R]$.

$$\begin{aligned} II &\leq R^s \sum_{j\geq j_R} 2^{-2(j-1)s} \int\limits_{\{2^{j-1}\leq |x'|<2^j\}} |u|^2 \qquad (13.117) \\ &= R^s\, 2^{2s} \sum_{j\geq j_R} 2^{-js} \left((2^{-j})^s \int\limits_{\{2^{j-1}\leq |x'|<2^j\}} |u|^2 \right) \\ &\leq R^s\, 2^{2s} \sum_{j\geq j_R} 2^{-js} \left(\sup_{\varrho>0} \varrho^{-s} \int\limits_{|x'|<\varrho} |u|^2 \right) \\ &\leq \left(\sup_{\varrho>0} \varrho^{-s} \int\limits_{|x'|<\varrho} |u|^2 \right) R^s\, 2^{2s} \frac{1}{2^{j_R s}} \frac{1}{1-2^{-s}} \\ &\leq \left(\sup_{\varrho>0} \varrho^{-s} \int\limits_{|x'|<\varrho} |u|^2 \right) \frac{2^{3s}}{1-2^{-s}}, \end{aligned}$$

since $2^{j_R} \geq \frac{R}{2}$. By (13.116), (13.117) we have

$$\begin{aligned} \int\limits_{\mathbf{R}^{n+\bar{m}}} \langle x'\rangle_R^{-2s} |u|^2 dx' dx'' &\leq \left(\sup_{\varrho>0} \varrho^{-s} \int\limits_{|x'|<\varrho} |u|^2 \right) \left(1 + \frac{2^{3s}}{1-2^{-s}} \right) \\ &\leq \left(1 + \frac{2^{4s}}{2^s - 1} \right) \sup_{\varrho>0} \varrho^{-s} \int\limits_{|x'|<\varrho} |u|^2. \end{aligned}$$

In particular, we get for $s = 1$

$$\|\langle x'\rangle_R^{-1} u\|_{L^2(\Omega)} \leq \sqrt{17} \|u\|_{X_1},$$

and for $s = 3$

$$\|\langle x'\rangle_R^{-3} u\|_{L^2(\Omega)} \leq 25 \|u\|_{X_3}. \qquad (13.118)$$

By a similar proof, we obtain

$$\|u\|_{X^*} \leq 16 \|\langle x'\rangle_R\, u\|_{L^2(\Omega)}. \qquad (13.119)$$

Finally, we notice that for any $\gamma > 0$ and $\varepsilon > 0$

$$\|\langle x'\rangle^{-\frac{\gamma}{2}-\varepsilon} u\|_{L^2(\Omega)} \leq C \sup_{R>0} \|\langle x'\rangle_R^{-\gamma}\, u\|_{L^2(\Omega)}, \qquad (13.120)$$

where $C > 0$ is a constant depending on γ and ε. To prove (13.120) we write

$$\begin{aligned} \int\limits_{\Omega} \langle x'\rangle^{-\gamma-2\varepsilon} |u|^2 &= \int\limits_{|x'|\leq 1} \cdots + \int\limits_{|x'|>1} \cdots \\ &\leq \int\limits_{|x'|\leq 1} |u|^2 + \sum_{j\geq 0} \int\limits_{\{2^j\leq |x'|<2^{j+1}\}} \langle x'\rangle^{-\gamma-2\varepsilon} |u|^2 \end{aligned}$$

$$\begin{aligned}
&\leq \sup_{R>0} \frac{1}{R^\gamma} \int_{|x'|\leq R} |u|^2 + \sum_{j\geq 0} 2^{-j(\gamma+2\varepsilon)} \int_{\{2^j\leq |x'|<2^{j+1}\}} |u|^2 \\
&\leq \sup_{R>0} \frac{1}{R^\gamma} \int_{|x'|\leq R} |u|^2 + \sum_{j\geq 0} 2^{-2\varepsilon j}\, 2^\gamma \frac{1}{2^{(j+1)\gamma}} \int_{\{|x'|\leq 2^{j+1}\}} |u|^2 \\
&\leq \left(\sup_{R>0} \frac{1}{R^\gamma} \int_{|x'|\leq R} |u|^2 \right) \left(1 + 2^\gamma \frac{1}{1-2^{-2\varepsilon}} \right) \\
&\leq \left(1 + \frac{2^\gamma}{1-2^{-2\varepsilon}} \right) 2^\gamma \sup_{R>0} \int_{|x'|\leq R} \langle x' \rangle_R^{-2\gamma} |u|^2,
\end{aligned}$$

the last inequality arising from

$$\frac{1}{R} \leq \frac{2}{R + \frac{|x'|^2}{R}} \qquad \text{for } |x'| \leq R.$$

Thus, (13.120) is proved with $C := \left(2^\gamma + \frac{2^{2\gamma}}{1-2^{-2\varepsilon}}\right)^{\frac{1}{2}}$.

Now we can state the main resolvent estimate.

Theorem 13.25 *Let Ω be a domain which is repulsive with respect to the x'-variables. Assume $n \geq 3$ and that the real-valued potential V satisfies (13.113). Then, for $\lambda, \varepsilon \in \mathbb{R}$ and $u \in W_0^{1,2}(\Omega)$ satisfying*

$$-\Delta u - (\lambda + i\varepsilon)u + Vu = f, \tag{13.121}$$

we have

$$\|\nabla_{x'} u\|_{X_1}^2 + \|u\|_{X_3}^2 + (|\lambda| + |\varepsilon|)\|u\|_{X_1}^2 \leq 5000\, n^2 \|f\|_{X^*}^2.$$

Sketch of the PROOF: (cp. [26]) We consider two real-valued piecewise smooth functions $\psi = \psi(x')$, $\phi = \phi(x')$, being independent of x'', such that

$$\nabla\psi, \Delta\psi, \nabla\Delta\psi, \phi, \nabla\phi \quad \text{are bounded for } |x'| \to \infty, \tag{13.122}$$

and

$$\nu \cdot \nabla\psi \leq 0 \quad \text{on } \partial\Omega. \tag{13.123}$$

Later on, ψ and ϕ will be given explicitly, the choice depending on the sign of λ. They will be functions only depending on $|x'|$. Then condition (13.123) turns for $\psi = h(|x'|)$ $h = h(r)$, into

$$0 \geq \nu(x', x'') \cdot \nabla\psi(x') = \nu \cdot (x', 0) \cdot |x'|^{-1} h_r(|x'|)$$

which is, since Ω is repulsive, equivalent to the condition that h_r or the radial derivative of ψ is non-negative,

$$x' \nabla_{x'} \psi \geq 0.$$

As in [9, 129] we will use a multiplier method. Multiplying the resolvent equation (13.121) by

$$(\Delta\psi - \phi)\bar{u} + 2\nabla\psi \cdot \nabla\bar{u}$$

and taking the real part, we obtain the identity

$$\begin{aligned}\nabla u \cdot (2\nabla^2\psi - \phi Id)\nabla\bar{u} + \frac{1}{2}\Delta(\phi - \Delta\psi)\,|u|^2 + \phi\lambda|u|^2 \\ -(\nabla V \cdot \nabla\psi + \phi V)|u|^2 + \operatorname{div}\operatorname{Re} Q_1 \\ = \operatorname{div}\operatorname{Re} Q + \operatorname{Re} f(2\nabla\psi \cdot \nabla\bar{u} + (\Delta\psi - \phi)\bar{u}) - 2\varepsilon \operatorname{Im}(\nabla\psi \cdot \nabla\bar{u}u),\end{aligned} \tag{13.124}$$

where

$$Q := (\Delta\psi)\bar{u}\nabla u - \frac{1}{2}(\nabla\Delta\psi)|u|^2 - (V - \lambda)(\nabla\psi)|u|^2 + \frac{1}{2}(\nabla\phi)|u|^2 - \phi\bar{u}\nabla u$$

and

$$Q_1 := \nabla\psi|\nabla u|^2 - 2\nabla u(\nabla\psi \cdot \nabla\bar{u}). \tag{13.125}$$

Finally, we shall integrate over Ω and we estimate the last term on the right-hand side of (13.124) as follows. Multiplying (13.121) by $\bar{u}$, we obtain

$$\operatorname{Im}\operatorname{div}((\nabla u)\bar{u}) + \varepsilon|u|^2 = -\operatorname{Im}(f\bar{u}) \tag{13.126}$$

and

$$\operatorname{Re}\operatorname{div}(-(\nabla u)\bar{u}) + |\nabla u|^2 = (\lambda - V)|u|^2 + \operatorname{Re}(f\bar{u}),$$

implying, with $\lambda^+ := \max\{\lambda, 0\}$,

$$\begin{aligned}|\varepsilon|\,|\nabla u|^2 &\leq |\varepsilon|\lambda^+|u|^2 + |\varepsilon|\operatorname{Re}(f\bar{u}) + \operatorname{Re}\operatorname{div}(|\varepsilon|\nabla u\bar{u}) \\ &= -s_\varepsilon\lambda^+\operatorname{Im}(f\bar{u}) + \operatorname{div}\left(s_\varepsilon \operatorname{Im}(\lambda^+ u\nabla\bar{u}) + \operatorname{Re}(|\varepsilon|(\nabla u)\bar{u})\right) + |\varepsilon|\operatorname{Re}(f\bar{u}),\end{aligned}$$

where we used the non-negativity of V and (13.126), and s_ε denotes the usual sign of ε. Hence

$$|\varepsilon||\nabla u|^2 \leq (\lambda^+ + |\varepsilon|)|f\bar{u}| + \operatorname{div}\left(s_\varepsilon \operatorname{Im}(\lambda^+ u\nabla\bar{u}) + \operatorname{Re}(|\varepsilon|\nabla u\bar{u})\right). \tag{13.127}$$

Using

$$2|\varepsilon u\nabla\bar{u}| \leq |\varepsilon|(\lambda^+ + |\varepsilon|)^{\frac{1}{2}}\,|u|^2 + |\varepsilon|(\lambda^+ + |\varepsilon|)^{-\frac{1}{2}}\,|\nabla u|^2$$

and (13.126), (13.127), we get

$$2|\varepsilon u\nabla\bar{u}| \leq 2\left(\sqrt{|\varepsilon|} + \sqrt{\lambda^+}\right)|f\bar{u}| + \operatorname{div} A,$$

where

$$A := \frac{|\varepsilon|\operatorname{Re}(\nabla u\bar{u}) + (s_\varepsilon 2\lambda^+ + |\varepsilon|)\operatorname{Im}(\nabla u\bar{u})}{(\lambda^+ + |\varepsilon|)^{\frac{1}{2}}}.$$

Inserting this into the basic equality (13.124), we obtain

$$\begin{aligned}\nabla u\cdot(2\nabla^2\psi-\phi Id)\nabla\bar u+\frac{1}{2}\Delta(\phi-\Delta\psi)\,|u|^2+\phi\lambda|u|^2\\ -(\nabla V\cdot\nabla\psi+\phi V)\,|u|^2+\operatorname{div}\operatorname{Re}Q_1\\ \le 2|f\nabla\psi\cdot\nabla\bar u|+|f(\Delta\psi-\phi)\bar u|+2\|\nabla\psi\|_{L^\infty}\Big(\sqrt{|\varepsilon|}+\sqrt{\lambda^+}\Big)|f\bar u|+\operatorname{div}\operatorname{Re}P,\end{aligned}\tag{13.128}$$

where $P:=Q+\|\nabla\psi\|_{L^\infty}A$.

The next goal is to estimate the integral over Ω of the right-hand side of (13.128). For this, we need an additional estimate, obtained by multiplying (13.121) by $\chi\bar u$, with choosing χ as a radial function of x' satisfying for some arbitrary, but fixed $R>0$

$$\chi(x')=\begin{cases}1, & \text{if } |x'|<R,\\ 0, & \text{if } |x'|>2R,\\ 2-\frac{|x'|}{R}, & \text{if } R\le|x'|\le 2R.\end{cases}$$

We get

$$s_\varepsilon\ \operatorname{Im}\Big(\operatorname{div}(\chi\nabla u\bar u)\Big)+|\varepsilon|\,|\chi|\,|u|^2=-s_\varepsilon\ \operatorname{Im}(\chi f\bar u)-s_\varepsilon\ \operatorname{Im}(\nabla\chi\cdot\nabla\bar u u).$$

Integrating over Ω, not producing boundary terms thanks to the Dirichlet boundary conditions, we arrive at

$$|\varepsilon|\int\limits_{\Omega\cap\{|x'|\le R\}}|u|^2\le\int\limits_{\Omega\cap\{|x'|\le 2R\}}|f\bar u|+\frac{1}{R}\int\limits_{\Omega\cap\{R\le|x'|\le 2R\}}|\nabla_{x'}u||u|$$

since χ only depends on x'. The right-hand side is estimated using Lemma 13.24 (ii), (iv), leading to

$$\frac{|\varepsilon|}{R}\int\limits_{\Omega\cap\{|x'|\le R\}}|u|^2\le 4\|f\|_{X^*}\|u\|_{X_3}+4\|\nabla_{x'}u\|_{X_1}\|u\|_{X_3}.$$

Hence, taking the supremum over R, we get

$$|\varepsilon|\,\|u\|_{X_1}^2\le 4(\|f\|_{X^*}+\|\nabla_{x'}u\|_{X_1})\|u\|_{X_3}.\tag{13.129}$$

Using Lemma 13.24 (i) and (13.129), we may estimate

$$\begin{aligned}2\Big(\sqrt{\lambda^+}+\sqrt{|\varepsilon|}\Big)\|f\bar u\|_{L^1} &\le 2\sqrt{\lambda^+}\|f\|_{X^*}\|u\|_{X_1}+4\|f\|_{X^*}(\|f\|_{X^*}+\|\nabla_{x'}u\|_{X_1})^{\frac12}\|u\|_{X_3}^{\frac12}\\ &\le \delta(\lambda^+\|u\|_{X_1}^2+\|\nabla_{x'}u\|_{X_1}^2+\|u\|_{X_3}^2)+5\,\delta^{-1}\|f\|_{X^*}^2,\end{aligned}$$

for all $\delta\in(0,1)$. This inequality will be used to estimate the third term on the right-hand side of (13.128).

Integration over Ω of the term $\operatorname{div} \operatorname{Re} P$ gives zero. This can be seen as follows. Let C_R denote the cylinder

$$C_R := \{(x', x'') \mid |x'| < R,\ x'' \in \mathbb{R}^{\bar{m}}\}.$$

We integrate $\operatorname{div} P$ on $\Omega \cap C_R$ and let $R \to \infty$. The boundary of $\Omega \cap C_R$ is the union of the two sets S_1 and S_2 given by

$$S_1 := \partial\Omega \cap C_R, \quad S_2 := \partial C_R \cap \Omega = \{(x', x'') \in \Omega \mid |x'| = R\}.$$

The surface integral over S_1 vanishes due to the boundary conditions. For the surface integral $\int_{S_2} \nu \cdot P$ we have, by assumption (13.122) on the boundedness of ψ, ϕ, and since $u \in W^{1,2}(\Omega)$,

$$\liminf_{R\to\infty} \int_{S_2} \nu \cdot P = 0. \tag{13.130}$$

This proves

$$\int_\Omega \operatorname{div} P = 0.$$

Concerning the first and the second term on the right-hand side of (13.128), we estimate their integrals using Lemma 13.24 (i) and (iii):

$$2\int_\Omega |f\nabla\psi \cdot \nabla\bar{u}| \le 2\|\nabla\psi\|_{L^\infty}\|f\|_{X^*}\|\nabla_{x'}u\|_{X_1}$$

and

$$\int_\Omega |f(\Delta\psi - \phi)\bar{u}| \le 2\|\,|x'|\,(\Delta\psi - \phi)\|_{L^\infty}\|f\|_{X^*}\|u\|_{X_3}.$$

Summarizing, the integral over Ω of the right-hand side of (13.128) is bounded by

$$C(\phi,\psi)\delta(\lambda^+\|u\|_{X_1}^2 + \|\nabla_{x'}u\|_{X_1}^2 + \|u\|_{X_3}^2) + C(\phi,\psi)\delta^{-1}\|f\|_{X^*}^2, \tag{13.131}$$

with

$$C(\phi,\psi) := 10\|\nabla\psi\|_{L^\infty} + 10\|\,|x'|(\Delta\psi - \phi)\|_{L^\infty}. \tag{13.132}$$

Now we consider the left-hand side of (13.128). The term in divergence form, $\operatorname{div} \operatorname{Re} Q_1$, with Q_1 given in (13.125), can be handled as above by integrating first on the cylinder C_R and then letting $R \to \infty$. The integral over $S_2 = \partial C_R \cap \Omega$ satisfies an analogous estimate to (13.130). On $S_1 = \partial\Omega \cap C_R$ we notice that $\nabla u = \frac{\partial u}{\partial\nu}\nu$ holds because of the Dirichlet boundary condition, implying

$$\nu \cdot Q_1 = -(\nu \cdot \nabla\psi)\left|\frac{\partial u}{\partial\nu}\right|^2.$$

Thus the integral over S_1 can be written as

$$I_R := -\int_{S_1} (\nu \cdot \nabla\psi)\left|\frac{\partial u}{\partial\nu}\right|^2.$$

Using assumption (13.123) on ψ, we obtain $I_R \geq 0$ for all R. Hence we can drop I_R in the sequel, and we get the basic inequality

$$\int_\Omega \left(\nabla u(2\nabla^2\psi - \phi Id)\nabla\bar{u} + \frac{1}{2}\left(\Delta(\phi - \Delta\psi)\right)|u|^2 + \phi\lambda|u|^2 - (\nabla V \cdot \nabla\psi + \phi V)|u|^2\right) \tag{13.133}$$
$$\leq C(\phi,\psi)\,\delta(\lambda^+\|u\|^2_{X_1} + \|\nabla_{x'}u\|^2_{X_1} + \|u\|^2_{X_3}) + C(\phi,\psi)\delta^{-1}\|f\|^2_{X^*}.$$

It remains to choose the functions ϕ, ψ in an appropriate way. For $\lambda > 0$ we make the following choice inspired by [9]:

$$\psi(x',x'') := \begin{cases} |x'|, & \text{if } |x'| \geq R, \\ \frac{R}{2} + \frac{|x'|^2}{2R}, & \text{if } |x'| < R, \end{cases} \qquad \phi(x',x'') := \begin{cases} 0, & \text{if } |x'| \geq R, \\ \frac{1}{R}, & \text{if } |x'| < R. \end{cases}$$

Then the assumptions (13.122), (13.123) are satisfied. We compute

$$\phi - \Delta\psi = \begin{cases} -\frac{n-1}{|x'|}, & \text{if } |x'| \geq R, \\ -\frac{n-1}{R}, & \text{if } |x'| < R, \end{cases}$$

and, in the distributional sense,

$$\Delta(\phi - \Delta\psi) = \frac{n-1}{R^2}\,\delta_{|x'|=R} + \begin{cases} \frac{\mu_n}{|x'|^3}, & \text{if } |x'| \geq R, \\ 0, & \text{if } |x'| < R, \end{cases}$$

where

$$\mu_n := (n-1)(n-3).$$

Moreover,

$$\|\nabla\psi\|_{L^\infty} = 1, \quad \|\,|x'|(\Delta\psi - \phi)\|_{L^\infty} = n-1,$$

implying

$$C(\phi,\psi) = 10n.$$

Denoting $x'_0 := \frac{x'}{|x'|}$, we have

$$(\nabla u)(2\nabla^2\psi - \phi)\nabla\bar{u} = \begin{cases} \frac{2}{R}|\nabla_{x'}u - (\nabla_{x'}u \cdot x'_0)x'_0|^2, & \text{if } |x'| \geq R, \\ \frac{1}{R}|\nabla_{x'}u|^2, & \text{if } |x'| < R. \end{cases}$$

The terms in (13.133) containing the potential V are non-negative thanks to assumption (13.113), hence we can drop them. Thus (13.133) implies

$$\frac{1}{R}\|\nabla_{x'}u\|^2_{L^2(\Omega\cap\{|x'|\leq R\})} + \frac{n-1}{2R^2}\left(\int_{\Omega\cap\{|x'|=R\}} |u|^2\right) + \frac{\lambda}{R}\|u\|^2_{L^2(\Omega\cap\{|x'|\leq R\})}$$
$$\leq 10n\delta(\lambda\|u\|^2_{X_1} + \|\nabla_{x'}u\|^2_{X_1} + \|u\|^2_{X_3}) + 10n\delta^{-1}\|f\|^2_{X^*},$$

and taking the supremum over $R > 0$ we obtain

$$\|\nabla_{x'} u\|_{X_1}^2 + \frac{n-1}{2}\|u\|_{X_2}^2 + \lambda\|u\|_{X_1}^2$$
$$\leq 10n\delta(\lambda\|u\|_{X_1}^2 + \|\nabla_{x'}u\|_{X_1}^2 + \|u\|_{X_3}^2) + 10n\delta^{-1}\|f\|_{X^*}^2.$$

Recalling Lemma 13.24 (v) and choosing $\delta := (20n)^{-1}$, we finally obtain for the case $\lambda > 0$

$$\|\nabla_{x'}u\|_{X_1}^2 + \|u\|_{X_2}^2 + \lambda\|u\|_{X_1}^2 \leq 400n^2\|f\|_{X^*}^2. \tag{13.134}$$

In the case $\lambda \leq 0$ we choose different weights. Following [25] we simply take $\phi = 0$ and

$$\psi(x', x'') := \int_0^{|x'|} \alpha(\tau)d\tau, \tag{13.135}$$

where

$$\alpha(\tau) := \begin{cases} \frac{1}{n} - \frac{1}{2n(n+2)}\frac{R^{n-1}}{\tau^{n-1}}, & \text{if } \tau \geq R, \\ \frac{1}{2n} + \frac{\tau}{2nR} - \frac{1}{2n(n+2)}\frac{\tau^3}{R^3}, & \text{if } \tau < R. \end{cases}$$

Then

$$\Delta\psi = \begin{cases} \frac{(n-1)}{n}\frac{1}{|x'|}, & \text{if } |x'| \geq R, \\ \frac{1}{2R} + \frac{n-1}{2n|x'|} - \frac{|x'|^2}{2nR^3}, & \text{if } |x'| < R, \end{cases}$$

$$\|\nabla\psi\|_{L^\infty} = \frac{1}{n}, \qquad \||x'|\Delta\psi\|_{L^\infty} \leq 1 - \frac{1}{n},$$

implying $C(\phi, \psi) \leq 10$.

For $n = 3$ we have

$$-\Delta^2\psi = \frac{1}{R^3}\chi_{\{|x'|<R\}} + \frac{4\pi}{3}\delta_0(x'),$$

where χ_Z is the characteristic function of the set Z, and $\delta_0(x')$ denotes the Dirac distribution in the variable x' (*Paul Adrian Maurice Dirac*, 8.8.1902 – 20.10.1984).

For $n \geq 4$ we have

$$-\Delta^2\psi = \Big(\frac{1}{R^3} + \frac{\mu_n}{2n|x'|^3}\Big)\chi_{\{|x'|<R\}} + \frac{\mu_n}{n|x'|^3}\chi_{\{|x'|\geq R\}} + \frac{n-3}{2nR^2}\delta_{|x'|\geq R}.$$

In all cases $n \geq 3$ we conclude

$$-\Delta^2\psi \geq \frac{1}{R^3}\chi_{\{|x'|<R\}}.$$

Proceeding as above we obtain

$$\frac{n-1}{n(n+2)}\|\nabla_{x'}u\|_{X_1}^2 + \|u\|_{X_3}^2 \leq 10\delta\Big(\|\nabla_{x'}u\|_{X_1}^2 + \|u\|_{X_3}^2\Big) + 10\delta^{-1}\|f\|_{X^*}^2,$$

and, choosing $\delta := (40n)^{-1}$, we conclude for $\lambda \leq 0$

$$\|\nabla_{x'} u\|_{X_1}^2 + \|u\|_{X_3}^2 \leq 800n^2 \|f\|_{X^*}^2. \tag{13.136}$$

Combining (13.134), (13.136) we obtain for all $\lambda \in \mathbb{R}$

$$\|\nabla_{x'} u\|_{X_1}^2 + \|u\|_{X_3}^2 + \lambda^+ \|u\|_{X_1}^2 \leq 800n^2 \|f\|_{X^*}^2. \tag{13.137}$$

As a last step, the factor λ^+ in (13.137) is improved to $|\lambda| + |\varepsilon|$. Recalling (13.129) and using (13.137), we get

$$|\varepsilon| \, \|u\|_{X_1}^2 \leq 4\Big(\|f\|_{X^*} + \|\nabla_{x'} u\|_{X_1}\Big) \|u\|_{X_3} \leq 3320n^2 \|f\|_{X^*}^2. \tag{13.138}$$

Assuming $\lambda^- := -\lambda \geq 0$ we multiply the resolvent equation (13.121) by $\bar{u}$ and take real parts, obtaining

$$|\nabla u|^2 + \lambda^- |u|^2 + V|u|^2 = \operatorname{Re}(f\bar{u}) + \frac{1}{2}\Delta(|u|^2).$$

Taking ψ as in (13.135), multiplying by $\Delta\psi$, using

$$\Delta\psi \geq \frac{1}{2R} \chi_{\{|x'|<R\}}, \quad \| \, |x'| \Delta\psi \|_{L^\infty} \leq 1,$$

and recalling Lemma 13.24 (iii), we get

$$\frac{1}{2R} \int_{\Omega \cap \{|x'|<R\}} \Big(|\nabla u|^2 + \lambda^- |u|^2\Big) \leq 2\|f\|_{X^*} \|u\|_{X_3}.$$

Taking the supremum over $R > 0$ gives

$$\|\nabla u\|_{X_1}^2 + \lambda^- \|u\|_{X_1}^2 \leq 4\|f\|_{X^*} \|u\|_{X_3} \leq 120n \|f\|_{X^*}^2, \tag{13.139}$$

where we used (13.137). Combining (13.137), (13.138) and (13.139), Theorem 13.25 is proved.

Q.E.D.

Using (13.118), Theorem 13.25 and (13.119) we conclude for z in the resolvent set of $H = -\Delta + V$, $u = \mathcal{R}(z)f = (H - z)^{-1}f$,

$$\|\langle x' \rangle_R^{-3} \mathcal{R}(z) f\|_{L^2(\Omega)}^2 \leq 729 \|u\|_{X_3}^2 \leq 3645000n^2 \|f\|_{X^*}^2 \leq 256 \cdot 3645000n^2 \|\langle x' \rangle_S f\|_{L^2(\Omega)}^2$$

for any $R, S > 0$, that is,

$$\|\langle x' \rangle_R^{-3} \mathcal{R}(z) f\|_{L^2(\Omega)} \leq c_n \|\langle x' \rangle_S f\|_{L^2(\Omega)} \tag{13.140}$$

with some positive constant depending only on n, denoted here and below by c_n. Thus, the operator

$$T_0 := \langle x' \rangle_R^{-3} \mathcal{R}(z) \langle x' \rangle_S^{-1} : L^2(\Omega) \to L^2(\Omega)$$

has a bounded adjoint operator T_0^*, implying that

$$T_1 := \langle x'\rangle_R^{-1}\mathcal{R}(z)\langle x'\rangle_S^{-3}$$

is continuous on $L^2(\Omega)$, hence

$$\|\langle x'\rangle_R^{-1}\mathcal{R}(z)f\|_{L^2(\Omega)} \le c_n\|\langle x'\rangle_S^3 f\|_{L^2(\Omega)}. \tag{13.141}$$

By (13.140), (13.141) we have $P := \mathcal{R}(z)$ as a bounded operator

$$P_\alpha : X_\alpha \to Y_\alpha,$$

for $\alpha = 0, 1$, where

$$X_0 := L^2_{\langle x'\rangle_S}, \quad Y_0 := L^2_{\langle x'\rangle_R^{-3}},$$

$$X_1 := L^2_{\langle x'\rangle_S^3}, \quad Y_1 := L^2_{\langle x'\rangle_R^{-1}}.$$

Denoting $L^2_\varrho := \{h \mid \varrho\, h \in L^2(\Omega)\}$, we obtain by complex interpolation that

$$P_\alpha : [X_0, X_1]_\alpha \to [Y_0, Y_1]_\alpha$$

is bounded for any $\alpha \in [0,1]$. Since (cp. [11])

$$[L^2_{\varrho_1}, L^2_{\varrho_2}]_\alpha = L^2_{\varrho_1^{1-\alpha}\,\varrho_2^\alpha}$$

we conclude the boundedness of

$$T_\alpha := \langle x'\rangle_R^{-3+2\alpha}\mathcal{R}(z)\langle x'\rangle_S^{-1-2\alpha} : L^2(\Omega) \to L^2(\Omega)$$

for any $\alpha \in [0,1]$, in particular for $\alpha = \frac{1}{2}$

$$\|\langle x'\rangle_R^{-2}\mathcal{R}(z)\langle x'\rangle_S^{-2} f\|_{L^2(\Omega)} \le c_n\|f\|_{L^2(\Omega)}. \tag{13.142}$$

Lemma 13.26 *Let $\gamma > 0$. If a bounded linear operator $\mathcal{A}$ satisfies for all $R, S > 0$ the estimate*

$$\|\langle x'\rangle_R^{-\gamma}\mathcal{A}\,\langle x'\rangle_S^{-\gamma} u\|_{L^2(\Omega)} \le C_0\|u\|_{L^2(\Omega)}, \tag{13.143}$$

with a positive constant C_0 being independent of R, S, u, then it also satisfies, for all $\varepsilon > 0$, the estimate

$$\|\langle x'\rangle^{-\frac{\gamma}{2}-\varepsilon}\mathcal{A}\,\langle x'\rangle^{-\frac{\gamma}{2}-\varepsilon} u\|_{L^2(\Omega)} \le C_0 C(\gamma,\varepsilon)\|u\|_{L^2(\Omega)},$$

where $C(\gamma, \varepsilon) > 0$ denotes a constant depending at most on γ and ε.

PROOF: Decomposing

$$v := \langle x' \rangle^{-\frac{\gamma}{2}-\varepsilon} u = v_0 + \sum_{j=1}^{\infty} v_j,$$

where v_j has support in $\{2^{j-1} \leq |x'| < 2^j\}$ for $j \geq 1$, and v_0 has support in $\{|x'| < 1\}$, we get

$$\mathcal{A}v = \mathcal{A}v_0 + \sum_{j=1}^{\infty} \mathcal{A}v_j,$$

and, applying (13.143) to v_j with $S := 2^j$,

$$\begin{aligned} \|\langle x' \rangle_R^{-\gamma} \mathcal{A}v\|_{L^2(\Omega)} &\leq \|\langle x' \rangle_R^{-\gamma} \mathcal{A}v_0\|_{L^2(\Omega)} + \sum_{j=1}^{\infty} \|\langle x' \rangle_R^{-\gamma} \mathcal{A}v_j\|_{L^2(\Omega)} \\ &\leq C_0 \|\langle x' \rangle^{\gamma} v_0\|_{L^2(\Omega)} + C_0 \sum_{j=1}^{\infty} \|\langle x' \rangle_{2^j}^{\gamma} v_j\|_{L^2(\Omega)}. \end{aligned}$$

Since, for $j \geq 1$ and $2^{j-1} \leq |x'| \leq 2^j$,

$$\begin{aligned} \langle x' \rangle_{2^j}^{2\gamma} &= \left(2^j + \frac{|x'|^2}{2^j}\right)^{\gamma} \leq 2^{\gamma}(2^{j\gamma} + 2^{j\gamma}) = 2^{\gamma+1} 2^{j\gamma} \\ &\leq 2^{\gamma+1}\, 2^{\gamma} |x'|^{\gamma} \leq 2^{2\gamma+1}\, 2^{2\varepsilon}\, 2^{-2\varepsilon j} |x'|^{\gamma+2\varepsilon}, \end{aligned}$$

we get

$$\begin{aligned} \|\langle x' \rangle_R^{-\gamma} \mathcal{A}v\|_{L^2(\Omega)} &\leq C_0 \|\langle x' \rangle^{\gamma} v_0\|_{L^2(\Omega)} + C_0 2^{2(\gamma+\varepsilon)+1} \sum_{j=1}^{\infty} \||x'|^{\frac{\gamma}{2}+\varepsilon} v_j\|_{L^2(\Omega)} \\ &\leq C_0 C(\gamma, \varepsilon) \|\langle x' \rangle^{\frac{\gamma}{2}+\varepsilon} v\|_{L^2(\Omega)}. \end{aligned} \tag{13.144}$$

Using (13.120) we obtain

$$\|\langle x' \rangle^{-\frac{\gamma}{2}-\varepsilon} \mathcal{A}v\|_{L^2(\Omega)} \leq C(\gamma, \varepsilon) \sup_{R>0} \|\langle x \rangle_R^{-\gamma} \mathcal{A}v\|_{L^2(\Omega)}. \tag{13.145}$$

A combination of (13.145), (13.144) completes the proof.

Q.E.D.

Applying Lemma 13.26 to (13.142) we have

$$\|\langle x' \rangle^{-1-\varepsilon} \mathcal{R}(z) \langle x' \rangle^{-1-\varepsilon} f\|_{L^2(\Omega)} \leq c_{n,\varepsilon} \|f\|_{L^2(\Omega)}, \tag{13.146}$$

where $c_{n,\varepsilon} > 0$ depends at most on n and ε. Similarly, one gets

$$\|\langle x' \rangle^{-\frac{1}{2}-\varepsilon} \nabla_{x'} \mathcal{R}(z) \langle x \rangle^{-\frac{1}{2}-\varepsilon} f\|_{L^2(\Omega)} \leq c_{n,\varepsilon} \|f\|_{L^2(\Omega)}$$

and

$$|z|^{\frac{1}{2}} \|\langle x' \rangle^{-\frac{1}{2}-\varepsilon} \mathcal{R}(z) \langle x \rangle^{-\frac{1}{2}-\varepsilon} f\|_{L^2(\Omega)} \leq c_{n,\varepsilon} \|f\|_{L^2(\Omega)}. \tag{13.147}$$

The concept of smoothing introduced by Kato [78] in scattering theory appearing the following theorem turned out to be useful for dispersive equations as revealed in [156], cp. [154], [128].

Theorem 13.27 *Let K be a self-adjoint operator in a Hilbert space $\mathcal{H}$. Let $\mathcal{R}(z) := (K-z)^{-1}$ for $z \in \mathbb{C} \setminus \mathbb{R}$ be the resolvent operator, let A be a densely defined closed operator from $D(A) \subset \mathcal{H}$ to another Hilbert space $\mathcal{H}_1$ with $D(K) \subset D(A)$. Assume that there is $c_0 > 0$ such that for all $f \in D(A^*) \subset \mathcal{H}_1$ with $\mathcal{R}(z)A^*f \in D(A)$ one has*

$$\sup_{z \notin \mathbf{R}} \|A\mathcal{R}(z)A^*f\|_{\mathcal{H}_1} \leq c_0^2 \|f\|_{\mathcal{H}_1}. \tag{13.148}$$

Then we have for all $f \in \mathcal{H}$

$$\|Ae^{itK}f\|_{L_t^2\mathcal{H}_1} \leq c_0 \|f\|_{\mathcal{H}_1}.$$

Choosing

$$\mathcal{H} := \mathcal{H}_1 := L^2(\Omega), \quad K := H = -\Delta + V, \quad A := \langle x' \rangle^{-1-\varepsilon} \text{ (multiplication operator)},$$

the estimate (13.146) gives condition (13.148) and hence the claim (13.149) in the following

Theorem 13.28 *Let Ω be a domain which is repulsive with respect to the x'-variables. Assume $n \geq 3$ and that the real-valued potential satisfies (13.113). Assume also that $H = -\Delta + V$ with Dirichlet boundary conditions is self-adjoint in $L^2(\Omega)$. Then, for any $\varepsilon > 0$, we have the smoothing estimates*

$$\|\langle x' \rangle^{-1-\varepsilon} e^{itH} f\|_{L_t^2 L^2(\Omega)} \leq c_{n,\varepsilon} \|f\|_{L^2(\Omega)}, \tag{13.149}$$

$$\|\langle x' \rangle^{-\frac{1}{2}-\varepsilon} \nabla_{x'} e^{itH} f\|_{L_t^2 L^2(\Omega)} \leq c_{n,\varepsilon} \||D_{x'}|^{\frac{1}{2}} f\|_{L^2(\Omega)}. \tag{13.150}$$

Here

$$|D_{x'}|^{\frac{1}{2}} g(x', x'') := (2\pi)^{-\frac{n}{2}} \int_{\mathbf{R}^n} |\xi'|^{\frac{1}{2}} \hat{g}(\xi', x'') e^{i\xi' x'} d\xi',$$

where $\hat{g}(\cdot, x'')$ denotes the Fourier transform of $x' \mapsto g(x', x'')$.
PROOF: For (13.150) see [26].

Q.E.D.
If we consider for $m \geq 0$

$$H_m := H + m = -\Delta + V + m \tag{13.151}$$

and the associated wave equation ($m = 0$) or Klein-Gordon equation ($m > 0$) with potential V,

$$u_{tt} + H_m u = 0, \tag{13.152}$$

then we have the following theorem. It is obtained from an adaptation of the above results observing that $Z := (u, u_t)'$ satisfies for $Z_0 := (u_0, u_1)' := (f, \sqrt{H} f)'$

$$Z(t) = \begin{pmatrix} e^{it\sqrt{H_m}} f \\ i\sqrt{H_m}\, e^{it\sqrt{H_m}} f \end{pmatrix} = e^{itK} Z_0, \quad \text{with} \quad K := \begin{pmatrix} 0 & -i \\ iH & 0 \end{pmatrix}.$$

Theorem 13.29 *Let the assumptions of Theorem 13.28 be fulfilled and H_m as in (13.151). Then the flow associated to wave or Klein-Gordon equations (13.152) satisfies*

$$\|\langle x'\rangle^{-\frac{1}{2}-\varepsilon}\, e^{it\sqrt{H_m}} f\|_{L^2_t L^2(\Omega)} \leq c_{n,\varepsilon}\|f\|_{L^2(\Omega)}.$$

PROOF: We choose

$$\mathcal{H} := D(\sqrt{H}) \times L^2(\Omega), \qquad \mathcal{H}_1 := L^2(\Omega), \qquad H := -\Delta + V(x', x'')$$

and $A : \mathcal{H} \to L^2(\Omega)$ defined by

$$A\begin{pmatrix} f \\ g \end{pmatrix} := \langle x\rangle^{-1/2-\varepsilon} H^{1/2} f, \quad \text{implying} \quad A^* g = \begin{pmatrix} H^{-1/2}\langle x\rangle^{-1/2-\varepsilon} g \\ 0 \end{pmatrix}.$$

Then the resolvent $R_K(z) := (K - z)^{-1}$ can be written in terms of the resolvent $\mathcal{R}(z) = (H - z)^{-1}$ as

$$R_K(z) = \begin{pmatrix} z\mathcal{R}(z^2) & -i\mathcal{R}(z^2) \\ -iH\mathcal{R}(z^2) & z\mathcal{R}(z^2) \end{pmatrix}.$$

Thus we see that, in order to apply the theory of Kato in Theorem 13.27, we need to prove that the following operator is bounded on $L^2(\Omega)$, uniformly in $z \notin \mathbb{R}$:

$$AR_K(z)A^* = \langle x\rangle^{-1/2-\varepsilon} z\mathcal{R}(z^2)\langle x\rangle^{-1/2-\varepsilon}.$$

This is precisely what is expressed by estimate (13.147).

Q.E.D.

Finally and without proof (cp. [26]) we state Strichartz estimates for simpler waveguides which are compactly supported perturbations of flat waveguides. We assume that there is $M > 0$ and a bounded domain $\omega \subset \mathbb{R}^{\overline{m}}$ such that

$$\Omega \cap \{(x', x'') \,|\, |x'| > M\} = (\mathbb{R}^n \times \omega) \cap \{(x', x'') \,|\, |x'| > M\}. \tag{13.153}$$

We have

Theorem 13.30 *In addition to the assumptions of Theorem 13.28 let (13.153) be satisfied. Then we have for all $f \in W_0^{1,2}(\Omega)$*

$$\|e^{itH} f\|_{L^2_t L^{\frac{2n}{n-2}}_{x'} L^2_{x''}} \leq c_{n,\varepsilon}\left(1 + \|\langle x'\rangle^{1+\varepsilon} V\|_{L^n_x L^2_{x''}}\right)\left(\|f\|_{L^2(\Omega)} + \|\,|D_{x'}|^{\frac{1}{2}} f\|_{L^2(\Omega)}\right).$$

It seems natural to apply these estimates to investigate the existence of global small solutions for nonlinear Schrödinger wave or Klein-Gordon equations on non-flat waveguides.

We abandon giving further references in this daily expanding field with recalling the last words of T. Fontane's father Briest, [32, p. 354] (*Theodor Fontane,* 30.12.1819 – 20.9.1898):

"...das ist ein *zu* weites Feld."

Appendix

To assure a more or less self-contained presentation we have compiled some of the basic results which were used in the previous chapters. The proofs are sketched (at least in the Appendices B,C).

A Interpolation

First, we state some general definitions and results on interpolation spaces. For details and proofs we refer the reader to the books of Bergh & Löfström [11] and H. Triebel [181]; sketches of the proofs are here given for the relevant applications:

Definition A.1 *(X_0, X_1) is called an* <u>*interpolation couple*</u> $:\iff$ *X_0, X_1 are Banach spaces which are continuously imbedded into a topological Hausdorff space.*

(*Felix Hausdorff*, 8.11.1868 – 26.1.1942)

Lemma A.2 *Let (X_0, X_1) be an interpolation couple. Then*

(i) *$X_0 \cap X_1$ is a Banach space with norm*

$$\|x\|_{X_0\cap X_1} := \max\{\|x\|_{X_0},\ \|x\|_{X_1}\},$$

(ii) *$X_0 + X_1$ is a Banach space with norm*

$$\|x\|_{X_0+X_1} := \inf\{\|x_0\|_{X_0} + \|x_1\|_{X_1} \ \mid\ x_0 \in X_0, x_1 \in X_1,\ x_0 + x_1 = x\}.$$

The proof of Lemma A.2 is straightforward.

Definition A.3 *Let (X_0, X_1) and (Y_0, Y_1) be interpolation couples and let X and Y be Banach spaces.*

(i) *X is called an* <u>*intermediate space*</u> *between X_0 and X_1 $:\iff$ $X_0 \cap X_1 \hookrightarrow X \hookrightarrow X_0 + X_1$ (continuous imbedding),*

(ii) *X, Y are called* <u>*interpolation spaces*</u> *for $(X_0, X_1), (Y_0, Y_1)$ $:\iff$ X is an intermediate space between X_0 and X_1, Y is an intermediate space between Y_0 and Y_1 and*

$$T : X_j \longrightarrow Y_j \ \text{ is continuous for } j = 0, 1 \ \Rightarrow T : X \longrightarrow Y \ \text{ is continuous.}$$

By an *interpolation method* two interpolation couples are attached to interpolation spaces. We consider the so-called *complex interpolation method.*
Let (X_0, X_1) be an interpolation couple,

$$Z := \{z \in \mathbb{C} \ \mid\ 0 \le \operatorname{Re} z \le 1\},$$

and $\mathcal{I}(X_0, X_1) := \Big\{ f : Z \longrightarrow X_0 + X_1 \ \big| \ f$ is continuous and bounded in Z, analytic in the interior of Z; $t \longmapsto f(j+it)$ maps $\mathbb{R}$ into X_j, continuously, and tends to zero as $|t| \to \infty,\ j = 0, 1 \Big\}$.
Then the following theorem can be proved:

Theorem A.4 *$\mathcal{I}(X_0, X_1)$ is a Banach space with norm*

$$\|f\|_{\mathcal{I}(X_0,X_1)} := \max\{\sup_{\eta\in\mathbf{R}} \|f(i\eta)\|_{X_0}, \sup_{\eta\in\mathbf{R}} \|f(1+i\eta)\|_{X_1}\}.$$

For $\theta \in [0, 1]$ let

$$[X_0, X_1]_\theta := \{f(\theta) \mid f \in \mathcal{I}(X_0, X_1)\}$$

and for $x \in [X_0, X_1]_\theta$ let

$$|x|_\theta := \inf\{\|f\|_{\mathcal{I}(X_0,X_1)} \mid f \in \mathcal{I}(X_0, X_1),\ f(\theta) = x\}.$$

$[X_0, X_1]_\theta$ has the following properties.

Theorem A.5

(i) *$[X_0, X_1]_\theta$ with norm $|\cdot|_\theta$ is an intermediate space between X_0 and X_1,*

(ii) *$X_0 \cap X_1$ is dense in $[X_0, X_1]_\theta$.*

In this abstract setting we finally quote the following general interpolation theorem.

Theorem A.6 *Let (X_0, X_1), (Y_0, Y_1) be interpolation couples, let*

$$T : X_0 + X_1 \longrightarrow Y_0 + Y_1$$

be linear with

$$T_{/X_0} : X_0 \longrightarrow Y_0 \ \text{ is bounded with norm } M_0,$$

$$T_{/X_1} : X_1 \longrightarrow Y_1 \ \text{ is bounded with norm } M_1.$$

Then we have for all $\theta \in (0, 1)$

$$T_{/[X_0,X_1]_\theta} : [X_0, X_1]_\theta \longrightarrow [Y_0, Y_1]_\theta \ \text{ is bounded with norm } M_\theta$$

and

$$M_\theta \le M_0^{1-\theta} M_1^\theta.$$

Now we turn to specific applications, the first being the interpolation theorem of Riesz & Thorin (*Marcel Riesz,* 16.11.1886 – 4.9.1969):

Theorem A.7 *Let $\theta \in (0,1)$, $p_0, p_1 \in [1,\infty]$. Then*

$$[L^{p_0}, L^{p_1}]_\theta = L^{p_\theta},$$

where

$$\frac{1}{p_\theta} = \frac{1-\theta}{p_0} + \frac{\theta}{p_1}.$$

PROOF: Without loss of generality we assume $p_0 \neq p_1$. It suffices to show

$$|a|_\theta = \|a\|_{p_\theta}$$

for all real-valued continuous functions a with compact support because the set C_0^0 of those functions is dense in L^{p_θ} and also in $L^{p_0} \cap L^{p_1}$ (and hence in $[L^{p_0}, L^{p_1}]_\theta$ according to Theorem A.5).

Step 1:

CLAIM: $|a|_\theta \leq \|a\|_{p_\theta}$.

PROOF: Without loss of generality we assume $\|a\|_{p_\theta} = 1$. For $z \in Z, x \in \mathbb{R}^n$, and $\varepsilon > 0$ let

$$p(z) := \left(\frac{1-z}{p_0} + \frac{z}{p_1}\right)^{-1},$$

$$f_\varepsilon(x,z) := \begin{cases} e^{\varepsilon(z^2-\theta^2)}|a(x)|^{p_\theta/p(z)}a(x)/|a(x)| & \text{if} \quad a(x) \neq 0 \\ 0 & \text{if} \quad a(x) = 0. \end{cases}$$

Then

$$f_\varepsilon \in \mathcal{I}(L^{p_0}, L^{p_1}), \quad f_\varepsilon(\cdot,\theta) = a.$$

Since

$$\|f_\varepsilon(\cdot, it)\|_{p_0} \leq 1, \qquad \|f_\varepsilon(\cdot, 1+it)\|_{p_1} \leq e^\varepsilon, \quad \text{for} \quad t \in \mathbb{R}$$

we conclude from Hadamard's Three-Line-Theorem

$$\|f_\varepsilon\|_{\mathcal{I}(L^{p_0}, L^{p_1})} \leq e^\varepsilon.$$

Letting ε tend to zero we get

$$|a|_\theta \leq 1.$$

(Q.E.D.)

Step 2:

CLAIM: $|a|_\theta \geq \|a\|_{p_\theta}$.

PROOF: Without loss of generality we assume $|a|_\theta = 1$. It has to be shown that

$$\|a\|_{p_\theta} = \sup\{|\langle\langle a, b\rangle\rangle| \mid b \in C_0^0, \|b\|_{p'_\theta} = 1\} = 1,$$

where

$$\langle\langle a, b\rangle\rangle := \int_{\mathbb{R}^n} a(x)b(x)\, dx$$

denotes the dual product. Here $a \in L^{p_\theta}, b \in L^{p'_\theta}$, and p'_θ is the dual Hölder exponent,

$$\frac{1}{p_\theta} + \frac{1}{p'_\theta} = 1.$$

Let $p'(z)$ be the dual Hölder exponent to $p(z)$ and let $\tilde{f}_\varepsilon(x, z)$ be defined for $b, p'_\theta, p'(z)$ in the same way as $f_\varepsilon(x, z)$ is defined for $a, p_\theta, p(z)$ above.
By the definition of $|\cdot|_{p_\theta}$ we obtain

$$\forall \delta > 0 \quad \exists f^\delta \in \mathcal{I}(L^{p_0}, L^{p_1}): \quad f^\delta(\theta) = a, \ \|f^\delta\|_{\mathcal{I}(L^{p_0}, L^{p_1})} \leq 1 + \delta.$$

For $z \in Z$ let

$$F_\varepsilon^\delta(z) := \langle\langle f^\delta(z), \tilde{f}_\varepsilon(\cdot, z)\rangle\rangle.$$

Then F_ε^δ is continuous and bounded in Z, analytic in the interior of Z and we have for sufficiently small δ

$$|F_\varepsilon^\delta(it)| \leq \mathrm{e}^\varepsilon, \quad |F_\varepsilon^\delta(1 + it)| \leq \mathrm{e}^{2\varepsilon},$$

whence

$$|F_\varepsilon^\delta(\theta + it)| \leq \mathrm{e}^{2\varepsilon}$$

follows by Hadamard's Three-Line-Theorem. In particular we get

$$|\langle\langle a, b\rangle\rangle| \leq |F_\varepsilon^\delta(\theta)| \leq \mathrm{e}^{2\varepsilon}.$$

Letting ε tend to zero we conclude

$$\|a\|_{p_\theta} \leq 1.$$

(Q.E.D.)

This completes the proof of Theorem A.7.

Q.E.D.

Remark: The interpolation theorem in L^p-spaces holds in more general measure spaces (U, μ) replacing ($\mathbb{R}^n$, Lebesgue measure) (*Henri Lebesgue*, 28.6.1875 – 26.7.1941).
In order to interpolate in Sobolev spaces we introduce the so-called Besov spaces and Bessel potential spaces. Among other features these spaces provide an interpretation of fractional derivatives.

Let $\mathcal{S}$ denote the usual space of C^∞-functions of rapid decrease and let $\mathcal{S}'$ denote its (topological) dual space, i.e. the space of tempered distributions.
Let $\varphi \in \mathcal{S}$ with

$$\operatorname{supp} \varphi \subset \{\xi \in \mathbb{R}^n \mid \frac{1}{2} \leq |\xi| \leq 2\}$$

and

$$\varphi(\xi) > 0 \quad \text{if} \quad \frac{1}{2} < |\xi| < 2,$$

$$\sum_{k=-\infty}^{\infty} \varphi(2^{-k}\xi) = 1 \quad \text{if} \quad \xi \neq 0.$$

(See [11, p. 136] for the existence of such a φ.)
Let φ_k and ψ be defined by

$$\begin{aligned} \mathcal{F}\varphi_k(\xi) &= \varphi(2^{-k}\xi), \quad k \text{ an integer}, \\ \mathcal{F}\psi(\xi) &= 1 - \sum_{k=1}^{\infty} \varphi(2^{-k}\xi), \end{aligned}$$

where $\mathcal{F}$ denotes the Fourier transform.

Definition A.8 *Let* $1 \leq p \leq q \leq \infty, s \geq 0$.

(i) *The* *Besov space* B^s_{pq} *is defined by*

$$B^s_{pq} := \{f \in \mathcal{S}' \mid \|f\|_{B^s_{pq}} < \infty\},$$

where

$$\|f\|_{B^s_{pq}} := \left\{ \|\psi * f\|_p^q + \sum_{k=1}^{\infty} (2^{sk}\|\varphi_k * f\|_p)^q \right\}^{1/q},$$

(with the usual convention for $q = \infty$*;* $*$ *denotes convolution).*

(ii) *The* *Bessel potential space* H^s_p *is defined by*

$$H^s_p := \{f \in \mathcal{S}' \mid \|f\|_{H^s_p} < \infty\},$$

where

$$\|f\|_{H^s_p} := \|\mathcal{F}^{-1}(1 + |\cdot|^2)^{s/2}\mathcal{F}f\|_p.$$

The spaces B^s_{pq} and H^s_p are Banach spaces with respect to the norms $\|\cdot\|_{B^s_{pq}}$ and $\|\cdot\|_{H^s_p}$ respectively. The spaces B^s_{pq} are independent of the choice of the special function φ.
For $m \in \mathbb{N}_0$ we have (cf. [11, pp. 141,152])

$$B^m_{22} = H^m_2 = W^{m,2}, \tag{A.1}$$

$$H^m_p = W^{m,p} \quad \text{if} \quad 1 < p < \infty, \tag{A.2}$$

$$H^{s+\varepsilon}_p \hookrightarrow B^{s+\varepsilon}_{p\max\{p,2\}} \quad \text{if} \quad s \geq 0, \varepsilon > 0, 1 < p < \infty. \tag{A.3}$$

Theorem A.9

(i) *If* $m \in \mathbb{N}_0$, $1 \le p < \infty$, *then*

$$B^m_{p1} \hookrightarrow W^{m,p}.$$

(ii) *If* $s \ge 0$, $1 < p < \infty$, $1 \le q \le \infty$, $\varepsilon > 0$, *then*

$$H^{s+\varepsilon}_p \hookrightarrow B^s_{pq}.$$

(iii) *If* $s_0 \neq s_1$, $1 \le p_0, q_0, p_1, q_1 \le \infty$, $\theta \in (0,1)$, *then*

$$[B^{s_0}_{p_0q_0}, B^{s}_{p_1q_1}]_\theta = B^{s_\theta}_{p_\theta q_\theta},$$

where

$$s_\theta := (1-\theta)s_0 + \theta s_1,$$
$$p_\theta := \left(\frac{1-\theta}{p_0} + \frac{\theta}{p_1}\right)^{-1},$$
$$q_\theta := \left(\frac{1-\theta}{q_0} + \frac{\theta}{q_1}\right)^{-1}.$$

PROOF: <u>(i)</u>: For $1 \le p, q < \infty$, $m \in \mathbb{N}$ there is the following norm on B^m_{pq} which is equivalent to $\|\cdot\|_{B^m_{pq}}$:

$$\sum_{|\alpha|\le m} \|\nabla^\alpha \cdot\|_{B^0_{pq}}$$

(see [181, p. 59]).
Moreover we have, defining $\varphi_0 := \psi$,

$$\|f\|_p = \|\sum_{k=0}^\infty \varphi_k * f\|_p \le \sum_{k=0}^\infty \|\varphi_k * f\|_p = \|f\|_{B^0_{p1}}.$$

(Q.E.D.)

<u>(ii)</u>: The imbedding

$$B^{s+\varepsilon}_{p\max\{p,2\}} \hookrightarrow B^{s+\varepsilon}_{p\infty}$$

is obvious.
The inequalities

$$\begin{aligned} \|f\|^q_{B^s_{pq}} &= \sum_{k=0}^\infty (2^{(s+\varepsilon-\varepsilon)k}\|\varphi_k * f\|_p)^q \\ &\le \sup_{k\in\mathbb{N}_0} (2^{(s+\varepsilon)k}\|\varphi_k * f\|_p)^q \underbrace{\sum_{k=0}^\infty 2^{-\varepsilon kq}}_{=:\, c_{\varepsilon,q}} \\ &\le c_{\varepsilon,q}\|f\|^q_{B^{s+\varepsilon}_{p\infty}} \end{aligned}$$

yield the imbedding

$$B^{s+\varepsilon}_{p\infty} \hookrightarrow B^s_{pq}$$

and hence the assertion follows with the help of (A.3).

(Q.E.D.)

(iii): The assertion (iii) is reduced to the statement

$$[\ell^{s_0}_{q_0}(L^{p_0}), \ell^{s_1}_{q_1}(L^{p_1})]_\theta = \ell^{s_\theta}_{q_\theta}(L^{p_\theta})$$

(see [11, p. 153]). Here $\ell^s_q(A)$ is defined for a given a Banach space A with norm $\|\cdot\|_A$ as follows:

$$\ell^s_q(A) := \{a = (a_k)_{k\in\mathbb{N}_0} \mid a_k \in A,\ \|a\|_{\ell^s_q(A)} < \infty\},$$

where

$$\|a\|_{\ell^s_q(A)} := \left\{\sum_{k=0}^{\infty} (2^{ks}\|a_k\|_A)^q\right\}^{1/q}.$$

Using the equivalence of ℓ_q and $L^q(d\mu)$, μ a pure point measure, Theorem A.7 yields the assertion.

Q.E.D.

Finally we present the interpolation theorem which was used in Chapter 2.

Theorem A.10 *Let the linear operator T satisfy*

$$T : W^{n,1} \longrightarrow L^\infty, \quad \textit{bounded with norm } M_0,$$

$$T : L^2 \longrightarrow L^2, \quad \textit{bounded with norm } M_1.$$

Let $1 < p < 2 < q < \infty$, $1/p + 1/q = 1$, $\theta := 2/q$, $N \in \mathbb{N}$ with $N > n(1-\theta)$. Then there is a constant $c = c(p,n)$ such that

$$T : W^{N,p} \longrightarrow L^q, \quad \textit{with norm } M,$$

and

$$M \le c\, M_0^{1-\theta} M_1^\theta.$$

Proof: According to (A.1) and Theorem A.9, (i), T maps as follows:

$$T : B^n_{11} \longrightarrow L^\infty, \quad \text{bounded with norm } cM_0 \quad (c = c(n))$$

and

$$T : B^0_{22} \longrightarrow L^2, \quad \text{bounded with norm } M_1.$$

Theorem A.6 then implies

$$T : [B^n_{11}, B^0_{22}]_\theta \longrightarrow [L^\infty, L^2]_\theta, \quad \text{bounded with norm } M,$$

and

$$M \leq (c\, M_0)^{1-\theta} M_1^\theta.$$

By Theorem A.7 we know

$$[L^\infty, L^2]_\theta = L^{2/\theta} = L^q.$$

Moreover we conclude from Theorem A.9, (iii), (ii):

$$[B^n_{11}, B^0_{22}]_\theta = B^{(1-\theta)n}_{\frac{2}{2-\theta}\frac{2}{2-\theta}} = B^{(1-\theta)n}_{pp}$$

and from (A.2) and (A.3):

$$W^{N,p} = H^N_p = H^{(1-\theta)n+\varepsilon}_p \hookrightarrow B^{(1-\theta)n}_{pp}$$

for $\varepsilon := N - (1-\theta)n > 0$. This completes the proof.

Q.E.D.

B The Theorem of Cauchy–Kowalevsky

The proof of the local existence theorem of Cauchy–Kowalevsky follows the presentation as in the book of F. John [71]. (Among the various spellings of the name of Sophie von Kowalevsky we chose that one which is used in her first paper in Crelle's Journal, volume 80 from 1874 (*August Leopold Crelle,* 11.3.1780 – 6.10.1855).)
The following initial value problem shall be solved (locally):

$$\partial_t u_j = \sum_{i=1}^n \sum_{k=1}^N a^i_{jk}(t,x,u)\partial_i u_k + b_j(t,x,u), \quad j = 1,\dots,N,$$

$$u(0,x) = u_0(x).$$

Here $u = (u_1,\dots,u_N) = u(t,x)$ is the unknown vector-valued function of $t \in \mathbb{R}$ and $x \in \mathbb{R}^n$. a^i_{jk}, b_j are real-analytic functions of their arguments, $i = 1,\dots,n; j,k = 1,\dots,N \in \mathbb{N}$, and u_0 is real-analytic in x.
Without loss of generality we may assume that $u_0 = 0$ (otherwise consider $\tilde{u} := u - u_0$) and that a^i_{jk} and b_j do not depend on t (otherwise introduce u_{N+1} with $\partial_t u_{N+1} = 1$, $u_{N+1}(0,x) = 0$). With these simplifications the following theorem will be proved.

Theorem B.1 *Let a^i_{jk} and b_j be real-analytic functions of $z = (x,u)$ in a neighbourhood of zero in $\mathbb{R}^{n+N}, i = 1,\dots,n;\ j,k = 1,\dots,N$. Then the system of differential equations*

$$\partial_t u_j = \sum_{i=1}^n \sum_{k=1}^N a^i_{jk}(z)\partial_i u_k + b_j(z), \quad j = 1,\dots,N, \tag{B.1}$$

with initial conditions

$$u(0,x) = 0, \quad x \in \mathbb{R}^n, \tag{B.2}$$

has a solution u in a neighbourhood of zero in $\mathbb{R}^{1+n}$ which is real-analytic there. The solution is unique in the class of real-analytic functions.

PROOF: Without loss of generality we assume $n = 1$ and we write a_{jk} instead of a^1_{jk}. The proof uses the fact that the coefficients $c^i_{\ell k}$ in the Taylor expansion of a solution u,

$$u_i(t,x) = \sum_{\ell,k=0}^{\infty} c^i_{\ell k} t^\ell x^k, \tag{B.3}$$

are necessarily determined by the differential equations (B.1) (*Brook Taylor*, 18.8.1685 – 29.12.1731). It is shown then that the series with these coefficients converges. For this purpose a majorant will be constructed. We have

$$c^i_{\ell k} = \frac{1}{\ell! k!} \left. \frac{\partial^{k+\ell} u_i(t,x)}{\partial t^\ell \partial x^k} \right|_{t=0, x=0}.$$

Then we get successively

$$\left. \frac{\partial^m u_i}{\partial x^m} \right|_{t=0} = 0 \qquad \text{from (B.2)},$$

$$\left. \frac{\partial u_i}{\partial t} \right|_{t=0}, \qquad \text{using (B.1)},$$

which yields

$$\left. \frac{\partial^2 u_i}{\partial t \partial x} \right|_{t=0}.$$

Differentiation of the differential equations (B.1) with respect to t then yields

$$\left. \frac{\partial^2 u}{\partial t^2} \right|_{t=0}$$

and so on, thus determining $c^i_{\ell k}$ for all $i = 1, \dots, N$ and $\ell, k \in \mathbb{N}_0$.
If

$$a_{jk}(z) = \sum_{|\alpha|=0}^{\infty} g^{jk}_\alpha z^\alpha$$

and

$$b_j(z) = \sum_{|\alpha|=0}^{\infty} h^j_\alpha z^\alpha$$

for all $|z| \le r$ for some fixed $r > 0$, then

$$c^i_{\ell k} = P^i_{\ell k} \left((g^{jm}_\alpha)_{\alpha jm}, (h^j_\alpha)_{\alpha j} \right)$$

where $P^i_{\ell k}$ is a polynomial with nonnegative coefficients. This is obvious by the definition of $c^i_{\ell k}$. The chain rule and the rule for differentiating products only contribute positive coefficients.

Now we construct majorant coefficients $C^i_{\ell k}$. Then u defined by (B.3) will automatically be the (unique) solution of (B.1), (B.2). We show that there is a (local) solution $v = (v_1, \ldots, v_N)$,

$$v_i(t,x) = \sum_{\ell,k=0}^{\infty} C^i_{\ell k} t^\ell x^k$$

of the initial value problem

$$\partial_t v_j = \sum_{k=1}^{N} A_{jk}(z) \frac{\partial}{\partial x} v_k + B_j(z),$$

$$v_j(0,x) = 0,$$

where A_{jk} and B_j have to be determined such that

$$A_{jk}(z) = \sum_{|\alpha|=0}^{\infty} G^{jk}_\alpha z^\alpha, \quad B_j(z) = \sum_{|\alpha|=0}^{\infty} H^j_\alpha z^\alpha$$

with the property

$$|g^{jk}_\alpha| \le G^{jk}_\alpha, \quad |h^j_\alpha| \le H^j_\alpha.$$

Then it follows

$$\begin{aligned} C^i_{\ell k} &= P^i_{\ell k}\left((G^{jm}_\alpha)_{\alpha jm}, (H^j_\alpha)_{\alpha j}\right) \ge \left|P^i_{\ell k}((|g^{jm}_\alpha|)_{\alpha jm}, (|h^j_\alpha|)_{\alpha j})\right| \\ &\ge |c^i_{\ell k}|. \end{aligned}$$

Hence we have found the desired majorant.

Now it only remains to determine A_{jk}, B_j and v appropriately.

With

$$M_1 := \max_{j,k=1,\ldots,N;|z|=r} |a_{jk}(z)|$$

the estimates

$$|g^{jk}_\alpha| \le \frac{M_1}{r^{|\alpha|}} \le \frac{M_1}{r^{|\alpha|}} \frac{|\alpha|!}{\alpha!} =: G^{jk}_\alpha,$$

hold, analogously with

$$\begin{aligned} M_2 &:= \max_{j=1,\ldots,N;|z|=r} |b_j(z)| : \\ |h^j_\alpha| &\le \frac{M_2}{r^{|\alpha|}} \le \frac{M_2}{r^{|\alpha|}} \frac{|\alpha|!}{\alpha!} =: H^j_\alpha. \end{aligned}$$

Without loss of generality we assume

$$M_1 = M_2 =: M.$$

Let

$$A_{jk}(z) := \sum_{|\alpha|=0}^{\infty} G_{\alpha}^{jk} z^{\alpha} = M \sum_{|\alpha|=0}^{\infty} \frac{|\alpha|!}{\alpha!} \left(\frac{z}{r}\right)^{\alpha}$$
$$= M \frac{1}{1 - \frac{z_1 + \ldots + z_{N+1}}{r}} \quad \text{if} \quad |z_1| + \ldots + |z_{N+1}| < r,$$

and analogously

$$B_j(z) := \sum_{|\alpha|=0}^{\infty} H_{\alpha}^{j} z^{\alpha} = M \frac{1}{1 - \frac{z_1 + \ldots + z_{N+1}}{r}}.$$

Since the chosen coefficients A_{jk} and B_j are independent of j and k we make the following ansatz for v:

$$v_i(t,x) = w(t,x), \quad i = 1, \ldots, N.$$

Then we have to solve

$$\partial_t w = \frac{Mr}{r - x - Nw} \left(1 + N \frac{\partial}{\partial x} w\right),$$
$$w(0,x) = 0 \qquad (|x| + N|w| < r).$$

This is explicitly solvable by

$$w(t,x) := \frac{1}{2N} \left(r - x - \sqrt{(r-x)^2 - 4MNrt}\right)$$

in a neighbourhood of zero in $\mathbb{R}^{1+n}$, e.g. where

$$|x| < \frac{r}{2}, \quad t < \frac{r}{16MN} =: T(M,r) \quad \text{for } N, n \text{ fixed,} \tag{B.4}$$

holds and there also u is analytic.

Q.E.D.

For the application in Chapter 3 the following remarks on the linear case are important: Let

$$Lu := A^0(t,x)\partial_t u + \sum_{j=1}^{n} A^j(t,x)\partial_j u + B(t,x)u = 0,$$
$$u(0,x) = P(x), \; x \in \mathbb{R}^n,$$

with analytic $N \times N$-matrices $A^0, A^1, \ldots, A^n, B$ in a cylinder $Z := \{(t,x) \,|\, 0 \le t \le T, |x| \le r\} \equiv Z_T^r$, for some $T, r > 0$.
Let A^0 be positive definite in Z and P be a polynomial. Then v defined by

$$v := u - P$$

should solve

$$\partial_t v = -\sum_{j=1}^{n}(A^0)^{-1}A^j\partial_j v - (A^0)^{-1}Bv - (A^0)^{-1}LP,$$

$$v(t=0) = 0.$$

According to Theorem B.1 there exists a solution v in $Z^{r/2}_{T(M,r)}$ where $T(M,r)$ is given in (B.4).
If P is replaced by $P_1 := P/c$ with a sufficiently large constant $c > 0$, then M, computed from $(A^0)^{-1}A^j, (A^0)^{-1}B, j = 1, \ldots, n$, is already a corresponding bound for $(A^0)^{-1}LP_1$. Hence the solution U corresponding to the initial value P_1 exists in $Z^{r/2}_{T(M,r)}$ with

$$M = M(((A^0)^{-1}A^j)_j, (A^0)^{-1}B).$$

But then also $u = cU$ exists there and this implies that $T(M,r)$ does not depend on the special polynomial P.

C A local existence theorem for hyperbolic-parabolic systems

In this appendix we present a local existence theorem for quasilinear hyperbolic-parabolic coupled systems, essentially taken from the paper of S. Kawashima [84, Chapter II], together with sketches of the proof.
We consider the initial value problem for a system of quasilinear differential equations of the form

$$A_1^0(u,v)u_t \ + \sum_{j=1}^{n} A_{11}^j(u,v)\partial_j u \ = \ f_1(u,v,\nabla v), \tag{C.1}$$

$$A_2^0(u,v)v_t \ - \sum_{j,k=1}^{n} B_2^{jk}(u,v,\nabla v)\partial_j\partial_k v \ = \ f_2(u,v,\nabla u,\nabla v), \tag{C.2}$$

where $t \geq 0, x \in \mathbb{R}^n$, $n \in \mathbb{N}$. $u = u(t,x)$ and $v = v(t,x)$ are vectors with m' and m'' components, respectively, $m', m'' \in \mathbb{N}_0$, one being different from zero. The pair $(u,v)(t,x)$ takes its values in an open convex set $\mathcal{U} \in \mathbb{R}^m$ $(m := m' + m'' \geq 1)$. A_1^0 and A_{11}^j $(j = 1, \ldots, n)$ (resp. A_2^0 and B_2^{jk} $(j,k = 1, \ldots, n)$) are square matrices of order m' (resp. m''), and f_1 (resp. f_2) is a $\mathbb{R}^{m'}$-valued (resp. $\mathbb{R}^{m''}$-valued) function.
The initial data are prescribed at $t = 0$ by

$$(u,v)(0,x) \ = \ (u_0,v_0)(x). \tag{C.3}$$

We assume that the system (C.1), (C.2) is symmetric hyperbolic-parabolic in the following sense:

Condition C1: The functions $A_1^0(u,v), A_2^0(u,v)$ and $A_{11}^j(u,v)$ $(j=1,\ldots,n)$ are sufficiently smooth in $(u,v)\in\mathcal{U}$ and $B_2^{jk}(u,v,\xi)$ $(j,k=1,\ldots,n)$ is sufficiently smooth in $(u,v,\xi)\in\tilde{\mathcal{U}}:=\mathcal{U}\times\mathbb{R}^{nm''}$, and

(i) *$A_1^0(u,v)$ and $A_2^0(u,v)$ are real symmetric and positive definite for $(u,v)\in\mathcal{U}$,*

(ii) *$A_{11}^j(u,v)$ is real-symmetric for $(u,v)\in\mathcal{U}$,*

(iii) *$B_2^{jk}(u,v,\xi)$ is real-symmetric and satisfies*

$$B_2^{jk}(u,v,\xi) = B_2^{kj}(u,v,\xi)$$

for $(u,v,\xi)\in\tilde{\mathcal{U}}$;
$\sum_{j,k=1}^{n} B_2^{jk}(u,v,\xi)\omega_j\omega_k$ is (real-symmetric and) positive definite for all $(u,v,\xi)\in\tilde{\mathcal{U}}$ and $\omega=(\omega_1,\ldots,\omega_n)\in\mathbb{R}^n$ with $|\omega|=1$.

Let $\eta\in\mathbb{R}^{nm'}$ denote a vector corresponding to ∇u.

Condition C2: The functions $f_1(u,v,\xi)$ and $f_2(u,v,\eta,\xi)$ are sufficiently smooth in $(u,v,\xi)\in\mathcal{U}\times\mathbb{R}^{nm'}$ and $(u,v,\eta,\xi)\in\mathcal{U}\times\mathbb{R}^{nm}$, respectively, and

$$f_1(\overline{u},\overline{v},0) = f_2(\overline{u},\overline{v},0,0) = 0$$

for some constant state $(\overline{u},\overline{v})\in\mathcal{U}$.

Remark: The results in this appendix hold in particular for $m''=0$ (symmetric hyperbolic systems, cf. Chapter 5) and for $m'=0$ (parabolic systems).
First we study solutions of the linearized equations

$$A_1^0(u,v)\tilde{u}_t + \sum_{j=1}^{n} A_{11}^j(u,v)\partial_j\tilde{u} = f_1, \tag{C.4}$$

$$A_2^0(u,v)\tilde{v}_t - \sum_{j,k=1}^{n} B_2^{jk}(u,v,\nabla v)\partial_j\partial_k\tilde{v} = f_2. \tag{C.5}$$

Let $Q_T := [0,T]\times\mathbb{R}^n$ $(T>0$ arbitrary but fixed), $\ell,s\in\mathbb{N}_0$ with $0\le\ell\le s$.

Theorem C.1 *Assume Condition C1 and $s\ge[\frac{n}{2}]+3$. Let (u,v) satisfy*

$$u-\overline{u}\in L^\infty([0,T],W^{s,2}),\ u_t\in L^\infty([0,T],W^{s-1,2}), \tag{C.6}$$

$$v-\overline{v}\in L^\infty([0,T],W^{s,2}),\ v_t\in L^\infty([0,T],W^{s-2,2})\cap L^2([0,T],W^{s-1,2}), \tag{C.7}$$

$$\forall (t,x)\in Q_T : (u,v,\nabla v)(t,x)\in\mathcal{U}_1, \tag{C.8}$$

where $\mathcal{U}_1$ is a bounded, open, convex set in $\mathbb{R}^{m+nm''}$ satisfying $\overline{\mathcal{U}}_1 \subset \mathcal{U} \times \mathbb{R}^{nm}$. Let

$$M := \operatorname*{ess\,sup}_{0\le t\le T} \|(u-\overline{u}, v-\overline{v})(t)\|_{s,2},$$

$$M_1 := \left(\int_0^T \|\partial_t(u,v)(t)\|_{s-1,2}^2\,dt\right)^{1/2}.$$

Let $0 \le \ell \le s$ be an integer and let f_1, f_2 satisfy:

$$f_1 \in L^\infty([0,T], W^{\ell-1,2}) \cap L^2([0,T], W^{\ell,2}), \tag{C.9}$$

$$f_2 \in L^\infty([0,T], W^{\ell-1,2}).$$

(i) *Assume that $\tilde{u}$ is a solution of (C.4) satisfying*

$$\tilde{u} \in L^\infty([0,T], W^{\ell,2}),\ \tilde{u}_t \in L^\infty([0,T], W^{\ell-1,2}). \tag{C.10}$$

Then we have $\tilde{u} \in C^0([0,T], W^{\ell,2})$. Furthermore there exist constants $C_1 = C_1(\mathcal{U}_1) > 1$ and $C_2 = C_2(\mathcal{U}_1, M) > 0$ such that the following energy inequality holds for $t \in [0,T]$:

$$\|\tilde{u}(t)\|_{\ell,2}^2 \le C_1^2 \left\{ \|\tilde{u}(0)\|_{\ell,2}^2 + C_2 t \int_0^t \|f_1(r)\|_{\ell,2}^2\,dr \right\} \exp\{C_2(Mt + M_1 t^{1/2})\}. \tag{C.11}$$

(ii) *Assume that $\tilde{v}$ is a solution of (C.5) satisfying*

$$\tilde{v} \in L^\infty([0,T], W^{\ell,2}),\ \tilde{v}_t \in L^\infty([0,T], W^{\ell-2,2}).$$

Then we have $\tilde{v} \in C^0([0,T], W^{\ell,2}) \cap L^2([0,T], W^{\ell+1,2})$ and the following energy inequality holds for $t \in [0,T]$ (with the constants C_1, C_2 from (i)):

$$\|\tilde{v}(t)\|_{\ell,2}^2 + \int_0^t \|\tilde{v}(r)\|_{\ell+1,2}^2 dr \tag{C.12}$$

$$\le C_1^2 \left\{ \|\tilde{v}(0)\|_{\ell,2}^2 + C_2 \int_0^t \|f_2(r)\|_{\ell-1,2}^2\,dr \right\} \exp\{C_2(t + M_1 t^{1/2})\}.$$

PROOF: (i):

1. Assume

$$v - \overline{v} \in L^\infty([0,T], W^{s,2}),\ v_t \in L^\infty([0,T], W^{s-1,2}), \tag{C.13}$$

$$f_1 \in L^\infty([0,T], W^{\ell,2}), \tag{C.14}$$

$$\tilde{u} \in L^\infty([0,T], W^{\ell+1,2}),\ \partial_t\tilde{u} \in L^\infty([0,T], W^{\ell,2}). \tag{C.15}$$

(Cf. the conditions (C.7), (C.9), (C.10).)
Applying ∇^k to the system (C.4), integrating by parts as usual, and summing up for $0 \le k \le \ell$, we arrive at

$$\frac{1}{2}\frac{d}{dt}(E_1(\tilde{u}))^2 - C_{01}(M + \|\partial_t(u,v)\|_{s-1,2})(E_1(\tilde{u}))^2 \le C_{02}\|f_1\|_{\ell,2}E_1(\tilde{u}) + C_{02}M(E_1(\tilde{u}))^2,$$

where

$$E_1(\tilde{u})(t) := \left(\sum_{k=0}^{\ell}\langle A_1^0(u,v)\nabla^k\tilde{u}, \nabla^k\tilde{u}\rangle(t)\right)^{1/2}$$

and

$$C_{01} = C_{01}(\mathcal{U}_1),\ C_{02} = C_{02}(\mathcal{U}_1, M)$$

are positive constants. Gronwall's inequality, Lemma 4.1, now yields (C.11).

2. Let $(u,v), f_1, \tilde{u}$ satisfy (C.6), (C.7), (C.14), (C.15).
This case is reduced to the situation in case 1 by using the Friedrichs mollifier $j_\delta, \delta > 0$ (cf. Chapter 4). Consider $v_\delta := j_\delta * v$ and let δ tend to zero.

3. Let $(u,v), f_1, \tilde{u}$ satisfy the assumptions of the theorem. Apply $j_\delta *$ to the system (C.4) and thus reduce it to the situation in case 2, then let δ tend to zero.

4. $\tilde{u} \in C^0([0,T], W^{\ell,2})$ follows by considering the system (C.4) for $\tilde{u}_\delta - \tilde{u}_{\delta'}$ instead of $\tilde{u}$, $u_\delta := j_\delta * u$. We have $\tilde{u}_\delta \in C^0([0,T], W^{\ell,2})$. Then let δ, δ' tend to zero.

(ii) is proved analogously. First assume that $(u,v), f_2, \tilde{v}$ satisfy (C.6), (C.13) and

$$f_2 \in L^\infty([0,T], W^{\ell,2}),$$
$$\tilde{v} \in L^\infty([0,T], W^{\ell+2,2}),\ \tilde{v}_t \in L^\infty([0,T], W^{\ell,2}),$$

then regularize.

Q.E.D.

An existence result for the system (C.4), (C.5) is given by the following theorem.

Theorem C.2 *Assume Condition C1 and $s \ge [\frac{n}{2}] + 3$. Let (u,v) satisfy*

$$u - \overline{u} \in C^0([0,T], W^{s,2}),\ u_t \in C^0([0,T], W^{s-1,2}), \tag{C.16}$$
$$v - \overline{v} \in C^0([0,T], W^{s,2}),\ v_t \in C^0([0,T], W^{s-2,2}) \cap L^2([0,T], W^{s-1,2}),$$
$$(C.8).$$

(i) *Let $1 \le \ell \le s$ be an integer and let f_1 satisfy*

$$f_1 \in C^0([0,T], W^{\ell-1,2}) \cap L^2([0,T], W^{\ell,2}).$$

If the prescribed initial data satisfy $\tilde{u}(0) \in W^{\ell,2}$, then the system (C.4) has a unique solution $\tilde{u} \in C^0([0,T], W^{\ell,2}) \cap C^1([0,T], W^{\ell-1,2})$ satisfying the estimate (C.11).

(ii) *Let $2 \le \ell \le s$ be an integer and let f_2 satisfy*

$$f_2 \in C^0([0,T], W^{\ell-1,2}).$$

If the prescribed initial data satisfy $\tilde{v}(0) \in W^{\ell,2}$, then the system (C.5) has a unique solution $\tilde{v} \in C^0([0,T], W^{\ell,2}) \cap C^1([0,T], W^{\ell-2,2}) \cap L^2([0,T], W^{\ell+1,2})$ satisfying the estimate (C.12).

PROOF: (i): The system (C.4) is written in the form

$$\tilde{u}_t + \tilde{A}_1(t)\tilde{u}(t) = \tilde{f}_1(t),\ t \in [0,T],$$

where

$$\begin{aligned}
\tilde{A}_1(t) &:= \sum_{j=1}^n \left(A_1^0(u,v)(t)\right)^{-1} A_{11}^j(u,v)(t)\partial_j,\\
\tilde{f}_1(t) &:= \left(A_1^0(u,v)(t)\right)^{-1} f_1(t).
\end{aligned}$$

Then the results of T. Kato on linear evolution equations can be applied, see [79] (with $S(t) \equiv S := (1-\Delta)^{s/2}$ there, cf. also [80]).

(ii): (By induction.) Let $\ell = 2$: $\tilde{v}$ satisfies

$$\tilde{v}_t + \tilde{A}_2(t)\tilde{v}(t) = \tilde{f}_2(t),$$

where

$$\begin{aligned}
\tilde{A}_2(t) &:= -\sum_{j,k=1}^n \left(A_2^0(u,v)(t)\right)^{-1} B_2^{jk}(u,v,\nabla v)(t),\\
\tilde{f}_2(t) &:= \left(A_2^0(u,v)(t)\right)^{-1} f_2(t).
\end{aligned}$$

Then the results from [79] can be used again (with $S(t) = \tilde{A}_2(t) + \beta,\ \beta > 1$ sufficiently large).

Q.E.D.

Now we consider the linearized equations arising from (C.1), (C.2):

$$A_1^0(u,v)\tilde{u}_t + \sum_{j=1}^n A_{11}^j(u,v)\partial_j\tilde{u} = f_1(u,v,\nabla v), \tag{C.17}$$

$$A_2^0(u,v)\tilde{v}_t - \sum_{j,k=1}^n B_2^{jk}(u,v,\nabla v)\partial_j\partial_k\tilde{v} = f_2(u,v,\nabla u,\nabla v), \tag{C.18}$$

with initial data

$$(\tilde{u},\tilde{v})(0,x) = (u,v)(0,x) = (u_0,v_0)(x). \tag{C.19}$$

Let $s \geq [\frac{n}{2}] + 3$, $T > 0$ and let $X_T^s(\mathcal{U}_1, M, M_1)$ be the set of functions (u, v) satisfying (C.16),

$$v - \overline{v} \in C^0([0,T], W^{s,2}) \cap L^2([0,T], W^{s+1,2}),$$

$$v_t \in C^0([0,T], W^{s-2,2}) \cap L^2([0,T], W^{s-1,2}),$$

$$(C.8),$$

$$\forall t \in [0,T] \quad : \quad \sup_{0 \leq r \leq t} \|(u - \overline{u}, v - \overline{v})(r)\|_{s,2}^2 + \int_0^t \|(v - \overline{v})(r)\|_{s+1,2}^2 \, dr \leq M^2, \quad \text{(C.20)}$$

$$\forall t \in [0,T] \quad : \quad \int_0^t \|\partial_t (u,v)(r)\|_{s-1,2}^2 \, dr \leq M_1^2.$$

The following existence result holds:

Theorem C.3 *Assume conditions C1 and C2. Let* $s \geq [\frac{n}{2}]+3$ *and* $(u_0 - \overline{u}, v_0 - \overline{v}) \in W^{s,2}$ *satisfy*

$$\forall x \in \mathbb{R}^n \, : \, (u_0, v_0, \nabla v_0)(x) \in \mathcal{U}_0, \qquad \text{(C.21)}$$

where $\mathcal{U}_0$ *is a bounded, open, convex set in* $\mathbb{R}^{m+nm''}$ *satisfying* $\overline{\mathcal{U}_0} \subset \mathcal{U} \times \mathbb{R}^{nm''}$.
Then there exist a positive constant T_0*, only depending on* $\mathcal{U}_0$, $\|(u_0 - \overline{u}, v_0 - \overline{v})\|_{s,2}$ *and* d_1*, where* d_1 *is an arbitrary positive number being smaller than the distance from* $\mathcal{U}_0$ *to the boundary of* $\mathcal{U} \times \mathbb{R}^{nm''}$*, such that if* $(u,v) \in X_{T_0}^s(\mathcal{U}_1, M, M_1)$ *then the problem (C.17), (C.18), (C.19) has a unique solution* $(\tilde{u}, \tilde{v}) \in X_{T_0}^s(\mathcal{U}_1, M, M_1)$.
Here

$$\left.\begin{aligned} \mathcal{U}_1 &:= d_1\text{-neighbourhood of } \mathcal{U}_0, \\ M &:= 2C_1 \|(u_0 - \overline{u}, v_0 - \overline{v})\|_{s,2}, \ M_1 := 2C_3 M, \end{aligned}\right\} \qquad \text{(C.22)}$$

where $C_1 = C_1(\mathcal{U}_1)$ *from Theorem C.1, and* $C_3 = C_3(\mathcal{U}_1, M)$ *is given from the valid relation*

$$\forall 0 \leq t \leq T : \int_0^t \|\partial_t (\tilde{u}, \tilde{v})(r)\|_{s-1,2}^2 \, dr \leq C_3^2 (\tilde{M}^2 + (\tilde{M}^2 + M^2)t), \qquad \text{(C.23)}$$

$$(\tilde{M} = \tilde{M}(\tilde{u}, \tilde{v}) \quad \text{as} \quad M = M(u,v) \quad \text{in (C.20)}).$$

Proof: The proof follows from Theorems C.1, C.2.

Q.E.D.

Remarks: 1. (C.23) follows from the differential equations (C.17), (C.18) and from the inequality

$$\|f_1(u, v, \nabla v)\|_{s-1,2} + \|f_2(u, v, \nabla u, \nabla v)\|_{s-1,2} \leq C(\mathcal{U}_1, M) M,$$

(which is obtained using Condition C2).
2. The proof requires at least $T_0 \leq 3/2$.
The solution of the nonlinear system (C.1), (C.2), (C.3) can now be constructed by successive approximation in the following way. Let L be the operator which maps (u, v) to $(\tilde{u}, \tilde{v})$ according to Theorem C.3. A sequence $(u^n, v^n)_{n \in \mathbb{N}_0}$ is defined by

$$\begin{aligned} (u^0, v^0)(t, x) &:= (\overline{u}, \overline{v}), \\ (u^{n+1}, v^{n+1}) &:= L(u^n, v^n). \end{aligned}$$

Consider the differential equations (C.1), (C.2) for $(u^{n+1} - \overline{u}, v^{n+1} - \overline{v})$ instead of (u, v). Using the preceding theorems it can be shown that $(u^n - \overline{u}, v^n - \overline{v})_n$ is a Cauchy sequence in $C^0([0, T_1], W^{s-1,2})$ and for a subsequence we have the following:
$(v^{n'} - v)_{n'}$ tends to zero weakly in $L^2([0, T_1], W^{s+1,2})$, $(u^n(t) - \overline{u}, v^n(t) - \overline{v})_n$ tends to zero weakly in $W^{s,2}$ for each $t \in [0, T_1]$ if $T_1 \in (0, T_0]$ is sufficiently small. The limit (u, v) is the desired solution of (C.1), (C.2), (C.3).
Thus, the following local existence theorem holds.

Theorem C.4 *Assume the Conditions C1, C2, $s \geq [\frac{n}{2}] + 3, (u_0 - \overline{u}, v_0 - \overline{v}) \in W^{s,2}$ and (C.21). Then there is a positive constant T_1, only depending on $\mathcal{U}_0$, d_1 and on $\|(u_0 - \overline{u}, v_0 - \overline{v})\|_{s,2}$, such that the quasilinear symmetric hyperbolic-parabolic initial value problem (C.1), (C.2), (C.3) has a unique solution $(u, v) \in X^s_{T_1}(\mathcal{U}_1, M, M_1)$, where $\mathcal{U}_1, M, M_1$, are defined by (C.22).*
In particular the solution satisfies

$$u - \overline{u} \in C^0([0, T_1], W^{s,2}) \cap C^1([0, T_1], W^{s-1,2}),$$

$$v - \overline{v} \in C^0([0, T_1], W^{s,2}) \cap C^1([0, T_1], W^{s-2,2}) \cap L^2([0, T_1], W^{s+1,2}),$$

$$\begin{aligned} &\sup_{0 \leq r \leq t} \|(u - \overline{u}, v - \overline{v})(r)\|^2_{s,2} + \int_0^t \left(\|(u - \overline{u})(r)\|^2_{s,2} + \|(v - \overline{v})(r)\|^2_{s+1,2} \right) dr \\ &\qquad \leq C_4^2 \|(u_0 - \overline{u}, v_0 - \overline{v})\|^2_{s,2}, \quad t \in [0, T_1], \end{aligned}$$

where $C_4 > 1$ is a constant which only depends on $\mathcal{U}_0, d_1$ and $\|(u_0 - \overline{u}, v_0 - \overline{v})\|_{s,2}$ (uniformly bounded for fixed T_1 and all (u_0, v_0) with $\|(u_0 - \overline{u}, v_0 - \overline{v})\|_{s,2} \leq 1$).

References

[1] Adams, R.A.: *Sobolev spaces.* Academic Press, New York et al. (1975).

[2] Agmon, S. *Lectures on elliptic boundary value problems.* Van Nostrand, New York (1975).

[3] Agmon, S.: Spectral properties of Schrödinger operators and scattering theory. *Ann. Scuola Norm. Sup. Pisa, Ser. 4,* **2** (1975), 151–218.

[4] Alber, H.-D., Cooper, J.: Quasilinear hyperbolic 2×2 systems with a free, damping boundary condition. *J. reine angew. Math.* **406** (1990), 10–43.

[5] Alsholm. P., Schmidt, G.: Spectral and scattering theory for Schrödinger operators. *Arch. Rat. Mech. Anal.* **40** (1971), 281–311.

[6] Alt, H.W.: *Lineare Funktionalanalysis.* Springer-Verlag, Berlin et al. (1985).

[7] Arnold, V.I.: *Geometrical methods in the theory of ordinary differential equations.* Grundlehren math. Wiss. **250**. Springer-Verlag, New York et al. (1988).

[8] Ball, J.M.: Convexity conditions and existence theorems in nonlinear elasticity. *Arch. Rational Mech. Anal.* **63** (1977), 337–403.

[9] Barceló, J.A., Ruiz, A., Vega, L., Vilela, M.C.: Weak dispersive estimates for Schrödinger equations with long range potentials. *Comm. PDE* **34** (2009), 74–105.

[10] Bemelmans, J., Hildebrandt, S., von Wahl, W.: Partielle Differentialgleichungen und Variationsrechnung. In: *Ein Jahrhundert Mathematik 1890 – 1990, Festschrift zum Jubiläum der DMV*; eds.: G. Fischer et al., Friedr. Vieweg & Sohn, Braunschweig/Wiesbaden (1990), 149–230.

[11] Bergh, J., Löfström, J.: *Interpolation spaces, an introduction.* Grundlehren math. Wiss. **223**. Springer-Verlag, Berlin et al. (1976).

[12] Bier, M.: *Globale Existenz kleiner Lösungen nichtlinearer Wellengleichungen in Wellenleitern.* Diplomarbeit (master thesis). University of Stuttgart (2002).

[13] Bloom, C.O.: Estimates for solutions of reduced hyperbolic equations of the second order with a large parameter. *J. Math. Anal. Appl.* **44** (1973), 310–332.

[14] Bosch, V.: *Schrödingergleichungen in Wellenleitern.* Diplomarbeit (master thesis). University of Konstanz (2009).

[15] Carlson, D.E.: Linear thermoelasticity. *Handbuch der Physik* **VIa/2** (1972), 297–346.

[16] Chen, Y.: Initial boundary-value problems of nonlinear wave equations in an exterior domain. *J. London Math. Soc.* **40** (1989), 519–534.

[17] Chow, S.-N., Hale, J.K.: *Methods of bifurcation theory.* Grundlehren math. Wiss. **251**. Springer-Verlag, New York et al. (1982).

[18] Christodoulou, D.: Global solutions of nonlinear hyperbolic equations for small initial data. *Comm. Pure Appl. Math.* **39** (1986), 267–282.

[19] Christodoulou, D., Klainerman, S.: Asymptotic properties of linear field equations in Minkowski space. *Comm. Pure Appl. Math.* **43** (1990), 137–199.

[20] Chrzęszczyk, A.: Some existence results in dynamical thermoelasticity. Part I. Nonlinear case. *Arch. Mech.* **39** (1987), 605–617.

[21] Ciarlet, P.G.: *Élasticité tridimensionelle.* Collection Recherches en Mathématiques Appliquées **1**. Masson, Paris et al. (1986).

[22] Constantin, P.: Decay estimates for Schrödinger equations. *Comm. Math. Phys.* **127** (1990), 101–108.

[23] Courant, R., Hilbert, D.: *Methods of mathematical physics II.* J. Wiley & Sons, New York et al. (1989).

[24] Dafermos, C.M., Hsiao, L.: Development of singularities in solutions of the equations of nonlinear thermoelasticity. *Quart. Appl. Math.* **44** (1986), 463–474.

[25] D'Ancona, P., Fanelli, L.: Smoothing estimates for the Schrödinger equation with unbounded potentials. *J. Differential Equations* **246** (2009), 4552–4567.

[26] D'Ancona, P., Racke, R.: Evolution equations on non-flat waveguides. *Arch. Rational Mech. Anal.* **206** (2012), 81–110.

[27] Dreher, M.: Decay estimates of solutions to wave equations in conical sets. *Commun. PDE* **32** (2007), 477–501.

[28] Duff, G.F.D.: The Cauchy problem for elastic waves in an anisotropic medium. *Philos. Trans. Roy. Soc. London, Ser. A.,* **252** (1960), 249–273.

[29] Dunford, N., Schwartz, J.T.: *Linear operators, vol. II.* John Wiley & Sons, New York et al. (1963).

[30] Eidus, D.: Asymptotical expansions of solutions of linear parabolic equations as $t \to \infty$. *J. Math. Anal. Appl.* **130** (1988), 155–170.

[31] Feireisl, E.: Forced vibrations in one-dimensional nonlinear thermoelasticity as a local coercive-like problem. *Comment. Math. Univ. Carolinae* **31** (1990), 243–255.

[32] Fontane, T.: *Effi Briest.* Insel TB **138**. Insel Verlag, Frankfurt a. Main (1989). (Reprint. Orig.: Verlag F. Fontane & Co., Berlin (October 1895, dated 1896)).

[33] Friedman, A.: *Partial differential equations.* Holt, Rinehart and Winston, New York et al. (1969).

[34] Friedrichs, K.O.: The identity of weak and strong extensions of differential operators. *Trans. Amer. Math. Soc.* **55** (1944), 132–151.

[35] ——— Symmetric hyperbolic linear differential equations. *Comm. Pure Appl. Math.* **7** (1954), 345–392.

[36] Gagliardo, E.: Ulteriori proprietà di alcune classi di funzioni in più variabili. *Ricerce Mat.* **8** (1959), 24–51.

[37] Giaquinta, M.: *Multiple integrals in the calculus of variations and nonlinear elliptic systems.* Annals of mathematics studies **105**. Princeton University Press, Princeton (1983).

[38] Ginibre, J., Velo, G.: On a class of nonlinear Schrödinger equations, I: The Cauchy problem, general case; II: Scattering theory, general case. *J. Functional Analysis* **32** (1979), 1–32; 33–71.

[39] Glassey, R.T.: Blow-up theorems for nonlinear wave equations. *Math. Z.* **132** (1973), 183–203.

[40] Greenberg, J.M., Li, T.-T.: The effect of boundary damping for the quasilinear wave equation. *J. Differential Equations* **52** (1984), 66–75.

[41] Günther, M.: Zum Einbettungssatz von J. Nash. *Math. Nachr.* **144** (1989), 165–187.

[42] Gurtin, M.E.: The linear theory of elasticity. *Handbuch der Physik* **VIa/2** (1972), 1–295.

[43] Hanouzet, B., Joly, J.-L.: Explosion pour des problèmes hyperboliques sémi-linéaires avec second membre non compatible. *C. R. Acad. Sc. Paris, Sér. I,* **301** (1985), 581–584.

[44] Hörmander, L.: Estimates for translation invariant operators in L^p spaces. *Acta Math.* **104** (1960), 93–140.

[45] ———— The boundary problems of physical geodesy. *Arch. Rational Mech. Anal.* **62** (1976), 1–52.

[46] ———— *The analysis of linear partial differential operators I.* Grundlehren math. Wiss. **256**. Springer-Verlag, Berlin et al. (1983).

[47] ———— On Sobolev spaces associated with some Lie algebras. *Report* **4**, Institut Mittag–Leffler, Djursholm (1985).

[48] ———— *Non-linear hyperbolic differential equations. Lectures 1986–1987.* University of Lund (1988).

[49] Hrusa, W.J., Messaoudi, S.A.: On formation of singularities in one-dimensional nonlinear thermoelasticity. *Arch. Rational Mech. Anal.* **111** (1990), 135–151.

[50] Hrusa, W.J., Tarabek, M.A.: On smooth solutions of the Cauchy problem in one-dimensional nonlinear thermoelasticity. *Quart. Appl. Math.* **47** (1989), 631–644.

[51] Hughes, T.J.R., Kato, T., Marsden, J.E.: Well-posed quasi-linear second-order hyperbolic systems with applications to nonlinear elastodynamics and general relativity. *Arch. Rational Mech. Anal.* **63** (1977), 273–294.

[52] Ikebe, T.: Eigenfunction expansions associated with the Schrödinger operators and their applications to scattering theory. *Arch. Rational Mech. Anal.* **5** (1960), 1–34.

[53] Ikehata, M., Nakamura, G.: Decaying and nondecaying properties of the local energy of an elastic wave outside an obstacle. *Japan J. Appl. Math.* **6** (1989), 83–95.

[54] Iwashita, H.: L_q–L_r estimates for solutions of the nonstationary Stokes equations in an exterior domain and the Navier–Stokes initial value problems in L_q spaces. *Math. Ann.* **285** (1989), 265–288.

[55] Iwashita, H., Shibata, Y.: On the analyticity of spectral functions for some exterior boundary value problems. *Glasnik Mat.* **23** (1988), 291–313.

[56] Jiang, S.: Global existence of smooth solutions in one-dimensional nonlinear thermoelasticity. *Proc. Roy. Soc. Edinburgh* **115A** (1990), 257–274.

[57] ———— Far field behavior of solutions to the equations of nonlinear 1-d-thermoelasticity. *Appl. Anal.* **36** (1990), 25–35.

[58] ———— Rapidly decreasing behaviour of solutions in nonlinear 3-D-thermoelasticity. *Bull. Austral. Math. Soc.* **43** (1991), 89–99.

[59] ———— Local energy decay for the damped plate equation in exterior domains. *Quart. Appl. Math.* **50** (1992), 257–272.

[60] ———— On global smooth solutions to the one-dimensional equations of nonlinear inhomogeneous thermoelasticity. *Nonlinear Analysis, T.M.A.*, **20** (1993), 1245–1256,

[61] ———— Global solutions of the Neumann problem in one-dimensional nonlinear thermoelasticity. *Nonlinear Analysis, T.M.A.*, **19** (1992), 107–121.

[62] Jiang, S., Racke, R.: On some quasilinear hyperbolic-parabolic initial boundary value problems. *Math. Meth. Appl. Sci.* **12** (1990), 315–339.

[63] Jiang, S., Racke, R.: *Evolution Equations in Thermoelasticity.* Monogr. Surv. Pure Appl. Math. **112**. Chapman & Hall/CRC, Boca Raton (2000).

[64] John, F.: Formation of singularities in one-dimensional nonlinear wave propagation. *Comm. Pure Appl. Math.* **27** (1974), 377–405.

[65] ———— Delayed singularity formation in solutions of nonlinear wave equations in higher dimensions. *Comm. Pure Appl. Math.* **29** (1976), 649–681.

[66] ———— Finite amplitude waves in a homogeneous isotropic elastic solid. *Comm. Pure Appl. Math.* **30** (1977), 421–446.

[67] ———— Blow-up of solutions of nonlinear wave equations in three space dimensions. *Manuscripta math.* **28** (1979), 235–268.

[68] ——— Blow-up for quasilinear wave equations in three space dimensions. *Comm. Pure Appl. Math.* **34** (1981), 29–51.

[69] ——— Formation of singularities in elastic waves. In: "Trends and applications of pure mathematics to mechanics". Proceedings Palaiseau (1983); eds.: P.G. Ciarlet, M. Roseau. *Lec. Notes Phys.* **195** (1984), 194–210.

[70] ——— Blow-up of radial solutions of $u_{tt} = c^2(u_t)\Delta u$ in three space dimensions. *Mat. Apl. Comp.* **4** (1985), 3–18.

[71] ——— *Partial differential equations.* Appl. Math. Sci **1**. Springer-Verlag, New York et al. (1986).

[72] ——— Existence for large times of strict solutions of nonlinear wave equations in three space dimensions for small initial data. *Comm. Pure Appl. Math.* **40** (1987), 79–109.

[73] ——— Almost global existence of elastic waves of finite amplitude arising from small initial disturbances. *Comm. Pure Appl. Math.* **41** (1988), 615–666.

[74] ——— *Nonlinear wave equations, formation of singularities.* University Lecture Series **2**. Amer. Math. Soc., Providence, Rhode Island (1990).

[75] John, F., Klainerman, S.: Almost global existence to nonlinear wave equations in three space dimensions. *Comm. Pure Appl. Math.* **37** (1984), 443–455.

[76] Jones, D.S.: The eigenvalues of $\nabla^2 u + \lambda u = 0$ when the boundary conditions are given on semi-infinite domains. *Proc. Cambridge Phil. Soc.* **49** (1953), 668–684.

[77] Kasuga, T.: On Sobolev–Friedrichs' generalisation of derivatives. *Proc. Jap. Acad.* **33** (1957), 596–599.

[78] Kato, T.: Wave operators and similarity for some non-selfadjoint operators. *Math. Annalen* **162** (1966), 258–279.

[79] Kato, T.: Linear evolution equations of "hyperbolic" type. *J. Fac. Sci. Univ. Tokyo* **17** (1970), 241–258.

[80] ——— Linear evolution equations of "hyperbolic" type II. *J. Math. Soc. Japan* **25** (1973), 648–666.

[81] ——— The Cauchy problem for quasi-linear symmetric hyperbolic systems. *Arch. Rational Mech. Anal.* **58** (1975), 181–205.

[82] ——— Quasi-linear equations of evolution, with applications to partial differential equations. In: "Spectral theory and differential equations"; ed.: W.N. Everitt. *Lec. Notes Math.* **448** (1975), 24–70.

[83] ——— *Abstract differential equations and nonlinear mixed problems.* Lezioni Fermiane. Accad. Naz. D. Linc. Scuola Norm. Sup. Pisa (1985).

[84] Kawashima, S.: *Systems of a hyperbolic-parabolic composite type, with applications to the equations of magnetohydrodynamics.* Thesis, Kyoto University (1983).

[85] Kawashima, S., Okada, M.: Smooth global solutions for the one-dimensional equations in magnetohydrodynamics. *Proc. Jap. Acad., Ser. A,* **58** (1982), 384–387.

[86] Klainerman, S.: Global existence for nonlinear wave equations. *Comm. Pure Appl. Math.* **33** (1980), 43–101.

[87] ———— Long-time behavior of solutions to nonlinear evolution equations. *Arch. Rational Mech. Anal.* **78** (1982), 73–98.

[88] ———— Uniform decay estimates and the Lorentz invariance of the classical wave equation. *Comm. Pure Appl. Math.* **38** (1985), 321–332.

[89] ———— Global existence of small amplitude solutions to nonlinear Klein–Gordon equations in four space-time dimensions. *Comm. Pure Appl. Math.* **38** (1985), 631–641.

[90] ———— The null condition and global existence to nonlinear wave equations. *Lec. Appl. Math.* **23** (1986), 293–326.

[91] ———— Remarks on the global Sobolev inequalities in the Minkowski space $\mathbb{R}^{n+1}$. *Comm. Pure Appl. Math.* **40** (1987), 111–117.

[92] ———— Remark on the asymptotic behavior of the Klein Gordon equation in $\mathbb{R}^{n+1}$. *Comm. Pure Appl. Math.* **46** (1993), 137–144.

[93] Klainerman S., Majda, A.: Formation of singularities for wave equations including the nonlinear vibrating string. *Comm. Pure Appl. Math.* **33** (1980), 241–263.

[94] Klainerman, S., Ponce, G.: Global, small amplitude solutions to nonlinear evolution equations. *Comm. Pure Appl. Math.* **36** (1983), 133–141.

[95] Klaus, F.: *Globale klassische Lösungen für das Cauchy-Problem divergenzfreier, quasilinearer Maxwell-Gleichungen mit kubischer Nichtlinearität für kleine Daten.* Diplomarbeit (master thesis), Bonn (1989).

[96] Kozono, H., Yanagisawa, T.: L^r-variational inequality for vector fields and the Helmholtz-Weyl decomposition in bounded domains. *Indiana University Math. J.* **58** (2009), 1853–1920.

[97] Lax, P.D.: *Hyperbolic systems of conservation laws and the mathematical theory of shock waves.* SIAM Reg. Conf. Appl. Math., Philadelphia (1973).

[98] Leis, R.: *Initial boundary value problems in mathematical physics.* B.G. Teubner-Verlag, Stuttgart; John Wiley & Sons, Chichester et al. (1986).

[99] Lesky, P.: Resonance phenomena for a class of partial differential equations of higher order in cylindrical waveguides. Math. Meth. Appl. Sci. **11** (1989), 697–723.

[100] ———— Local existence for solutions of fully nonlinear wave equations. Math. Meth. Appl. Sci. **14** (1991), 483–508.

[101] Lesky, P.H., Racke, R.: Nonlinear wave equations in infinite waveguides. *Comm. PDE* **28** (2003), 1265–1301.

[102] ———— Elastic and electro-magnetic waves in infinite waveguides. *J. Differential Equations* **244** (2008), 945–971.

[103] Levine, H. A.: The role of critical exponents in blowup theorems. *SIAM Review* **32** (1990), 262–288.

[104] Li, T.-T., Chen, Y.: Initial value problems for nonlinear wave equations. *Comm. PDE* **13** (1988), 383–422.

[105] Li, T.-T., Yu, X.: Dureé de vie des solutions régulières pour les équations des ondes non linéaires. *C. R. Acad. Sc. Paris, Sér. I,* **309** (1989), 469–472.

[106] Li, T.-T., Zhou, Y.: Nonlinear stability for two space dimensional wave equations with higher order perturbations. *UMPA Prépublication* **49**, École Normale Supérieure de Lyon (1991).

[107] Liess, O.: *Global existence for the nonlinear equations of crystal optics.* "Journées Équations aux Dérivés Partielles" (Saint Jean de Monts (1989)), Exp. No. **V**, 11 pp., École Polytech., Palaiseau (1989).

[108] ———— Decay estimates for the solutions of the system of crystal optics. *Asymptotic Analysis* **4** (1991), 61–95.

[109] Lindblad, H.: Blow-up for solutions of $\Box u = |u|^p$ with small initial data. *Comm. PDE* **15** (1990), 757–821.

[110] ———— On the lifespan of solutions of nonlinear wave equations with small initial data. *Comm. Pure Appl. Math.* **43** (1990), 445–472.

[111] Littman, W.: Fourier transforms of surface-carried measures and differentiability of surface averages. *Bull. Amer. Math. Soc.* **69** (1963), 766–770.

[112] Liu, T.-P.: Development of singularities in the nonlinear waves for quasi-linear hyperbolic partial differential equations. *J. Differential Equations* **33** (1979), 92–111.

[113] Majda, A.: Outgoing solutions for perturbations of $-\Delta$ with applications to spectral and scattering theory. *J. Differential Equations* **16** (1974), 515–547.

[114] ———— *Compressible fluid flow and systems of conservation laws in several space variables.* Appl. Math. Sci. **53**. Springer-Verlag, New York et al. (1984).

[115] Marzuola, J.L., Metcalfe, J., Tataru, D.: Quasilinear Schrödinger equations I: small data and quadratic interactions. *Adv. Math.* **231** (2012), 1151–1172.

[116] ———— Quasilinear Schrödinger equations II: small data and cubic nonlinearities. *Kyoto J. Math.* (to appear).

[117] Matsumura, A.: *Initial value problems for some quasilinear partial differential equations in mathematical physics.* Thesis, Kyoto University (1980).

[118] ———— An energy method for the equations of motion of compressible viscous heat-conductive fluids. *MRC Technical Summary Report* **2194**, University of Wisconsin—Madison (1981).

[119] Matsumura, A., Nishida, T.: The initial value problem for the equations of motion of compressible viscous and heat-conductive fluids. *Proc. Jap. Acad., Ser. A,* **55** (1979), 337–341.

[120] ———— Initial boundary value problems for the equations of motion of compressible viscous and heat conductive fluids. *Comm. Math. Physics* **89** (1983), 445–464.

[121] Mehra, M.L.: Zur asymptotischen Verteilung der Eigenwerte des Maxwellschen Randwertproblems. *Bonner Math. Schriften* **109** (1978).

[122] Melrose, R.B.: Singularities and energy decay in acoustical scattering. *Duke Math. J.* **46** (1979), 43–59.

[123] Metcalfe, J., Sogge, C.D., Stewart, A.: Nonlinear hyperbolic equations in infinite homogeneous waveguides. *Comm. PDE* **30** (2005), 643–661.

[124] Metcalfe, J., Perry, J.: Global solutions to quasilinear wave equations in homogeneous waveguides with Neumann boundary conditions. *Comm. Pure Appl. Anal.* **11** (2012), 547–556.

[125] Meyers, N.G., Serrin, J.: $H = W$. *Proc. Nat. Acad. Sci. USA* **51** (1964), 1055–1056.

[126] Miller, G.F., Musgrave, J.P.: On the propagation of elastic waves in aeolotropic media. III. Media of cubic symmetry. *Proc. Roy. Soc. London, Ser. A,* **236** (1956), 352–383.

[127] Mochizuki, K.: Spectral and scattering theory for symmetric hyperbolic systems in an exterior domain. *Publ. RIMS, Kyoto University* **5** (1969), 219–258.

[128] ———— Uniform resolvent estimates for magnetic Schrödinger operators and smoothing effects for related evolution equations. *Publ. RIMS Kyoto Univ.* **46** (2010), 741–754.

[129] Morawetz, C.S.: Time decay for the nonlinear Klein-Gordon equation. *Proc. Roy. Soc. Ser. A* **306** (1968), 291–296.

[130] Morawetz, C.S., Ludwig, D.: An inequality for the reduced wave operator and the justification of geometrical optics. *Comm. Pure Appl. Math.* **21** (1968), 187–203.

[131] Morawetz, C.S., Ralston, J., Strauss, W.: Decay of solutions of the wave equation outside nontrapping obstacles. *Comm. Pure Appl. Math.* **30** (1977), 447–508.

[132] Morgenröther, K., Werner, P.: Resonances and standing waves. *Math. Methods Appl. Sci.* **9** (1987), 105–126.

[133] Moser, J.: A rapidly convergent iteration method and non-linear partial differential equations. *Ann. Scuola Norm. Sup. Pisa, Ser. 3,* **20** (1966), 265–315 and 499–535.

[134] Mukoyama, T.: On the Cauchy problem for quasilinear hyperbolic-parabolic coupled systems in higher dimensional spaces. *Tsukuba J. Math.* **13** (1989), 363–386.

[135] Nash, J.: The embedding problem for Riemannian manifolds. *Ann. Math.* **63** (1956), 20–63.

[136] Nirenberg, L.: On elliptic partial differential equations. *Ann. Scuola Norm. Sup. Pisa, Ser. 3,* **13** (1959), 115–162.

[137] Payne, L.E., Sattinger, D.H.: Saddle points and instability of nonlinear hyperbolic equations. *Israel J. Math.* **22** (1975), 273–303.

[138] Pecher, H.: L^p-Abschätzungen und klassische Lösungen für nichtlineare Wellengleichungen. I. *Math. Z.* **150** (1976), 159–183.

[139] ———— Scattering for semilinear wave equations with small data in three space dimensions. *Math. Z.* **198** (1988), 277–289.

[140] ———— Global smooth solutions to a class of semilinear wave equations with strong nonlinearities. *Manuscripta math.* **69** (1990), 71–92.

[141] Ponce, G.: Global existence of small solutions to a class of nonlinear evolution equations. *Nonlinear Analysis, T.M.A.,* **9** (1985), 399–418.

[142] Ponce, G., Racke, R.: Global existence of small solutions to the initial value problem for nonlinear thermoelasticity. *J. Differential Equations* **87** (1990), 70–83.

[143] Racke, R.: On the Cauchy problem in nonlinear 3-d-thermoelasticity. *Math. Z.* **203** (1990), 649–682.

[144] ———— Blow-up in nonlinear three-dimensional thermoelasticity. *Math. Meth. Appl. Sci.* **12** (1990), 267–273.

[145] ———— L^p–L^q-estimates for solutions to the equations of linear thermoelasticity in exterior domains. *Asymptotic Analysis* **3** (1990), 105–132.

[146] ———— Decay rates for solutions of damped systems and generalized Fourier transforms. *J. reine angew. Math.* **412** (1990), 1–19.

[147] ———— Non-homogeneous, non-linear damped wave equations in unbounded domains. *Math. Meth. Appl. Sci.* **13** (1990), 481–491.

[148] ———— Anfangswertprobleme bei nichtlinearen Evolutionsgleichungen. *SFB 256 Vorlesungsreihe* **13**, Universität Bonn (1990).

[149] ———— Remarks on the formation of singularities for nonlinear wave equations in bounded domains. *SFB 256 Preprint* **153**, Universität Bonn (1991).

[150] ———— *Lectures on nonlinear evolution equations. Initial value problems.* Aspects of Mathematics **E19**. Friedr. Vieweg & Sohn: Braunschweig/Wiesbaden, 1992.

[151] Racke, R., Shibata, Y.: Global smooth solutions and asymptotic stability in one-dimensional nonlinear thermoelasticity. *Arch. Rational Mech. Anal.* **116** (1991), 1–34.

[152] Racke, R., Shibata, Y., Zheng, S.: Global solvability and exponential stability in one-dimensional nonlinear thermoelasticity. *Quart. Appl. Math.* **51** (1993), 751–763.

[153] Racke, R., Zheng, S.: Global existence of solutions to a fully nonlinear fourth-order parabolic equation in exterior domains. *Nonlinear Analysis, T.M.A.,* **17** (1991), 1027–1038, and (Erratum) **18** (1992), 401.

[154] Reed, M., Simon, B.: *Methods of modern mathematical physics. IV. Analysis of operators.* Academic Press, New York (1978).

[155] Rellich, F.: Das Eigenwertproblem von $\Delta u + \lambda u = 0$ in Halbröhren. Studies and essays presented to R. Courant, New York (1948).

[156] Rodnianski. I., Schlag, W.: Time decay for solutions of Schrödinger equations with rough and time-dependent potentials. *Inventiones Math.* **155** (2004), 451–513.

[157] Schauder, J.: Das Anfangswertproblem einer quasilinearen hyperbolischen Differentialgleichung zweiter Ordnung in beliebiger Anzahl von unabhängigen Veränderlichen. *Fund. Math.* **24** (1935), 213–246.

[158] Shatah, J.: Global existence of small solutions to nonlinear evolution equations. *J. Differential Equations* **46** (1982), 409–425.

[159] ———— Normal forms and quadratic nonlinear Klein–Gordon equations. *Comm. Pure Appl. Math.* **38** (1985), 685–696.

[160] Shenk, N.A.: Eigenfunction expansions and scattering theory for the wave equation in an exterior region. *Arch. Rational Mech. Anal.* **21** (1966), 120–150.

[161] Shenk, N.A., Thoe, D.: Eigenfunction expansions and scattering theory for perturbations of $-\Delta$. *J. Math. Anal. Appl.* **36** (1971), 313–351.

[162] Shibata, Y.: On the global existence of classical solutions of mixed problem for some second order non-linear hyperbolic operators with dissipative term in the interior domain. *Funkcial. Ekvac.* **25** (1982), 303–345.

[163] ———— On the global existence of classical solutions of second order fully nonlinear hyperbolic equations with first order dissipation in the exterior domain. *Tsukuba J. Math.* **7** (1) (1983), 1–68.

[164] ———— Neumann problem for one-dimensional nonlinear thermoelasticity. *Banach Center Publications* **27** (1992), 457-480.

[165] Shibata, Y., Tsutsumi, Y.: Global existence theorem of nonlinear wave equations in the exterior domain. In: "Recent Topics in Nonlinear PDE"; eds.: M. Mimura, T. Nishida. North Holland Math. Studies 98; *Lec. Notes in Num. Appl. Anal.* **6** (1983), 155–195.

[166] ———— On a global existence theorem of Neumann problem for some quasilinear hyperbolic equations. In: "Recent Topics in Nonlinear PDE II"; eds.: K. Masuda, M. Mimura. North Holland Math. Studies 128; *Lec. Notes in Num. Appl. Anal.* **8** (1985), 175–228.

[167] ———— On a global existence theorem of small amplitude solutions for nonlinear wave equations in an exterior domain. *Math. Z.* **191** (1986), 165–199.

[168] ———— Local existence of solution for the initial boundary value problem of fully nonlinear wave equation. *Nonlinear Analysis, T.M.A.*, **11** (1987), 335–365.

[169] Shibata, Y., Zheng, S.: On some nonlinear hyperbolic systems with damping boundary conditions. *Nonlinear Analysis, T.M.A.*, **17** (1991), 233-266.

[170] Shizuta, Y., Kawashima, S.: Systems of equations of hyperbolic-parabolic type with applications to the discrete Boltzmann equation. *Hokkaido Math. J.* **14** (1985), 249–275.

[171] Sideris, T.C.: Global behavior of solutions to nonlinear wave equations in three dimensions. *Comm. PDE* **8** (1983), 1291–1323.

[172] ———— Formation of singularities in solutions to nonlinear hyperbolic equations. *Arch. Rational Mech. Anal.* **86** (1984), 369–381.

[173] ———— Decay estimates for the three-dimensional inhomogeneous Klein–Gordon equation and applications. *Comm. PDE* **14** (1989), 1421–1455.

[174] Simader, C.G.; Sohr, H. *The Dirichlet problem for the Laplacian in bounded and unbounded domains: a new approach to weak, strong and (2+k)-solutions in Sobolev-type spaces.* Pitman Research Notes in Mathematics Series 360; Addison Wesley Longman: Harlow, 1996.

[175] Slemrod, M.: Global existence, uniqueness, and asymptotic stability of classical smooth solutions in one-dimensional non-linear thermoelasticity. *Arch. Rational Mech. Anal.* **76** (1981), 97–133.

[176] Smoller, J.: *Shock waves and reaction-diffusion equations.* Grundlehren math. Wiss. **258**. Springer-Verlag, New York et al. (1983).

[177] Stoth, M.: Globale klassische Lösungen der quasilinearen Elastizitätsgleichungen für kubisch elastische Medien im $\mathbb{R}^2$. *SFB 256 Preprint* **157**, Universität Bonn (1991).

[178] Strauss, W.: Nonlinear scattering theory at low energy. *J. Functional Analysis* **41** (1981), 110–133.

[179] ———— *Nonlinear wave equations.* CBMS Series **73**, Amer. Math. Soc., Providence, Rhode Island (1989).

[180] Struwe, M.: Globally regular solutions to the u^5 Klein–Gordon equation. *Ann. Scuola Norm. Sup. Pisa, Ser. 4,* **15** (1988), 495–513.

[181] Triebel, H.: *Theory of function spaces.* Monographs Math. **78**. Birkhäuser Verlag, Basel et al. (1983).

[182] Tsutsumi, Y.: Global solutions of the nonlinear Schrödinger equation in exterior domains. *Comm. PDE* **8** (1983), 1337–1374.

[183] ———— Local energy decay of solutions to the free Schrödinger equation in exterior domains. *J. Fac. Sci. Univ. Tokyo, Sect. IA, Math.,* **31** (1984), 97–108.

[184] Vainberg, B.R.: On the short wave asymptotic behaviour of solutions of stationary problems and the asymptotic behaviour as $t \to \infty$ of solutions of non-stationary problems. *Russian Math. Surveys* **30** (1975), 1–58.

[185] ———— *Asymptotic methods in equations of mathematical physics.* Gordon and Breach Science Publishers. New York et al. (1989).

[186] von Wahl, W.: Klassische Lösungen nichtlinearer Wellengleichungen im Großen. *Math. Z.* **112** (1969), 241–279.

[187] ———— L^p-decay rates for homogeneous wave-equations. *Math. Z.* **120** (1971), 93–106.

[188] ———— On nonlinear evolution equations in a Banach space and on nonlinear vibrations of the clamped plate. *Bayreuther Math. Schriften* **7** (1981), 93 pp.

[189] ———— Corrections to my paper: "On nonlinear evolution equations in a Banach space and on nonlinear vibrations of the clamped plate", Bayr. Math. Schriften, Heft 7. *Bayreuther Math. Schriften* **20** (1985), 205–209.

[190] Werner, P.: Selfadjoint extensions of the Laplace operator with respect to electric and magnetic boundary conditions. *J. Math. Anal. Appl.* **70** (1979), 131-160.

[191] ———— Regularity properties of the Laplace operator with respect to electric and magnetic boundary conditions. *J. Math. Anal. Appl.* **87** (1982), 560-602.

[192] ———— Spectral properties of the Laplace operator with respect to electric and magnetic boundary conditions. *J. Math. Anal. Appl.* **92** (1983), 1-65.

[193] ———— Resonance phenomena in cylindrical waveguides. *J. Math. Anal. Appl.* **121** (1987), 173–214.

[194] ———— Aperiodic electromagnetic waves in cylindrical waveguides. *J. Math. Anal. Appl.* **121** (1987), 215–272.

[195] ———— Resonanzphänomene in akustischen und elektromagnetischen Wellenleitern. *Z. angew. Math. Mech.* **67** (1987), T43–T54.

[196] Weyl, H.: Das asymptotische Verteilungsgesetz der Eigenschwingungen eines beliebig gestalteten elastischen Körpers. *Rend. Circolo Mat. Palermo* **39** (1915), 1-50.

[197] Wilcox, C.H.: *Scattering theory for the d'Alembert equation in exterior domains.* Lec. Notes Math. **442** (1975).

[198] Willems, F.: Globale Existenz von Lösungen zum Anfangswertproblem der nichtlinearen Plattengleichung. *SFB 256 Preprint* **176**, Universität Bonn (1991).

[199] Witsch, K.-J.: On a free boundary value problem of physical geodesy, II (existence). *Math. Meth. Appl. Sci.* **8** (1986), 1–22.

[200] Witsch, K.-J.: Examples of embedded eigenvalues for the Dirichlet-Laplacian in domains with infinite boundaries. *Math. Methods Appl. Sci.* **12** (1990), 177–182.

[201] Yamamoto, K.: Characterization of a convex obstacle by singularities of the scattering kernel. *J. Differential Equations* **64** (1986), 283–293.

[202] Zheng, S.: Global solutions and applications to a class of quasilinear hyperbolic-parabolic coupled systems. *Sci. Sinica, Ser. A,* **27** (1984), 1274–1286.

[203] ———— Remarks on global existence for nonlinear parabolic equations. *Nonlinear Analysis, T.M.A.,* **10** (1986), 107–114.

[204] ———— Global solutions to the second initial boundary value problem for fully nonlinear parabolic equations. *Acta Math. Sinica* **3** (1987), 237–246.

[205] ———— Global existence of solutions to initial boundary value problem for nonlinear parabolic equation. *Acta Math. Sinica* **4** (1987), 361–374.

[206] Zheng, S., Chen, Y.: Global existence for nonlinear parabolic equations. *Chin. Ann. Math., Ser. B.,* **7** (1986), 57–73.

[207] ———— On blowing up of the solutions to initial boundary value problems of nonlinear evolution equations (in Chinese). *J. Fudan University (Natural Science)* **26** (1987), 19–27.

[208] Zheng, S., Shen, W.: Global solutions to the Cauchy problem of quasilinear hyperbolic parabolic coupled systems. *Sci. Sinica, Ser. A,* **30** (1987), 1133–1149.

Notation

a.e.	almost everywhere
$B(x_0, r)$	closed ball in $\mathbb{R}^n$ with center x_0 and radius $r > 0$
C^k	$C^k(\mathbb{R}^n)$
$C^k(D)$	k-times continuously differentiable functions from $D \subset \mathbb{R}^m$ into $\mathbb{R}$ (or $\mathbb{R}^j, \mathbb{C}, \mathbb{C}^j$); $k \in \mathbb{N}_0 \cup \{\infty\}$; $m, j \in \mathbb{N}$
C_b^k	$C_b^k(\mathbb{R}^n)$
$C_b^k(D)$	$C^k(D)$-functions with bounded derivatives up to order k, $k \in \mathbb{N}_0 \cup \{\infty\}$
$C^k(I, E)$	space of k-times strongly differentiable functions from an interval $I \subset \mathbb{R}$ into a Banach space E, $k \in \mathbb{N}_0 \cup \{\infty\}$
C_0^∞	$C_0^\infty(\mathbb{R}^n)$
$C_0^\infty(D)$	$C^\infty(D)$-functions with compact support
$C_w(I, E)$	space of weakly continuous functions from an interval $I \subset \mathbb{R}$ into a Banach space E
$\delta_{\alpha\beta}$	Kronecker delta
∂_j	$\partial/\partial x_j$
∂_t	$\partial/\partial t$ (also indicated by a subindex t)
D	$(\partial_t, \partial_1, \ldots, \partial_n)'$ (in Section 11.6: symbol for the displacement current)
D^α	$(\partial_t^{\alpha_0}, \partial_1^{\alpha_1}, \ldots, \partial_n^{\alpha_n})'$, $\alpha = (\alpha_0, \alpha_1, \ldots, \alpha_n) \in {\mathbb{N}_0}^{n+1}$, $n \in \mathbb{N}$
det	determinant
ess sup	essential supremum
$\exp \cdots$	$e^{\cdots}$
$\mathcal{F}$	Fourier transform
Im	imaginary part
$\langle \cdot, \cdot \rangle$	inner product in L^2
κ_s	Sobolev constant in the imbeddding $W^{s,2} \hookrightarrow C^0$, $s \in \mathbb{N}$
$Lip(I, E)$	space of Lipschitz continuous functions from an interval $I \subset \mathbb{R}$ into a Banach space E
log	natural logarithm
L^p	$L^p(\mathbb{R}^n)$
$L^p(I, E)$	space of strongly measurable functions from $I \subset \mathbb{R}$ into a Banach space E, the p-th powers of which are integrable (essentially bounded if $p = \infty$), $1 \leq p \leq \infty$
L_{loc}^p	set of functions being locally in L^p
$L^p(\Omega)$	$W^{0,p}(\Omega)$
∇	$(\partial_1, \ldots, \partial_n)'$
∇^α	$(\partial_1^{\alpha_1}, \ldots, \partial_n^{\alpha_n})'$, $\alpha = (\alpha_1, \ldots, \alpha_n) \in {\mathbb{N}_0}^n$, $n \in \mathbb{N}$
$\|\cdot\|_E$	norm in the Banach space E
$\|\cdot\|_{m,p}$	norm in $W^{m,p}$

$\|\cdot\|_p$	norm in L^p
$\|u\|_{s,T}$	$\sup_{0\le t\le T}\|u(t)\|_{s,2}$
$\|D^m u\|_p$	$\left(\sum_{\|\alpha\|=m}\|D^\alpha u\|_p^p\right)^{1/p}$,
$\|\overline{D}^m u\|_p$	$\left(\sum_{0\le\|\alpha\|\le m}\|D^\alpha u\|_p^p\right)^{1/p}$,
$\|\nabla^m u\|_p$	$\left(\sum_{\|\alpha\|=m}\|\nabla^\alpha u\|_p^p\right)^{1/p}$,
$\|\overline{\nabla}^m u\|_p$	$\left(\sum_{0\le\|\alpha\|\le m}\|\nabla^\alpha u\|_p^p\right)^{1/p}$, $m\in\mathbb{N}_0$, $1\le p<\infty$ ($p=\infty$ as usual).
Re	real part
S^{n-1}	unit sphere in $\mathbb{R}^n$
supp	support
$W^{m,p}$	$W^{m,p}(\mathbb{R}^n)$
$W^{m,p}(\Omega)$	usual Sobolev space, $m\in\mathbb{N}_0$, $1\le p\le\infty$, (see R.A. Adams [1]); several copies are denoted with the same symbol
$W_0^{1,2}(\Omega)$	usual Sobolev space generalizing zero boundary values, see [1]
$\hookrightarrow$	continuous imbedding
$\rightharpoonup$	weak convergence
$[x]$	largest integer which is less than or equal to x, $x\in\mathbb{R}$
$'$	used for transposition, e.g. for the divergence ∇', for one-dimensional derivatives, for indexing, and, in Chapter 13, for denoting parts of space variables

Index

MIX
Papier aus verantwortungsvollen Quellen
Paper from responsible sources
FSC® C105338

If you have any concerns about our products,
you can contact us on
ProductSafety@springernature.com

In case Publisher is established outside the EU,
the EU authorized representative is:
Springer Nature Customer Service Center GmbH
Europaplatz 3, 69115 Heidelberg, Germany

Printed by Libri Plureos GmbH
in Hamburg, Germany